北京理工大学"双一流"建设精品出版工程

Tactical Missile Guidance and Control System Design

林德福◎著

北京理工大学出版社
BEIJING INSTITUTE OF TECHNOLOGY PRESS

版权专有　侵权必究

图书在版编目（CIP）数据

战术导弹制导控制系统设计 = Tactical Missile Guidance and Control System Design：英文/林德福著 . —北京：北京理工大学出版社，2020.12
　　ISBN 978 – 7 – 5682 – 9307 – 5

Ⅰ. ①战… Ⅱ. ①林… Ⅲ. ①战术导弹 – 导弹制导 – 控制系统设计 – 英文 Ⅳ. ①TJ765

中国版本图书馆 CIP 数据核字（2020）第 244011 号

出版发行 / 北京理工大学出版社有限责任公司
社　　址 / 北京市海淀区中关村南大街 5 号
邮　　编 / 100081
电　　话 / （010）68914775（总编室）
　　　　　（010）82562903（教材售后服务热线）
　　　　　（010）68948351（其他图书服务热线）
网　　址 / http：//www.bitpress.com.cn
经　　销 / 全国各地新华书店
印　　刷 / 三河市华骏印务包装有限公司
开　　本 / 787 毫米 × 1092 毫米　1/16
印　　张 / 18.25　　　　　　　　　　　　　　　　　责任编辑 / 梁铜华
字　　数 / 426 千字　　　　　　　　　　　　　　　　文案编辑 / 梁铜华
版　　次 / 2020 年 12 月第 1 版　2020 年 12 月第 1 次印刷　责任校对 / 周瑞红
定　　价 / 62.00 元　　　　　　　　　　　　　　　　责任印制 / 李志强

图书出现印装质量问题，请拨打售后服务热线，本社负责调换

Preface

This book is pitched as an introductory text for tactical missiles, covering the missile systems and design processes that must be considered. The main contents of this book are the summary and extension of the authors' 20 years' research works in the field of missile guidance and control.

We start from Chapter 1 with the basics of missile guidance by exposing the reader to critical concepts such as lateral acceleration as the preferred guidance command, the interplay between guidance, navigation and control within the context of missile systems. Chapter 2 discusses the missile dynamics with detailed aerodynamic models. Chapter 3 is intended to present another brief overview of several pertinent missile sub-systems and sensors, supposedly covering actuator dynamics, rate gyro dynamics, accelerometers and integrated navigation systems. Chapter 4 presents a discussion of the radar guidance system and attempts to extract this system's performance characteristics with the greatest impact on guidance loop design and overall system performance. Chapter 5 describes commonly used seekers and the parasitic loop and real seeker model. Details of how to design a proper missile autopilot are presented in Chapter 6, including the most well-known two-loop and three-loop autopilot. Chapter 7 introduces the line-of-sight guidance. The optimality of proportional navigation and its variants are discussed in Chapter 8. In the final chapter, we introduce some modern optimal guidance laws, including optimal trajectory shaping guidance and gravity-turn-assisted guidance.

The book can benefit researchers, engineers and graduate students in the field of Guidance, Navigation and Control. It is our hope for this book to serve as a useful step for permitting further advances in the field of missile guidance and control. The authors have carefully reviewed the content of this book before the printing stage. However, it does not mean that this book is completely free from any possible errors. Consequently, the authors would be grateful to readers who will call out attention on mistakes as they might discover.

Finally, the authors would like to thank colleagues from the Institute of Autonomous UAV Control, Beijing Institute of Technology for providing valuable and constructive comments. Without their support, the writing of the book would not have been a success.

Contents

1 The Overview of Missile Guidance and Control 001

2 Missile Mathematical Models 007
  - § 2.1 Symbols and Definitions 007
  - § 2.2 Euler Equations of the Rigid Body Motion 008
  - § 2.3 Configuration of the Control Surfaces 013
  - § 2.4 Missile Aerodynamic Derivatives and Dynamic Coefficients 014
  - § 2.5 Aerodynamic Transfer Functions of the Missile 019

3 Simplified Models of Missile Control Components 028

4 Guidance Radar 030
  - § 4.1 Introduction 030
  - § 4.2 Motion Characteristic of Line-of-sight 030
  - § 4.3 Control Loop of the Guidance Radar 034
  - § 4.4 Effect of the Receiver Thermal Noise on the Guidance Performance 041
  - § 4.5 Effect of Target Glint on the Guidance Performance 044
  - § 4.6 Effect of Other Disturbances on the Guidance Performance 045
    - 4.6.1 Effect of Disturbance Moment 045
    - 4.6.2 Effect of Target Maneuvers 047

5 Seekers 049
  - § 5.1 Definitions 049
  - § 5.2 Different Kinds of Seekers 050
    - 5.2.1 Dynamic Gyro Seeker 050
    - 5.2.2 Stabilized Platform Seeker 055
    - 5.2.3 Detector Strapdown Stabilized Optic Seeker 059
    - 5.2.4 Semi-strapdown Platform Seeker 060
    - 5.2.5 Strapdown Seeker 060

    5.2.6   Roll-pitch Seeker ·· 063
§ 5.3   Anti-disturbance Moment of the Seeker ···································· 065
§ 5.4   Body Motion Coupling and the Parasitic Loop ···························· 068
    5.4.1   Body Motion Coupling Dynamics Model ································ 068
    5.4.2   Guidance Parasitic Loop Introduced by Seeker-missile Coupling ·············· 071
§ 5.5   A Real Seeker Model ··· 074
    5.5.1   A Real Seeker Model ··· 074
    5.5.2   Testing Methods ··· 077
§ 5.6   Other Parasitic Loop Models ··· 078
    5.6.1   Parasitic Loop Model for a Phase Array Strapdown Seeker ················· 078
    5.6.2   Parasitic Loop Model Due to Radome Slope Error ······················· 079
    5.6.3   Beam Control Gain Error ΔK_B of the Phased Array Seeker and the Radome Slope Error R_{dom} Effect on the Seeker's Performance ······················ 081
    5.6.4   A Novel Online Estimation and Compensation Method for Strapdown Phased Array Seeker Disturbance Rejection Effect Using Extended State Kalman Filter ··· 083
§ 5.7   Platform Based Seeker Design ··· 090
    5.7.1   Stabilization Loop Design ··· 090
    5.7.2   Tracking Loop Design ·· 091

6　Missile Autopilot Design ·· 094

§ 6.1   Acceleration Autopilot ··· 094
    6.1.1   Two-loop Acceleration Autopilot ······································ 094
    6.1.2   Two-loop Acceleration Autopilot with PI Compensation ················· 098
    6.1.3   Three-loop Autopilot with Pseudo Angle of Attack Feedback ············ 101
    6.1.4   Classic Three-loop Autopilot ·· 107
    6.1.5   Discussion of Variable Acceleration Autopilot Structures ················ 111
    6.1.6   Hinge Moment Autopilot ··· 112
    6.1.7   Questions Concerning Acceleration Autopilot Design ···················· 115
    6.1.8   The Acceleration Autopilot Design Method ···························· 120
§ 6.2   Pitch/Yaw Attitude Autopilot ·· 131
§ 6.3   Flight Path Angle Autopilot ·· 134
§ 6.4   Roll Attitude Autopilot ··· 135
§ 6.5   BTT Missile Autopilot ··· 140
§ 6.6   Thrust Vector Control and Thruster Control ······························· 149
§ 6.7   Spinning Missile Control ·· 155
    6.7.1   Aerodynamic Coupling ··· 156
    6.7.2   Control Coupling ··· 156

7 Line of Sight Guidance Methods ... 160

§ 7.1 LOS Guidance System ... 160
§ 7.2 Required Acceleration for the Missile with LOS Guidance ... 161
§ 7.3 Analysis of the LOS Guidance Loop ... 166
§ 7.4 Lead Angle Guidance Method ... 173

8 Proportional Navigation and Extended Proportional Navigation Guidance Laws ... 176

§ 8.1 Proportional Navigation Guidance Law ... 176
  8.1.1 Proportional Navigation Guidance Law (PN) ... 176
  8.1.2 PN Analysis without Guidance System Lag ... 182
  8.1.3 PN Characteristics Including the Missile Guidance Dynamics ... 187
  8.1.4 Adjoint Method ... 199
§ 8.2 Optimal Proportional Navigation Guidance Laws (OPN) ... 203
  8.2.1 OPN1: Considering the Missile Guidance Dynamics ... 203
  8.2.2 OPN2: Considering the Constant Target Maneuver ... 208
  8.2.3 OPN3: Considering Both Constant Target Maneuvers and Missile Guidance Dynamics ... 211
  8.2.4 Estimation of Target Maneuver Acceleration ... 214
  8.2.5 Estimation of t_{go} ... 215
  8.2.6 Extended Guidance Law with Impact Angle Constraint ... 215
§ 8.3 Other Proportional Navigation Laws ... 218
  8.3.1 Proportional Navigation Law with Gravity Over-compensation ... 218
  8.3.2 Lead Angle Proportional Navigation Guidance Law ... 221
§ 8.4 Target Acceleration Estimation ... 223
§ 8.5 Optimum Trajectory Control System Design ... 236

9 Optimal Guidance for Trajectory Shaping ... 242

§ 9.1 Optimality of Error Dynamics in Missile Guidance ... 242
  9.1.1 Optimal Error Dynamics ... 243
  9.1.2 Analysis of Optimal Error Dynamics ... 245
§ 9.2 Optimal Predictor-corrector Guidance ... 247
  9.2.1 General Approach for Guidance Law Design ... 247
  9.2.2 Impact Angle Control ... 247
  9.2.3 Impact Time Control ... 250
§ 9.3 Gravity-turn-assisted Optimal Guidance Law ... 253
  9.3.1 Zero-control-effort Trajectory Considering Gravity ... 254
  9.3.2 Optimal Guidance Law Design and Analysis ... 257

  9.3.3 Characteristics Analysis by Simulations 260
§9.4 Three-dimensional Optimal Impact Time Guidance for Antiship Missiles 265
  9.4.1 Problem Formulation 266
  9.4.2 Three-dimensional Optimal Impact Time Guidance Law Design 267
  9.4.3 Analysis of Proposed Guidance Law 271
  9.4.4 Numerical Simulations 274

References 279

1

The Overview of Missile Guidance and Control

The purpose of missile control is to make the missile hit the target at the end of its flight. In order to achieve this goal, it is essential for the missile to constantly acquire the motion information of the target and of the missile itself in the course of the flight and adopt a tactic (that is a guidance law) to decide how to change the missile's velocity direction based on the current missile and target relative motion, allowing the missile to finally hit the target. As the relationship between the angular velocity $\dot{\theta}$ of the missile velocity vector and its normal acceleration a of the missile is as follows:

$$\dot{\theta} = \frac{a}{V} (V \text{ is the missile velocity}) \tag{1.1-1}$$

the command of a guidance law that is generated to change the missile velocity vector direction is usually the normal acceleration a_c of the missile. This missile and target interception control loop is quite different from the conventional tracking control loop; the former is a time-varying control system, and its analysis method is completely different from the general linear time-invariant time-domain and frequency-domain analysis method. So a special term (guidance loop) has historically been given to this particular missile control outer loop.

With the help of autopilots, the missile output acceleration a will follow the above guidance acceleration command a_c. Under the assumptions of small perturbation, linearization and constant system parameter value, this autopilot loop is a linear time-invariant system and so different traditional control theory design methods can all be applied. Therefore, the autopilot loop that acts as the guidance inner loop historically is still referred to as the control loop.

The missile position and velocity information needed in the guidance process are obtained by an inertial navigation or integrated inertial navigation system. The process of obtaining the missile position and orientation information is called navigation. It is noteworthy that the term navigation here does not refer to the historical definition of directing the course of a ship or an aircraft. Fig. 1.1-1 shows the relationships between the terms navigation, guidance and control in missile control loops.

It has been stated before that the task of a missile control system is to use missile normal acceleration to change the missile's velocity direction according to the guidance law command. For tactical missiles flying in the atmosphere, this normal acceleration is generated by normal aerodynamic forces. As we know, when the missile has an angle of attack with respect to its velocity vector, the corresponding lift will produce a normal acceleration. However, for the missile to maintain a steady angle of attack, this angle of attack induced aerodynamic moment must be

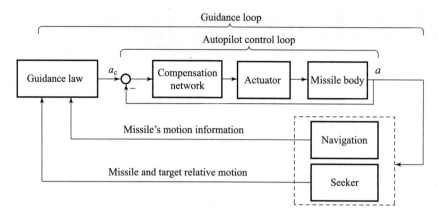

Fig. 1.1 – 1 Block diagram of the missile guidance and control loops

balanced by the control surface deflection induced control moment.

When the center of gravity of the missile is located in front of the center of pressure, the angle of attack generated aerodynamic moment will decrease the existing angle of attack and meanwhile, the x-axis of the missile body will try to coincide with the missile velocity axis. This type of aerodynamic layout is known as a statically stable aerodynamic configuration (Fig. 1.1 – 2). However, when the center of pressure of the missile is in front of its center of gravity, the existing angle of attack will continuously increase under the action of its corresponding destabilizing aerodynamic moment. Therefore, the missile is in a divergent state. This aerodynamic layout is called a statically unstable aerodynamic configuration (Fig. 1.1 – 3).

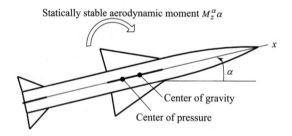

Fig. 1.1 – 2 Missile in a statically stable aerodynamic configuration

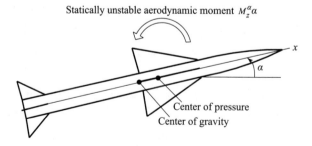

Fig. 1.1 – 3 Missile in a statically unstable aerodynamic configuration

In general, there are three types of aerodynamic configurations for the generation of a missile control moment:

1) Normal aerodynamic configuration

In this aerodynamic configuration, the missile actuator is arranged at the tail of the missile (Fig. 1.1 -4). The benefit of this configuration is that when the control moment is balanced by the angle of attack produced moment, the control surface incident angle is the difference between the control surface deflection angle and the angle of attack, which is the most efficient way of using the control deflection angle, thus allowing the use of a larger control surface deflection and a larger angle of attack for maneuvering. But the drawback is that the position of the actuator in this configuration coincides with the rear end motor, which places certain restrictions on the size of the actuator. In addition, when the missile is to maneuver, the control surface force is in the opposite direction to the angle of attack produced normal force, which will cause some total normal force loss. However, taking these advantages and disadvantages into account, this configuration is still the most commonly used aerodynamic configuration for tactical missiles.

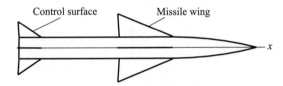

Fig. 1.1 -4 Normal aerodynamic configuration

2) Canard aerodynamic configuration

In this aerodynamic configuration, the actuator is positioned at the head of the missile (Fig. 1.1 - 5). The benefit of this arrangement is that the missile motor can be arranged independently, avoiding the need to contend for space with other subsystems. In addition, when the missile is to maneuver, the control surface force is in the same direction as the angle of attack produced normal force, thus achieving higher maneuvering force utilization efficiency. However, in this configuration, the actuator incident angle is the sum of the actuator deflection angle and the missile angle of attack. As the maximum allowed control surface incident angle is limited, a large angle of attack maneuvering cannot be achieved. Therefore, nowadays this aerodynamic configuration is less commonly seen in missile applications.

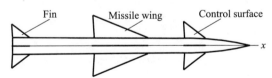

Fig. 1.1 -5 Canard aerodynamic configuration

3) Moving-wing scheme

With this aerodynamic configuration, the missile wing can be turned as a control surface (Fig. 1.1 -6), and the full center of pressure is positioned in front of the center of gravity, similar to the canard aerodynamic configuration but with a short control arm. However, the required lift for missile maneuvering is essentially provided by the wing deflection. This is because the wing has a very large lifting surface. For this reason, the angle of attack required for missile maneuvering could be small. Therefore, it is particularly suitable to be used when the missile turbine engine for cruising flight is not allowed to work at a large angle of attack. However, due to the higher power requirement for the wing actuator, its operating frequency bandwidth is limited, and so is the response speed of the related autopilot. For this reason, this aerodynamic configuration is rarely used nowadays in engineering practice.

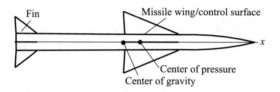

Fig. 1.1-6 Moving-wing aerodynamic configuration

Fig. 1.1 - 7 and Fig. 1.1 - 8 show the situations in which the control moment and the aerodynamic moment are in an equilibrium state when there is a steady state angle of attack for statically stable and statically unstable missiles.

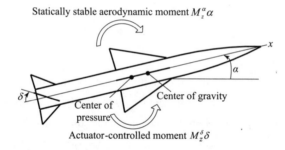

Fig. 1.1 -7 Moment equilibrium of a statically stable missile

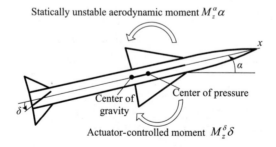

Fig. 1.1 -8 Moment equilibrium of a statically unstable missile

It is noteworthy that for a statically stable missile, the control moment generated by the actuator deflection angle δ will make the missile rotate in the required direction to produce an angle of attack. When the aerodynamic stabilizing moment that increases with the angle of attack increases to the same level as the control moment, the corresponding angle of attack will be at an equilibrium state. Therefore, missiles with sufficient static stability can also be designed without autopilot.

However, this type of aerodynamic feedback solution has less precise missile normal acceleration control compared with an acceleration autopilot solution. But for statically unstable missiles, a steady state angle of attack can only be generated through autopilot closed-loop control to maintain a required equilibrium between the control moment and aerodynamic moment generated by the angle of attack.

As mentioned above, a steady state angle of attack α is achieved when the control moment and the aerodynamic moment are in equilibrium, that is

$$\underset{(\text{Control moment})}{M_z^\delta \cdot \delta} = \underset{(\text{Aerodynamic moment})}{M_z^\alpha \cdot \alpha}$$

The transfer function with the actuator deflection angle δ as the input and the angle of attack α as the output, shown below, can be regarded as the object being controlled for the autopilot (Fig. 1.1-9).

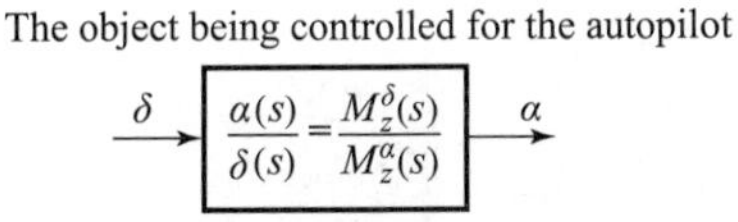

The object being controlled for the autopilot

$$\delta \rightarrow \boxed{\frac{\alpha(s)}{\delta(s)} = \frac{M_z^\delta(s)}{M_z^\alpha(s)}} \rightarrow \alpha$$

Fig. 1.1-9 The object being controlled for the autopilot

The missile's static stability is directly proportional to the distance between its center of gravity and its center of pressure, and this distance is small for missiles with low static stability. Therefore, when center of gravity or center of pressure of the missile with low static stability deviates from its designed value, the value of M_z^δ and the gain of the transfer function $\dfrac{M_z^\alpha}{M_z^\delta}$ from the actuator δ to the angle of attack α will change greatly from its designed value, which means that the open loop gain of the autopilot loop will also change greatly. This is unacceptable for a normally designed control loop. Therefore, to reduce the autopilot open loop gain change, the missile static stability is often taken at around 4% - 8%. For missiles that must have low static stability aerodynamic configurations for other considerations, the gain from δ to α could be stabilized by designing a pseudo angle of attack feedback loop. For a detailed discussion of this option, see the autopilot design section.

At present, a skid-to-turn (STT) control scheme is adopted in most tactical missiles. That is, in the Cartesian coordinate system, a missile pitch turn is achieved by the generation of angle of attack α, and a yaw turn is achieved by the generation of sideslip angle β, as shown in Fig. 1.1-10.

Such a control scheme has a very fast response, but it is necessary to be roll-stabilized. Clearly, STT is most suitable for aerodynamically symmetrical missiles.

Another control scheme is bank-to-turn (BTT). This scheme is generally used for surface-symmetrical missiles, especially when there is a big difference between the pitch and yaw lift surface areas of the missile. In this scheme, the missile must turn the main lift surface by an angle φ with the help of a roll control autopilot to have the missile angle of attack in the required maneuvering direction (Fig. 1.1-11).

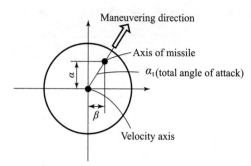

Fig. 1.1 – 10   Skid-to-turn (STT) polar diagram

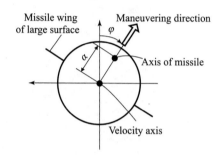

Fig. 1.1 – 11   Bank-to-turn (BTT) polar diagram

For the missile with BTT control, when the missile maneuvering direction needs to be changed, it is possible that the missile has to roll a large roll angle to a new direction, clearly this leads to a slow missile maneuvering response. For this reason, BTT control is more suitable for missile midcourse guidance phase.

2

Missile Mathematical Models

§ 2.1 Symbols and Definitions

The origin of the missile body coordinate system $Ox_b y_b z_b$ is defined at the center of gravity of the missile, and each axis is defined as follows (suppose that the missile is an axisymmetric or plane-symmetric rigid body, see Fig. 2.1 – 1)

Roll axis Ox_b: lies in the symmetry plane. Pointing forward is positive.

Yaw axis Oy_b: located in the symmetry plane of the missile body, with upwards as the positive direction.

Pitch axis Oz_b: forms the right-handed rectangular coordinate system together with axes Ox_b and Oy_b.

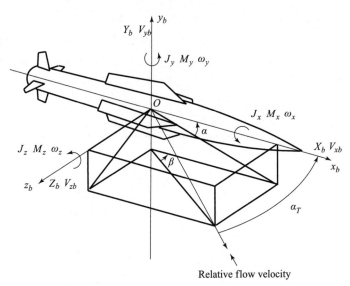

Fig. 2.1 – 1 Definitions of aerodynamic force, moment, etc., of the missile
NOTE: O is the center of gravity of the missile

Table 2.1 – 1 defines the symbols for aerodynamic forces, moments acting on the missile, linear velocities, and angular velocities, as well as moments of inertia (as shown in Fig. 2.1 – 1). The moment of inertia around each axis is defined as

$$J_x = \sum m_i(y_i^2 + z_i^2), \qquad (2.1-1)$$
$$J_y = \sum m_i(z_i^2 + x_i^2), \qquad (2.1-2)$$
$$J_z = \sum m_i(x_i^2 + y_i^2). \qquad (2.1-3)$$

The product of inertia around each axis is defined as

$$J_{yz} = \sum m_i y_i z_i, \qquad (2.1-4)$$
$$J_{zx} = \sum m_i z_i x_i, \qquad (2.1-5)$$
$$J_{xy} = \sum m_i x_i y_i. \qquad (2.1-6)$$

The plane $Ox_b y_b$ is the pitch plane and the plane $Ox_b z_b$ is the yaw plane. The relevant angles are defined as follows:

α —angle of attack in the pitch plane;
β —angle of attack in the yaw plane (angle of sideslip);
α_T —total angle of attack;
λ —angle of attack plane angle.

Therefore

$$\tan\alpha = \tan\alpha_T \cdot \cos\lambda, \qquad (2.1-7)$$
$$\tan\beta = \tan\alpha_T \cdot \sin\lambda. \qquad (2.1-8)$$

That is,

$$\alpha = \arctan(\tan\alpha_T \cdot \cos\lambda), \qquad (2.1-9)$$
$$\beta = \arctan(\tan\alpha_T \cdot \sin\lambda). \qquad (2.1-10)$$

The axial velocity of the missile body V_{xb} is a large but slowly varying variable, and its variation is usually less than a few percents per second. However, the angular velocity $\omega_x, \omega_y, \omega_z$ and velocity components V_{yb}, V_{zb} of the pitch and yaw axes are usually small. They can be positive or negative, and they can have large rates of changes.

Table 2.1-1 Definition of symbols

Item	Roll axis x_b	Yaw axis y_b	Pitch axis z_b
Angular velocity (missile body coordinate system)	ω_x	ω_y	ω_z
Velocity component (missile body coordinate system)	V_{xb}	V_{yb}	V_{zb}
Forces acting on the missile (missile body coordinate system)	X_b	Y_b	Z_b
Moments acting on the missile (missile body coordinate system)	M_x	M_y	M_z
Moments of inertia	J_x	J_y	J_z
Product of inertia	J_{yz}	J_{zx}	J_{xy}

§2.2 Euler Equations of the Rigid Body Motion

The six-degree-of-freedom model of a missile motion in space consists of six dynamic equations (three center of gravity motion dynamical equations and three rotational dynamical equations) and six kinematic equations (three center of gravity motion kinematic equations and three rotational kinematic equations).

The coordinate systems involved in the study of missile guidance and control problems include the earth coordinate system, the missile body coordinate system, the trajectory coordinate system and the velocity coordinate system. The x-axis of the last two coordinate systems coincides with the missile velocity vector. However, the y-axis of the trajectory coordinate system is in the vertical plane, and the y-axis of the velocity coordinate system is in the longitudinal symmetrical plane of the missile body. The transformation between the four coordinate systems can be accomplished by a series of rotations (Fig. 2.2 − 1). Detailed descriptions of these coordinate systems can be found in general flight dynamics textbooks.

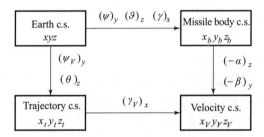

Fig. 2.2 − 1 Transformation from the earth coordinate system to other coordinate systems

NOTE: c. s. is the simple form of coordinate system.

For example, the rotation transformation from the earth coordinate system to the missile body coordinate system is shown in Fig. 2.2 − 2.

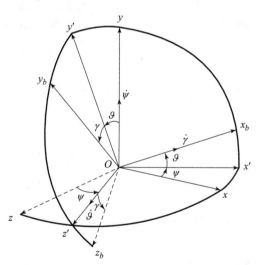

Fig. 2.2 − 2 Relationship between the earth coordinate system and the missile body coordinate system

In the study of coordinate transformation, it is necessary to know three basic coordinate system transformation matrixes about axes x, y, z:

Rotation matrix that does rotation about the x-axis by angle φ_x

$$L_x(\varphi_x) = \begin{bmatrix} 1 & 0 & 0 \\ 0 & \cos\varphi_x & \sin\varphi_x \\ 0 & -\sin\varphi_x & \cos\varphi_x \end{bmatrix}.$$

Rotation matrix that does rotation about the y-axis by angle φ_y

$$L_y(\varphi_y) = \begin{bmatrix} \cos\varphi_y & 0 & -\sin\varphi_y \\ 0 & 1 & 0 \\ \sin\varphi_y & 0 & \cos\varphi_y \end{bmatrix}.$$

Rotation matrix that does rotation about the z-axis by angle φ_z

$$L_z(\varphi_z) = \begin{bmatrix} \cos\varphi_z & \sin\varphi_z & 0 \\ -\sin\varphi_z & \cos\varphi_z & 0 \\ 0 & 0 & 1 \end{bmatrix}.$$

Define the following variables:

V ——missile body velocity;

ψ_V, θ ——missile flight path angle;

γ_V ——missile symmetrical plane deflection angle;

ψ, ϑ, γ ——missiles' yaw angle, pitch angle, roll angle;

α, β ——angle of attack, angle of sideslip;

V_x, V_y, V_z ——velocity component (earth coordinate system);

V_{xb}, V_{yb}, V_{zb} ——velocity component (missile body coordinate system);

ω_x, ω_y, ω_z ——missile angular velocity component (missile body coordinate system);

F_{xt}, F_{yt}, F_{zt} ——resultant force component acting on the missile (trajectory coordinate system);

F_{xb}, F_{yb}, F_{zb} ——resultant force component acting on the missile (missile body coordinate system).

The total force F acting on the missile consists of aerodynamic force $R = \begin{bmatrix} X_b \\ Y_b \\ Z_b \end{bmatrix}$ (missile body coordinate system), thrust $P = \begin{bmatrix} P \\ 0 \\ 0 \end{bmatrix}$ (missile body coordinate system), and gravity $G = \begin{bmatrix} 0 \\ -G \\ 0 \end{bmatrix}$ (earth coordinate system). The moment acting on the missile body is $M = \begin{bmatrix} M_x \\ M_y \\ M_z \end{bmatrix}$ (missile body coordinate system). The projections of related components in other coordinate systems are shown in Table 2.2-1.

Table 2.2-1 Related projections in different coordinate systems

Item	Earth coordinate system	Trajectory coordinate system	Missile body coordinate system
Aerodynamic force R		$L_x(-\gamma_v)L_y(-\beta)L_z(-\alpha)\begin{bmatrix} X_b \\ Y_b \\ Z_b \end{bmatrix}$	$\begin{bmatrix} X_b \\ Y_b \\ Z_b \end{bmatrix}$
Gravity G	$\begin{bmatrix} 0 \\ -G \\ 0 \end{bmatrix}$	$L_z(\theta)L_y(\psi_v)\begin{bmatrix} 0 \\ -G \\ 0 \end{bmatrix}$	$L_x(\gamma)L_z(\vartheta)L_y(\psi)\begin{bmatrix} 0 \\ -G \\ 0 \end{bmatrix}$
Thrust P		$L_x(-\gamma_v)L_y(-\beta)L_z(-\alpha)\begin{bmatrix} P \\ 0 \\ 0 \end{bmatrix}$	$\begin{bmatrix} P \\ 0 \\ 0 \end{bmatrix}$
Resultant force F ($F = R + G + P$)		$\begin{bmatrix} F_{xt} \\ F_{yt} \\ F_{zt} \end{bmatrix}$	$\begin{bmatrix} F_{xb} \\ F_{yb} \\ F_{zb} \end{bmatrix}$
Aerodynamic moment M			$\begin{bmatrix} M_x \\ M_y \\ M_z \end{bmatrix}$
Velocity V	$\begin{bmatrix} V_x \\ V_y \\ V_z \end{bmatrix} = L_y(-\psi)L_z(-\vartheta)L_x(-\gamma)\begin{bmatrix} V_{xb} \\ V_{yb} \\ V_{zb} \end{bmatrix}$		$\begin{bmatrix} V_{xb} \\ V_{yb} \\ V_{zb} \end{bmatrix}$

The six-degree-of-freedom missile model can be given in the trajectory or the missile body coordinate system. When the six-degree-of-freedom model is given in the trajectory coordinate system (Equation (2.2-1)), the state variables of the three dynamic translational and three-rotational equations are taken as the velocity V,θ,ψ_V (trajectory coordinate system) and the angular velocity $\omega_x,\omega_y,\omega_z$ (missile body coordinate system). The state variables of the six kinematic equations are respectively taken as the position component x,y,z (earth coordinate system) and the Euler angle ϑ,ψ,γ (missile body coordinate system). Other dependent derived parameters include α,β and γ_V.

$$m\dot{V} = F_{xt},$$
$$mV\dot{\theta} = F_{yt},$$
$$-mV\cos\theta\,\dot{\psi}_V = F_{zt},$$
$$J_x\dot{\omega}_x - (J_y - J_z)\omega_y\omega_z - J_{yz}(\omega_y^2 - \omega_z^2) - J_{zx}(\dot{\omega}_z + \omega_x\omega_y) - J_{xy}(\dot{\omega}_y - \omega_x\omega_z) = M_x,$$
$$J_y\dot{\omega}_y - (J_z - J_x)\omega_z\omega_x - J_{zx}(\omega_z^2 - \omega_x^2) - J_{xy}(\dot{\omega}_x + \omega_y\omega_z) - J_{yz}(\dot{\omega}_z - \omega_y\omega_x) = M_y,$$
$$J_z\dot{\omega}_z - (J_x - J_y)\omega_x\omega_y - J_{xy}(\omega_x^2 - \omega_y^2) - J_{yz}(\dot{\omega}_y + \omega_z\omega_x) - J_{zx}(\dot{\omega}_x - \omega_z\omega_y) = M_z,$$

Tactical Missile Guidance and Control System Design

$$\dot{x} = V\cos\theta\cos\psi_V,$$
$$\dot{y} = V\sin\theta,$$
$$\dot{z} = -V\cos\theta\sin\psi_V, \qquad (2.2-1)$$
$$\dot{\vartheta} = \omega_y\sin\gamma + \omega_z\cos\gamma,$$
$$\dot{\psi} = (\omega_y\cos\gamma - \omega_z\sin\gamma)/\cos\vartheta,$$
$$\dot{\gamma} = \omega_x - \tan\vartheta(\omega_y\cos\gamma - \omega_z\sin\gamma),$$
$$\sin\beta = \cos\theta[\cos\gamma\sin(\psi-\psi_V) + \sin\vartheta\sin\gamma\sin(\psi-\psi_V)] - \sin\theta\cos\vartheta\sin\gamma,$$
$$\sin\alpha = \{\cos\theta[\sin\vartheta\cos\gamma\cos(\psi-\psi_V) - \sin\gamma\sin(\psi-\psi_V)] - \sin\theta\cos\vartheta\cos\gamma\}/\cos\beta,$$
$$\sin\gamma_V = (\cos\alpha\sin\beta\sin\vartheta - \sin\alpha\sin\beta\cos\gamma\cos\vartheta + \cos\beta\sin\gamma\cos\vartheta)/\cos\theta.$$

When the six-degree-of-freedom model of the missile is given in the missile body coordinate system (Equation (2.2-2)), aside from the state variables of the three translational dynamic equations changing to the velocity components V_{xb}, V_{yb}, V_{zb} (missile body coordinate system), the remaining state variables are the same as the trajectory system. That is, the state variables of the three dynamic rotational equations are taken as the angular velocity components $\omega_x, \omega_y, \omega_z$ (missile body coordinate system). The state variables of the six kinematic equations are taken as the position components x, y, z (earth coordinate system) and the Euler angle ϑ, ψ, γ (missile body coordinate system), respectively. Other useful dependent derived parameters are $V_x, V_y, V_z, V, \theta, \psi_V, \alpha, \beta$ and γ_V.

$$m(\dot{V}_{xb} + V_{zb}\omega_y - V_{yb}\omega_z) = F_{xb} = X_b - G\sin\vartheta + P,$$
$$m(\dot{V}_{yb} + V_{xb}\omega_z - V_{zb}\omega_x) = F_{yb} = Y_b - G\cos\vartheta\cos\gamma,$$
$$m(\dot{V}_{zb} + V_{yb}\omega_x - V_{xb}\omega_y) = F_{zb} = Z_b + G\cos\vartheta\sin\gamma,$$
$$J_x\dot{\omega}_x - (J_y - J_z)\omega_y\omega_z - J_{yz}(\omega_y^2 - \omega_z^2) - J_{zx}(\dot{\omega}_z + \omega_x\omega_y) - J_{xy}(\dot{\omega}_y - \omega_x\omega_z) = M_x,$$
$$J_y\dot{\omega}_y - (J_z - J_x)\omega_z\omega_x - J_{zx}(\omega_z^2 - \omega_x^2) - J_{xy}(\dot{\omega}_x + \omega_y\omega_z) - J_{yz}(\dot{\omega}_z - \omega_y\omega_x) = M_y,$$
$$J_z\dot{\omega}_z - (J_x - J_y)\omega_x\omega_y - J_{xy}(\omega_x^2 - \omega_y^2) - J_{yz}(\dot{\omega}_y + \omega_z\omega_x) - J_{zx}(\dot{\omega}_x - \omega_z\omega_y) = M_z,$$
$$\dot{x} = \cos\psi\cos\vartheta V_{xb} - (\cos\psi\sin\vartheta\cos\gamma - \sin\psi\sin\gamma)V_{yb} + (\cos\psi\sin\vartheta\sin\gamma + \sin\psi\cos\gamma)V_{zb},$$
$$\dot{y} = \sin\vartheta V_{xb} + \cos\vartheta\cos\gamma V_{yb} - \cos\vartheta\sin\gamma V_{zb}, \qquad (2.2-2)$$
$$\dot{z} = -\sin\psi\cos\vartheta V_{xb} + (\sin\psi\sin\vartheta\cos\gamma + \cos\psi\sin\gamma)V_{yb} - (\sin\psi\sin\vartheta\sin\gamma - \cos\psi\cos\gamma)V_{zb},$$
$$\dot{\vartheta} = \omega_y\sin\gamma + \omega_z\cos\gamma,$$
$$\dot{\psi} = (\omega_y\cos\gamma - \omega_z\sin\gamma)/\cos\vartheta,$$
$$\dot{\gamma} = \omega_x - \tan\vartheta(\omega_y\cos\gamma - \omega_z\sin\gamma),$$
$$V = \sqrt{V_{xb}^2 + V_{yb}^2 + V_{zb}^2} = \sqrt{V_x^2 + V_y^2 + V_z^2} \text{ (Expressions of } V_x, V_y, V_z \text{ are shown in Table 2.2-1)},$$
$$\theta = \arctan(V_y/\sqrt{V_x^2 + V_z^2}),$$
$$\psi_V = \arctan(-V_z/V_x),$$
$$\alpha = \arctan(-V_{yb}/V_{xb}),$$
$$\beta = -\arcsin(V_{zb}/V),$$
$$\gamma_V = \arcsin[(\cos\alpha\sin\beta\sin\vartheta - \sin\alpha\sin\beta\cos\gamma\cos\vartheta + \cos\beta\sin\gamma\cos\vartheta)/\cos\theta].$$

Typically, aerodynamic force **R** and moment **M** are functions of Mach number Ma, angle of attack α, angle of sideslip β, three-channel control surface deflection angles $\delta_x, \delta_y, \delta_z$, and three angular velocities ω_x, ω_y, ω_z.

$$R = R(Ma, \alpha, \beta, \delta_x, \delta_y, \delta_z),$$
$$M = M(Ma, \alpha, \beta, \delta_x, \delta_y, \delta_z, \omega_x, \omega_y, \omega_z).$$

The exact expression of these functions and their reasonable simplification can be obtained through wind tunnel tests and test data analysis.

Missile guidance and control is achieved through control surface deflection δ_x, δ_y, δ_z commanded by guidance and control laws. Models with guidance and control will contain more equations. For example, the mathematical model of the entire system will also include the seeker dynamic mathematical model, autopilot mathematical model, command guidance radar mathematical model, the control surface servo mechanism mathematical model, etc.

The above dynamic equations can often be simplified in specific mathematical simulations. For example, for axisymmetric missiles, their cross inertia moments J_{xy}, J_{yz}, J_{zx} can be safely omitted. For three-channel control missiles, the related angular velocity ω_x, ω_y, ω_z are so small that their product $\omega_x\omega_y$, $\omega_y\omega_z$, $\omega_z\omega_x$, ω_x^2, ω_y^2, ω_z^2 can also be omitted. Furthermore, since the projections of the velocity vectors on the missile body coordinate system V_{yb}, V_{zb} are also of a small quantity, their product with the component of ω can also be omitted. Therefore, the dynamic equations represented in the missile body coordinate system can be simplified as

$$m\dot{V}_{xb} = F_{xb}, \tag{2.2-3}$$
$$m(\dot{V}_{yb} + V_{xb} \cdot \omega_z) = F_{yb}, \tag{2.2-4}$$
$$m(\dot{V}_{zb} - V_{xb} \cdot \omega_y) = F_{zb}, \tag{2.2-5}$$
$$J_x \cdot \dot{\omega}_x = M_x, \tag{2.2-6}$$
$$J_y \cdot \dot{\omega}_y = M_y, \tag{2.2-7}$$
$$J_z \cdot \dot{\omega}_z = M_z. \tag{2.2-8}$$

When presented in the trajectory coordinate system, the above translational dynamic equations can be described as

$$m\dot{V} = F_{xt}, \tag{2.2-9}$$
$$mV\dot{\theta} = F_{yt}, \tag{2.2-10}$$
$$mV\cos\theta\dot{\psi}_V = -F_{zt}. \tag{2.2-11}$$

§2.3　Configuration of the Control Surfaces

The sequential numbering of the control surface is shown in Fig. 2.3 – 1. The deflection angles δ_1, δ_2, δ_3, δ_4 generated by turning clockwise along each coordinate axis positive directions are defined as positive. The respective deflection angles are defined as follows

Roll control deflection angle: $\delta_x = \dfrac{1}{4}(\delta_1 + \delta_2 + \delta_3 + \delta_4)$.

(When only a pair of actuators is moved, there is $\delta_x = (\delta_1 + \delta_3)/2$ or $\delta_x = (\delta_2 + \delta_4)/2$)

Pitch control deflection angle: $\delta_z = \dfrac{1}{2}(\delta_1 - \delta_3)$.

Yaw control deflection angle: $\delta_y = \dfrac{1}{2}(\delta_4 - \delta_2)$.

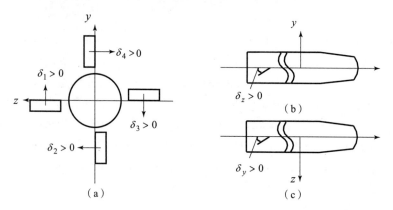

Fig. 2.3 – 1 Definition of control surface angles
(a) Front view; (b) Side view; (c) Top view

Readers can verify that:

A positive roll control deflection angle produces a negative moment around the x-axis. For normal control missiles, a positive pitch control deflection angle δ_z produces a negative pitch moment around the z-axis, and a positive yaw control deflection angle δ_y produces a negative yaw moment around the y-axis. A positive pitch control deflection angle δ_z produces a positive force along the y-axis, and a positive yaw control deflection angle δ_y produces a negative force along the z-axis.

For canard controlled missiles, the pitch and yaw moments produced by the same control channel actuator deflection directions are opposite to the normal control missile, but the force direction remains the same.

§ 2.4 Missile Aerodynamic Derivatives and Dynamic Coefficients

In order to conveniently use the most mature linear time-invariant system (LTI system) for missile control system design, some simplifications have to be made. For this purpose, we first linearize the assumed small-disturbance nonlinear dynamic equations to obtain their linearized time-varying differential equations. Then we assume that the time-varying parameters of the system change slowly in the system transient time and can be taken as constants. In this way the simplified linear time-invariant system can be designed by using various mature control theories (such as the frequency analysis design, root locus design, optimal control and robust control theory design, etc.). In the above-mentioned simplification process, an important task is to linearize the aerodynamic force and moment functions.

Suppose a certain fin-controlled missile flies at sea level with $Ma = 1.5$. The relation between the roll moment M_x, the roll actuator deflection angle δ_x and the total angle of attack α_T is shown in Fig. 2.4 – 1. It can be seen that here M_x is not a strict linear function of δ_x. It can also be seen that the roll actuator produced M_x moment is slightly reduced with the increase of the total angle of attack α_T.

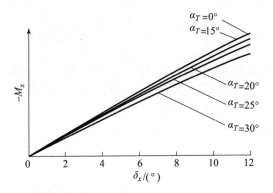

Fig. 2.4-1 Relationship between the roll moment M_x, the roll actuator deflection angle δ_x and the total angle of attack α_T

$M_x^{\delta_x}$ is defined as

$$M_x^{\delta_x} = \frac{\partial M_x}{\partial \delta_x}. \tag{2.4-1}$$

The moment increment ΔM_x caused by small increment $\Delta \delta_x$ is

$$\Delta M_x = M_x^{\delta_x} \cdot \Delta \delta_x. \tag{2.4-2}$$

Among these, the value of the roll moment derivative $M_x^{\delta_x}$ is closely related to the chosen flight condition (the set point on the trajectory). It is noteworthy that since the value of $\Delta \delta_x$ in most applications is only a few degrees, $M_x^{\delta_x}$ is always regarded as a constant value at the selected flight condition.

$M_x^{\omega_x}$ is a roll damping derivative whose dimension is: moment per unit roll angular velocity. Because this moment always prevents the rolling motion, its sign is always negative. For a given Mach number and flight altitude, $M_x^{\omega_x}$ is often considered as a constant. Aside from $M_x^{\delta_x}$ and $M_x^{\omega_x}$, there are no other important roll derivatives.

Let us now consider the aerodynamic derivatives related to the pitch and yaw. The lift Y caused by an angle of attack of the missile is usually expressed as

$$Y = \frac{1}{2}\rho V^2 S C_y. \tag{2.4-3}$$

where ρ is the atmosphere density; S is the characteristic area of the missile body, which is usually taken as the cross-sectional area of the missile body; $C_y(Ma, \alpha, \delta_z)$ is known as the lift coefficient, which is a function of the angle of attack and the actuator deflection angle δ_z for a given Mach number Ma. For symmetrically arranged missiles, $C_z(Ma, \beta, \delta_y)$ is related to the sideslip angle and the actuator deflection angle δ_y, and is equal to the lift coefficient C_y. The related derivatives are defined as follows

$$Y^\alpha = \frac{\partial Y}{\partial \alpha} = \frac{\partial C_y}{\partial \alpha} \cdot \frac{1}{2}\rho V^2 S = C_y^\alpha \cdot \frac{1}{2}\rho V^2 S, \tag{2.4-4}$$

$$Y^{\delta_z} = \frac{\partial Y}{\partial \delta_z} = \frac{\partial C_y}{\partial \delta_z} \cdot \frac{1}{2}\rho V^2 S = C_y^{\delta_z} \cdot \frac{1}{2}\rho V^2 S, \tag{2.4-5}$$

$$Z^\beta = \frac{\partial Z}{\partial \beta} = \frac{\partial C_z}{\partial \beta} \cdot \frac{1}{2}\rho V^2 S = C_z^\beta \cdot \frac{1}{2}\rho V^2 S, \qquad (2.4-6)$$

$$Z^{\delta_y} = \frac{\partial z}{\partial \delta_y} = \frac{\partial C_z}{\partial \delta_y} \cdot \frac{1}{2}\rho V^2 S = C_z^{\delta_y} \cdot \frac{1}{2}\rho V^2 S. \qquad (2.4-7)$$

When designing most wings and control surfaces, the lift generated by a small angle of attack should be proportional to the angle of attack; but the lift generated by slender missile bodies includes two parts: one is proportional to α, the other is proportional to α^2. This situation is common in fin-controlled supersonic missiles, see Fig. 2.4 – 2.

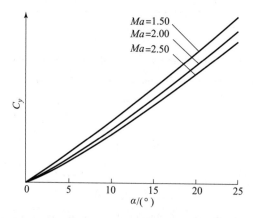

Fig. 2.4 – 2 Relationship of the lift coefficient, angle of attack and Mach number

It should be noted that if there is an angle of attack α and actuator deflection δ_z at the same time, the actual aerodynamic incident angle of the pitch actuator is $\alpha + \delta_z$; however, the total aerodynamic force increment is not $Y^\alpha \cdot \alpha + Y^{\delta_z} \cdot (\alpha + \delta_z)$, but $Y^\alpha \cdot \alpha + Y^{\delta_z} \cdot \delta_z$. This is because the lift, which is generated by the angle of attack α, has already been included in Y^α.

Take the expression of the pitch moment M_z as

$$M_z = \frac{1}{2}\rho V^2 SLm_z, \qquad (2.4-8)$$

where $m_z(Ma, \alpha, \delta_z, \omega_z)$ is called the pitch moment coefficient, L is the characteristic length of the missile (usually taken as the missile length), and the derivatives related to the pitch moment are defined as follows

$$M_z^\alpha = \frac{\partial M_z}{\partial \alpha} = \frac{1}{2}\rho V^2 SLm_z^\alpha, \quad m_z^\alpha = \frac{\partial m_z}{\partial \alpha}, \qquad (2.4-9)$$

$$M_z^{\delta_z} = \frac{\partial M_z}{\partial \delta_z} = \frac{1}{2}\rho V^2 SLm_z^{\delta_z}, \quad m_z^{\delta_z} = \frac{\partial m_z}{\partial \delta_z}, \qquad (2.4-10)$$

$$M_z^{\omega_z} = \frac{\partial M_z}{\partial \omega_z} = \frac{1}{2}\rho V^2 SLm_z^{\omega_z}, \quad m_z^{\omega_z} = \frac{\partial m_z}{\partial \omega_z}. \qquad (2.4-11)$$

where M_z^α is the product of the aerodynamic force derivative Y^α and the distance from the center of gravity to the center of pressure, and the distance between the center of gravity and the center of pressure is called the static stability. In essence, it is an aerodynamic anti-disturbance parameter which indicates the amount of the missile's static stability. If the center of pressure is behind the center of gravity, an aerodynamic restoration moment will be generated to help reduce the angle of attack and stabilize the missile body when there is an angle of attack disturbance. Conversely, if the center of pressure is in front of the center of gravity, then the angle of attack caused by any disturbance will get larger and larger. Such a missile body is called a statically unstable body.

In general, if the position of the center of gravity is approximately 50% of the missile length from the head of the missile, we expect that the center of pressure of the missile body will be placed approximately 52% – 55% of the missile length from the head of the missile. It can be seen that at subsonic and low supersonic speeds, the position of the center of pressure is farther ahead than that in the case of a high Mach number. In addition, at low speeds, the position of the center of pressure is greatly influenced by the angle of attack. This is mainly due to the fact that the center of pressure component of the missile body without wings or fins often moves backward as the angle of attack increases, but the center of pressure components of the control surface and the missile wing do not change much. Unfortunately, the position of the center of pressure is also a function of the angle of attack plane angle λ that characterizes the roll direction of the total angle of attack α_T plane.

$M_z^{\delta_z}$ is the moment derivative due to the pitch actuator deflection, which is equal to the actuator force derivative Y^{δ_z} multiplied by the distance from the center of gravity to the pitch actuator center of pressure ℓ_c. It is clear that if Y^{δ_z} is a constant, $M_z^{\delta_z}$ will only change when the center of gravity moves.

$M_z^{\omega_z}$ is the pitch damping moment derivative, which is equal to the aerodynamic moment produced by unit pitch angular velocity. This derivative is a small term not sensitive to the angle of attack.

To facilitate the control system design, the aerodynamic force derivative is usually divided by the product mV of the missile mass and speed, and the aerodynamic moment derivative is divided by its respective moment of inertia. Doing this we will have all the dynamic coefficients associated with the missile control system design which are given in Table 2.4 – 1 and Table 2.4 – 2. These are the dynamic coefficients a_α, a_δ, a_ω that are related to the missile pitch and yaw rotation, the dynamic coefficients b_α, b_δ which are related to the missile translational motion, and the roll relative rotational dynamic coefficients c_δ, c_ω.

Table 2.4 – 1 Symbols and dimensions of the main aerodynamic derivatives

The main aerodynamic derivatives			The main dynamic coefficients			
Symbols	Algebraic sign	Dimensions	Symbols	Algebraic sign	Dimensions	Physical meaning
M_z^α	– ve (Statically stable missile body); + ve (Statically unstable missile body)	N · m	$a_\alpha = \dfrac{-M_z^\alpha}{J_z}$	+ ve (Statically stable missile body); – ve (Statically unstable missile body)	s^{-2}	$-\dfrac{\partial \dot\omega_z}{\partial \alpha}$
$M_z^{\delta_z}$	– ve (Fin-controlled); + ve (Canard-controlled)	N · m	$a_\delta = \dfrac{-M_z^{\delta_z}}{J_z}$	+ ve (Fin-controlled); – ve (Canard-controlled)	s^{-2}	$-\dfrac{\partial \dot\omega_z}{\partial \delta_z}$
$M_z^{\omega_z}$	– ve	N · m · s	$a_\omega = \dfrac{-M_z^{\omega_z}}{J_z}$	+ ve	s^{-1}	$-\dfrac{\partial \dot\omega_z}{\partial \omega_z}$

Continued

The main aerodynamic derivatives			The main dynamic coefficients			
Symbols	Algebraic sign	Dimensions	Symbols	Algebraic sign	Dimensions	Physical meaning
Y^α	+ ve	N	$b_\alpha = \dfrac{P + Y^\alpha}{mV}$	+ ve	s^{-1}	$\dfrac{\partial \dot\theta}{\partial \alpha}$
Y^{δ_z}	+ ve	N	$b_\delta = \dfrac{Y^{\delta_z}}{mV}$	+ ve	s^{-1}	$\dfrac{\partial \dot\theta}{\partial \delta_z}$
$M_x^{\delta_x}$	− ve	N · m	$c_\delta = \dfrac{-M_x^{\delta_x}}{J_x}$	+ ve	s^{-2}	$-\dfrac{\partial \dot\omega_x}{\partial \delta_x}$
$M_x^{\omega_x}$	− ve	N · m · s	$c_\omega = \dfrac{-M_x^{\omega_x}}{J_x}$	+ ve	s^{-1}	$-\dfrac{\partial \dot\omega_x}{\partial \omega_x}$

Table 2.4−2   Main dynamic coefficient expressions

Symbols	Expressions	Notes
$a_\alpha = \dfrac{-M_z^\alpha}{J_z}$	$\dfrac{-m_z^\alpha qSL}{J_z} = \dfrac{-m_z^\alpha \rho V^2 SL}{2J_z}$	$a_\alpha > 0$, $m_z^\alpha < 0$, Statically stable missile body; $a_\alpha < 0$, $m_z^\alpha > 0$, Statically unstable missile body
$a_\delta = \dfrac{-M_z^{\delta_z}}{J_z}$	$\dfrac{-m_z^{\delta_z} qSL}{J_z} = \dfrac{-m_z^{\delta_z} \rho V^2 SL}{2J_z}$	$a_\delta > 0$, $m_z^{\delta_z} < 0$, Fin-controlled; $a_\delta < 0$, $m_z^{\delta_z} > 0$, Canard-controlled
$a_\omega = \dfrac{-M_z^{\omega_z}}{J_z}$	$\dfrac{-m_z^{\omega_z} qSL \cdot \dfrac{L}{V}}{J_z} = \dfrac{-m_z^{\omega_z} \rho VSL^2}{2J_z}$	$a_\omega > 0$, $m_z^{\omega_z} < 0$
$b_\alpha = \dfrac{P + Y^\alpha}{mV}$	$\dfrac{P + c_y^\alpha qS}{mV} = \dfrac{2P + c_y^\alpha \rho V^2 S}{2mV}$	$b_\alpha > 0$, $c_y^\alpha > 0$
$b_\delta = \dfrac{Y^{\delta_z}}{mV}$	$\dfrac{c_y^{\delta_z} qS}{mV} = \dfrac{c_y^{\delta_z} \rho VS}{2m}$	$b_\delta > 0$, $c_y^{\delta_z} > 0$
$c_\delta = \dfrac{-M_x^{\delta_x}}{J_x}$	$\dfrac{-m_x^{\delta_x} qSL}{J_x} = \dfrac{-m_x^{\delta_x} \rho V^2 SL}{2J_x}$	$c_\delta > 0$, $m_x^{\delta_x} < 0$
$c_\omega = \dfrac{-M_x^{\omega_x}}{J_x}$	$\dfrac{-m_x^{\omega_x} qSL}{J_x} = \dfrac{-m_x^{\omega_x} \rho V^2 SL}{2J_x}$	$c_\omega > 0$, $m_x^{\omega_x} < 0$

The physical meaning of the dynamic coefficients defined above is explained below. The first thing to remember is that in order to facilitate the use of these coefficients in the missile body transfer function study in the next section, we have deliberately defined the following:

(1) All aerodynamic coefficients (a_α, a_δ, a_ω, b_α, b_δ, c_δ, c_ω) of a statically stable and normally controlled missile are positive.

(2) For canard-controlled missiles, only $a_\delta < 0$, the others are positive.

(3) For statically unstable missiles, only $a_\alpha < 0$, the others are positive.

The physical meaning of a_α is the missile pitch angular acceleration produced by the unit angle of attack. Its unit is $(rad/s^2)/rad$, or s^{-2}. It reflects the level of the missile's static stability.

a_δ is the missile angular acceleration producted by unit control of surface deflection, and its unit is $(rad/s^2)/rad$ or s^{-2}. It reflects the control surface's efficiency in controlling missile rotation.

a_ω is the missile angular acceleration produced by unit missile angular velocity. Its unit is $(rad/s^2)/(rad/s)$ or s^{-1}. It reflects the amount of the missile aerodynamic damping.

b_α is the missile's velocity vector rotation angular velocity produced by unit angle of attack, and its unit is $(rad/s)/rad$ or s^{-1}. It is a very important aerodynamic derivative which characterizes the maneuvering ability of the missile by using the angle of attack producing normal force to rotate the velocity vector.

b_δ is the missile velocity vector rotation angular velocity generated by unit control surface deflection and its unit is $(rad/s)/rad$, or s^{-1}. Since its value is relatively small compared with b_α, it has limited contribution to missile velocity rotation. However, as the rotational control moment is equal to the product of the actuator force and the distance between the actuator's center of pressure and the missile center of gravity, it has a direct impact on a_δ value.

c_δ is the missile roll angle acceleration caused by unit roll actuator deflection, and its unit is $(rad/s^2)/rad$ or s^{-2}. It reflects the control efficiency of the missile's roll actuator.

c_ω is the roll angle acceleration of the missile caused by unit missile roll angular velocity, and it is expressed in unit of $(rad/s^2)/(rad/s)$ or s^{-1}. It reflects the size of the missile's roll damping.

§2.5 Aerodynamic Transfer Functions of the Missile

As mentioned in the previous section, the missile dynamic equations can be simplified to a set of linear time-invariant differential equations at a given trajectory set point by using small disturbance, linearization and constant parameter assumptions. With this simplification, the transfer function of the missile body as an object being controlled will be able to be studied as follows. Although the differential equations are discussed for small disturbance variables, for the sake of brevity, i.e., not making the symbol too complicated, the general practice is to omit the small disturbance symbol in front of both the state variables and the control variables in the differential equations.

For axisymmetric missiles, the pitch channel and the yaw channel are symmetric, so we only discuss the pitch channel transfer functions.

The simplified missile pitch channel differential equations are given below.

$$\ddot{\vartheta} = -a_\omega \cdot \dot{\vartheta} - a_\alpha \cdot \alpha - a_\delta \cdot \delta_z, \qquad (2.5-1)$$

$$\dot{\theta} = b_\alpha \cdot \alpha + b_\delta \cdot \delta_z, \qquad (2.5-2)$$

$$\alpha = \vartheta - \theta. \qquad (2.5-3)$$

This set of equations is a 3-state variable differential equation system. The system control variable is δ_z, the independent state variables are $\vartheta, \dot{\vartheta}$ and θ, and the derived dependent state variable is $\alpha = \vartheta - \theta$. So the equations above can be changed to

$$\frac{d\vartheta}{dt} = \dot{\vartheta}, \tag{2.5-4}$$

$$\frac{d\dot{\vartheta}}{dt} = -a_\omega \cdot \dot{\vartheta} - a_\alpha \cdot (\vartheta - \theta) - a_\delta \cdot \delta_z, \tag{2.5-5}$$

$$\frac{d\theta}{dt} = b_\alpha \cdot (\vartheta - \theta) + b_\delta \cdot \delta_z. \tag{2.5-6}$$

That is

$$\frac{d}{dt}\begin{bmatrix}\vartheta\\ \dot{\vartheta}\\ \theta\end{bmatrix} = \begin{bmatrix}0 & 1 & 0\\ -a_\alpha & -a_\omega & a_\alpha\\ b_\alpha & 0 & -b_\alpha\end{bmatrix}\begin{bmatrix}\vartheta\\ \dot{\vartheta}\\ \theta\end{bmatrix} + \begin{bmatrix}0\\ -a_\delta\\ b_\delta\end{bmatrix}\delta_z, \tag{2.5-7}$$

$$\alpha = \vartheta - \theta. \tag{2.5-8}$$

The important transfer functions of the pitch channel obtained from the pitch state equations are:

(1) The transfer function from the pitch actuator to the missile normal acceleration $a_y(s)/\delta_z(s)$. The transfer function of $a_y(s)/\delta_z(s)$ is derived with the help of the relation $a_y = V \cdot \dot{\theta}$.

$$\frac{a_y(s)}{\delta_z(s)} = -V \cdot \frac{-b_\delta s^2 - a_\omega b_\delta s + (a_\delta b_\alpha - a_\alpha b_\delta)}{s^2 + (a_\omega + b_\alpha)s + (a_\alpha + a_\omega b_\alpha)}, \tag{2.5-9}$$

$$\frac{a_y(s)}{\delta_z(s)} = \frac{k_a(A_2 s^2 + A_1 s + 1)}{T_m^2 s^2 + 2\mu_m T_m s + 1}, \tag{2.5-10}$$

where,

$$k_a = -\frac{V(a_\delta b_\alpha - a_\alpha b_\delta)}{a_\alpha + a_\omega b_\alpha}\ ((\text{m}\cdot\text{s}^{-2})/\text{rad}),\quad T_m = \frac{1}{\sqrt{a_\alpha + a_\omega b_\alpha}}\ (\text{s}),$$

$$\omega_m = \frac{1}{T_m} = \sqrt{a_\alpha + a_\omega b_\alpha}\ (1/\text{s}),\quad \mu_m = \frac{a_\omega + b_\alpha}{2\sqrt{a_\alpha + a_\omega b_\alpha}},$$

$$A_1 = -\frac{a_\omega b_\delta}{a_\delta b_\alpha - a_\alpha b_\delta}\ (\text{s}),\quad A_2 = -\frac{b_\delta}{a_\delta b_\alpha - a_\alpha b_\delta}\ (\text{s}^2).$$

This is a typical second-order oscillation transfer function and its undamped natural frequency is $\omega_m = \sqrt{a_\alpha + a_\omega b_\alpha}$. Since the term $a_\omega b_\alpha$ for a common missile is much smaller than a_α, it has

$$\omega_m \approx \sqrt{a_\alpha} = \sqrt{\frac{\text{Restoring moment produced by unit angle of attack}}{\text{Rotational inertia}}}$$

$$= \sqrt{\frac{-M_z^\alpha}{J_z}} = \sqrt{\frac{-m_z^\alpha \rho V^2 SL}{2J_z}} = \sqrt{\frac{c_y^\alpha S \rho V^2 x^*}{2J_z}}. \tag{2.5-11}$$

where, x^* is the distance from the center of pressure to the center of gravity of the missile. It characterizes the static stability level of the missile. The higher the missile's static stability, the greater the a_α and the missile's natural frequency.

For a rear controlled surface-to-air missile, the trajectory parameters for the chosen set point are as follows: The missile height from the ground is 1,500 m, the Mach number is 1.4, $V = 467$ m/s, and the dynamic coefficients of the missile at this set point are shown in Table 2.5-1.

2 Missile Mathematical Models

Table 2.5-1 The dynamic coefficients of the missile at the above chosen set point

a_α / s^{-2}	a_δ / s^{-2}	a_ω / s^{-1}	b_α / s^{-1}	b_δ / s^{-1}
321	534	2.89	2.74	0.42

Among these, the missile length is 2 m, the moment of inertia around the z-axis is $J_z = 12.8 \text{ kg} \cdot \text{m}^2$, and the mass of the missile is $m = 53 \text{ kg}$. Therefore, the static stability could be calculated as $x^* = -M_z^\alpha / Y^\alpha = a_\alpha J_z / b_\alpha mV = 321 \times 12.8 \div (2.74 \times 53 \times 467) = 61$ (mm), which is approximately 3% of the missile's length. With all the related missile dynamic coefficients from the known equation (2.5-9), we have

$$\frac{a_y(s)}{\delta_z(s)} = \frac{-1,886 \times (-3.16 \times 10^{-4} s^2 - 9.14 \times 10^{-4} s + 1)}{0.003 s^2 + 0.0171 s + 1} \quad (2.5-12)$$

Based on this transfer function, it is known that the undamped natural frequency of this second-order oscillation is $\omega_m = 18.1 \text{ rad/s} = 2.88 \text{ Hz}$, and its damping coefficient is $\mu_m = 0.155$. In addition, $k_a = -1,886 \text{ (m} \cdot \text{s}^{-2})/\text{rad}$, that is to say, a pitch actuator deflection angle of $5°$ will produce a normal acceleration $a_y = 165 \text{ m/s}^2 = 16.8 \text{ g}$. Fig. 2.5-1 shows the normal acceleration transient process generated by a $5°$ step pitch actuator deflection angle.

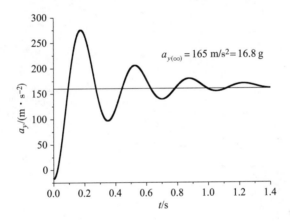

Fig. 2.5-1 Normal acceleration response generated by a $5°$ step pitch actuator deflection angle

The figure shows that the aerodynamic damping of a general missile body is very low (in this case, $\mu_m = 0.155$). The transient process of an uncontrolled missile body oscillates severely and its overshoot is very large. Only the artificial damping generated by an autopilot can help to improve the transient process.

If the center of gravity of the above fin-controlled missile is moved forward and its static stability increased by a factor of four, its steady-state gain will be reduced to about a quarter, its oscillation frequency will be doubled, and its damping coefficient will be halved. This shows that the missile static stability is a very important design parameter, and a larger static stability will lead to:

(a) Smaller steady state gain (poor maneuverability) ($k_a \propto 1/x^*$); good resistance to x^* change, that is, a better robustness.

(b) Higher short period oscillation frequency ($\omega_m \propto \sqrt{x^*}$).

(c) A smaller damping coefficient ($\mu_m \propto 1/\sqrt{x^*}$).

Similarly, a lower static stability will result in:

(a) Larger steady state gains (higher maneuverability); poor resistance to x^* change, that is, weak robustness.

(b) Lower short period oscillation frequency.

(c) Still lower but improved damping coefficient.

Therefore, when designing control systems, designers are very concerned about where the positions of center of pressure and center of gravity of the missile are for the chosen design.

In this case, the numerator transfer function is

$$(A_2 s^2 + A_1 s + 1) = -0.000,316 s^2 - 0.000,914 s + 1$$

that is

$$(s/59.3 + 1)(-s/56.4 + 1)$$

The Bode diagram of the individual two elements and the Bode diagram of the total second order numerator transfer function are separately given in Fig. 2.5 – 2.

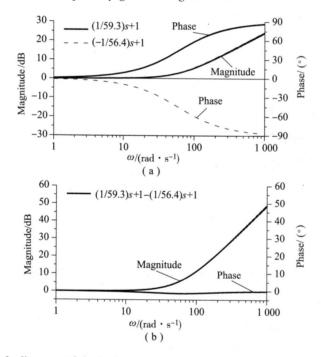

Fig. 2.5 – 2 Bode diagrams of the individual two elements and the numerator transfer function

The s^2 term coefficient A_2 of the transfer function numerator is given as:

$$|A_2| = \frac{b_\delta}{a_\delta b_\alpha - a_\alpha b_\delta} = \left(\frac{b_\delta}{b_\alpha}\right)\left(\frac{1}{a_\delta - \frac{a_\alpha b_\delta}{b_\alpha}}\right) = \left(\frac{b_\delta}{b_\alpha}\right)\left(\frac{1}{a_\delta}\right)\frac{1}{1 - \left(\frac{a_\alpha}{b_\alpha}\right)\left(\frac{b_\delta}{a_\delta}\right)}$$

and the following relations are known:

$\dfrac{a_\alpha}{b_\alpha} = x^*$, x^* is the distance from the center of gravity to the missile center of pressure.

$\dfrac{a_\delta}{b_\delta} = l_\delta$, l_δ is the distance from the center of gravity to the actuator center of pressure.

Therefore, $|A_2| = \left(\dfrac{b_\delta}{b_\alpha}\right)\left(\dfrac{1}{a_\delta}\right)\dfrac{1}{1 - \left(\dfrac{x^*}{l_\delta}\right)}$, since x^* is smaller than l_δ, the last item can be simplified as 1. Thus, $|A_2| \approx \left(\dfrac{b_\delta}{b_\alpha}\right)\left(\dfrac{1}{a_\delta}\right)$. Therefore, the cutoff frequency of the numerator transfer function can be given as $\omega^* = \sqrt{\dfrac{1}{|A_2|}} = \sqrt{\dfrac{b_\alpha}{b_\delta}}\sqrt{a_\delta}$, when a_α and a_δ are in the same magnitude order, the following relation can be obtained, $\omega^* = \sqrt{\dfrac{b_\alpha}{b_\delta}}\sqrt{a_\alpha} = \sqrt{\dfrac{b_\alpha}{b_\delta}}\omega_m$. Since the lift force generated by the angle of attack is greater than the lift force of the actuator, commonly we have $\dfrac{b_\alpha}{b_\delta} \approx 4 \sim 10$. Then, the cutoff frequency ω^* of the numerator transfer function of $\dfrac{a_y(s)}{\delta_z(s)}$ will be about $2 \sim 3.2$ times the natural frequency of the missile body ω_m. Here ω_m is determined by the missile transfer function denominator.

It is known that the s item coefficient A_1 of the numerator transfer function is $A_1 = -\dfrac{a_\omega b_\delta}{a_\delta b_\alpha - a_\alpha b_\delta}$. Because the missile body damping a_ω is very small, A_1 is approximately 0. The results of this are that the two first-order cutoff frequencies of the missile numerator transfer function are approximately the same, and the resulting phase is almost zero. The reason for this is that these two phases are equal in magnitude but opposite in sign. It is more important that the same magnitudes of their two components will lead to an increase in the magnitude of the transfer function $\dfrac{a_y(s)}{\delta_z(s)}$, making its crossover frequency move to a higher frequency. Its corresponding larger phase lag of the denominator at a higher frequency will make the autopilot design more difficult. However, the amplitudes of these two first order transfer functions must be added together.

Since the value of ω^* is higher than the natural frequency of the missile ω_m and the earlier missile autopilot design bandwidth is relatively low, the numerator transfer function was always omitted in past textbooks when the normal acceleration transfer function of the missile body is discussed. However, the bandwidth of the present autopilot design is quite high, so the numerator item of the normal acceleration transfer function of the missile body should not be omitted.

For axisymmetric missiles, the transfer functions a_z/δ_y and a_y/δ_z are essentially the same.

(2) Transfer function $\dot{\theta}(s)/\delta_z(s)$ from pitch actuator deflection angle δ_z to flight path angle angular velocity $\dot{\theta}$.

As $a_y(s) = V\dot{\theta}(s)$, so

$$\dfrac{\dot{\theta}(s)}{\delta_z(s)} = \dfrac{a_y(s)}{V} = \dfrac{k_{\dot{\theta}}(A_2 s^2 + A_1 s + 1)}{T_m^2 s^2 + 2\mu_m T_m s + 1}, \qquad (2.5-13)$$

where

$$k_{\dot{\theta}} = -\frac{a_\delta b_\alpha - a_\alpha b_\delta}{a_\alpha + a_\omega b_\alpha} \quad ((\text{rad} \cdot \text{s}^{-1})/\text{rad}).$$

It should be noted that all the transfer functions that use the actuator deflection angle δ_z as input share the same denominator and are with second-order oscillation characteristics.

(3) Transfer function $\dot{\vartheta}(s)/\delta_z(s)$ from actuator deflection angle δ_z to missile angular velocity $\dot{\vartheta}$.

We can infer that it can be derived from equation (2.5-4) ~ equation (2.5-6) that

$$\frac{\dot{\vartheta}(s)}{\delta_z(s)} = -\frac{a_\delta s + (a_\delta b_\alpha - a_\alpha b_\delta)}{s^2 + (a_\omega + b_\alpha)s + (a_\alpha + a_\omega b_\alpha)}, \tag{2.5-14}$$

$$\frac{\dot{\vartheta}(s)}{\delta_z(s)} = \frac{k_{\dot{\vartheta}}(T_\alpha s + 1)}{T_m^2 s^2 + 2\mu_m T_m s + 1}. \tag{2.5-15}$$

It is noteworthy that:

(a) The transfer function $\dot{\vartheta}(s)/\delta_z(s)$ and the transfer function $\dot{\theta}(s)/\delta_z(s)$ share the same gain, that is

$$k_{\dot{\vartheta}} = k_{\dot{\theta}}.$$

(b) We can learn from the previous derivation that $T_\alpha = \dfrac{a_\delta}{a_\delta b_\alpha - a_\alpha b_\delta} = \dfrac{1}{b_\alpha(1 - x^*/\ell_\delta)} \approx \dfrac{1}{b_\alpha}$.

Here the unit of T_α is second, which is called the angle of attack time constant. Since the physical meaning of b_α is the flight path angle angular velocity $\dot{\theta}$ produced by unit angle of attack α, the smaller the T_α, the higher the missile maneuverability. Because the air density and the value of b_α decrease with the increase in altitude, T_α will generally increase from a fraction of a second to several seconds with the increase in missile flight altitude.

(c) If the small term in the numerator of transfer function $a_y(s)/\delta_z(s)$ and $\dot{\theta}(s)/\delta_z(s)$ is omitted as $b_\delta \approx 0$, the relationships between the missile normal acceleration a_y, the flight path angle angular velocity $\dot{\theta}$ and the pitch angular velocity $\dot{\vartheta}$ are shown in Fig. 2.5-3.

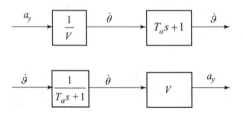

Fig. 2.5-3 The relationships between the normal acceleration, the flight path angle angular velocity and the pitch angle velocity

Since $\dot{\vartheta}(s) = \dfrac{1}{V}(T_\alpha s + 1)a_y(s)$, the angular velocity feedback in the inner loop of an acceleration autopilot design is clearly equivalent to a lead compensation for the normal acceleration.

(d) Since $\dot{\theta}(s) = \dfrac{1}{(T_\alpha s + 1)}\dot{\vartheta}(s)$ and $\theta(s) = \dfrac{1}{(T_\alpha s + 1)}\vartheta(s)$, $\dot{\theta}$ can be considered as the response of $\dot{\theta}$ to the pitch angular velocity $\dot{\vartheta}$ through dynamic lag $\dfrac{1}{(T_\alpha s + 1)}$ and the response of θ to the pitch angle ϑ through dynamic lag $\dfrac{1}{(T_\alpha s + 1)}$. Fig. 2.5-4 shows the time response curves of the missile body's angular velocity $\dot{\vartheta}$, flight path angle angular velocity $\dot{\theta}$, missile body attitude angle ϑ and the flight path angle θ of a fin-controlled missile with unit step pitch actuator deflection. As can be clearly seen from Fig. 2.5-4, the $\dot{\theta}$ response lags behind the $\dot{\vartheta}$ response, and the θ response lags behind the ϑ response.

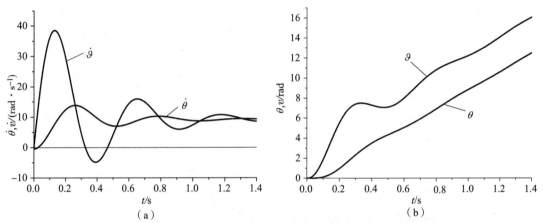

Fig. 2.5-4 The missile body pitch response to unit step actuator input

The physical mechanism of the flight path angle θ lagging behind the attitude angle ϑ is very simple. The change of the missile attitude produces an angle of attack. The normal acceleration generated by the angle of attack will produce a flight path angle angular velocity change and its integral is the flight path angle. From this process, it can be seen that the response of the flight path angle will lag behind the missile attitude, as in Fig. 2.5-5.

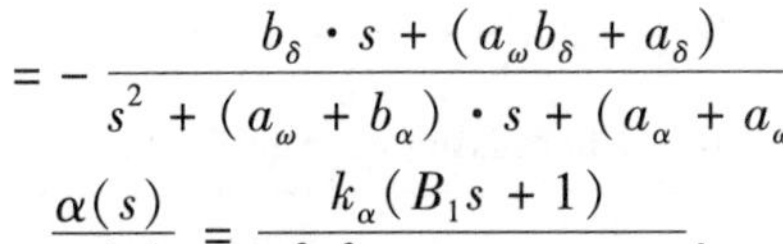

Fig. 2.5-5 Physical mechanism of the flight path angle θ lagging behind the attitude angle ϑ

(4) Transfer function from the pitch actuator deflection angle to the angle of attack $\alpha(s)/\delta_z(s)$.

$$\frac{\alpha(s)}{\delta_z(s)} = -\frac{b_\delta \cdot s + (a_\omega b_\delta + a_\delta)}{s^2 + (a_\omega + b_\alpha) \cdot s + (a_\alpha + a_\omega b_\alpha)}, \quad (2.5-16)$$

$$\frac{\alpha(s)}{\delta_z(s)} = \frac{k_\alpha(B_1 s + 1)}{T_m^2 s^2 + 2\mu_m T_m s + 1}, \quad (2.5-17)$$

where

$$k_\alpha = -\frac{a_\omega b_\delta + a_\delta}{a_\alpha + a_\omega b_\alpha}\,(\text{rad/rad}), \quad B_1 = \frac{b_\delta}{a_\omega b_\delta + a_\delta}\,(\text{s}).$$

Since the value of a_ω is very small, k_α can be expressed as $k_\alpha = -\dfrac{a_\delta}{a_\alpha}$. In missile design, $|k_\alpha| = \dfrac{a_\delta}{a_\alpha}$ is usually referred to as the control ratio of the missile, and it represents the steady angle of attack generated by the unit actuator deflection angle of the missile. It represents the actuator capability for producing angle of attack efficiency. Therefore, it seems that improving the control ratio of the missile can improve the efficiency of the control. However, due to the limitation of the size of the actuator surface, the value of a_δ is unlikely to increase significantly. Therefore, the control ratio can only be improved by reducing the missile's static stability margin, that is, reducing the distance x^* between the center of gravity and the center of pressure. When x^* is very small, small changes in the position of center of gravity or center of pressure can cause large changes in the value of a_α, that is, the control ratio. This could make k_α greatly deviate from the designed value, and the designed autopilot may fail to pass the robustness evaluation. An efficient solution is to add a pseudo angle of attack feedback loop to the missile body to construct a highly robust new virtual missile body. This design method will be described in Chapter 6.

(5) Transfer function of the roll channel $\dot{\gamma}(s)/\delta_x(s)$.

Roll channel dynamics is a second-order differential equation

$$\ddot{\gamma} = \frac{\mathrm{d}}{\mathrm{d}t}(\dot{\gamma}) = -c_\omega \cdot \dot{\gamma} - c_\delta \cdot \delta_x, \tag{2.5-18}$$

where, γ ——roll angle, rad;

$\dot{\gamma}$ ——roll angular velocity, rad/s;

$\ddot{\gamma}$ ——roll angular acceleration, rad/s^2.

Express it in the form of a state equation, take the state variables as γ and $\dot{\gamma}$, and the control variable as δ_x, that is

$$\frac{\mathrm{d}}{\mathrm{d}t}\begin{bmatrix}\gamma\\\dot{\gamma}\end{bmatrix} = \begin{bmatrix}0 & 1\\0 & -c_\omega\end{bmatrix}\begin{bmatrix}\gamma\\\dot{\gamma}\end{bmatrix} + \begin{bmatrix}0\\-c_\delta\end{bmatrix}\delta_x. \tag{2.5-19}$$

So, it can be derived that

$$\frac{\dot{\gamma}(s)}{\delta_x(s)} = \frac{-c_\delta}{s+c_\omega} = \frac{-c_\delta/c_\omega}{(1/c_\omega)s+1} = \frac{-c_\delta/c_\omega}{T_r s+1}, \tag{2.5-20}$$

$$\frac{\dot{\gamma}(s)}{\delta_x(s)} = \frac{k_r}{T_r s+1}. \tag{2.5-21}$$

where $k_r = -c_\delta/c_\omega$ is the steady state gain from δ_x to $\dot{\gamma}$, and $T_r = 1/c_\omega$ is the aerodynamic time constant for this transfer function.

The detailed expressions of all the above transfer functions are shown in Table 2.5-2.

Table 2.5-2 Expressions of aerodynamic transfer functions and their coefficients

Aerodynamic transfer function	Unit	Expression of the numerator related coefficients	Expression of the denominator related coefficient
$\dfrac{a_y(s)}{\delta_z(s)} = k_a \dfrac{A_2 s^2 + A_1 s + 1}{T_m^2 s^2 + 2\mu_m T_m s + 1}$	$(\mathrm{m}\cdot\mathrm{s}^{-2})/\mathrm{rad}$	$k_a = -V \cdot \dfrac{a_\delta b_\alpha - a_\alpha b_\delta}{a_\alpha + a_\omega b_\alpha}$ $A_1 = \dfrac{-a_\omega b_\delta}{a_\delta b_\alpha - a_\alpha b_\delta}$	$T_m = \dfrac{1}{\sqrt{a_\alpha + a_\omega b_\alpha}}$ $\mu_m = \dfrac{a_\omega + b_\alpha}{2\sqrt{a_\alpha + a_\omega b_\alpha}}$

Continued

Aerodynamic transfer function	Unit	Expression of the numerator related coefficients	Expression of the denominator related coefficient
$\dfrac{a_y(s)}{\delta_z(s)} = k_a \dfrac{A_2 s^2 + A_1 s + 1}{T_m^2 s^2 + 2\mu_m T_m s + 1}$	$(m \cdot s^{-2})/rad$	$A_2 = \dfrac{-b_\delta}{a_\delta b_\alpha - a_\alpha b_\delta}$	
$\dfrac{\dot\theta(s)}{\delta_z(s)} = k_{\dot\theta} \dfrac{A_2 s^2 + A_1 s + 1}{T_m^2 s^2 + 2\mu_m T_m s + 1}$	$(rad \cdot s^{-1})/rad$	$k_{\dot\theta} = -\dfrac{a_\delta b_\alpha - a_\alpha b_\delta}{a_\alpha + a_\omega b_\alpha}$ $A_1 = \dfrac{-a_\omega b_\delta}{a_\delta b_\alpha - a_\alpha b_\delta}$ $A_2 = \dfrac{-b_\delta}{a_\delta b_\alpha - a_\alpha b_\delta}$	
$\dfrac{\dot\vartheta(s)}{\delta_z(s)} = k_{\dot\vartheta} \dfrac{T_\alpha s + 1}{T_m^2 s^2 + 2\mu_m T_m s + 1}$	$(rad \cdot s^{-1})/rad$	$k_{\dot\vartheta} = -\dfrac{a_\delta b_\alpha - a_\alpha b_\delta}{a_\alpha + a_\omega b_\alpha}$ $T_\alpha = \dfrac{a_\delta}{a_\delta b_\alpha - a_\alpha b_\delta}$	
$\dfrac{\alpha(s)}{\delta_z(s)} = k_\alpha \dfrac{B_1 s + 1}{T_m^2 s^2 + 2\mu_m T_m s + 1}$	rad/rad	$k_\alpha = -\dfrac{a_\omega b_\delta + a_\delta}{a_\alpha + a_\omega b_\alpha}$ $B_1 = \dfrac{b_\delta}{a_\omega b_\delta + a_\delta}$	
$\dfrac{\dot\gamma(s)}{\delta_x(s)} = \dfrac{k_r}{T_r s + 1}$	$(rad \cdot s^{-1})/rad$	$k_r = -\dfrac{c_\delta}{c_\omega}$	$T_r = \dfrac{1}{c_\omega}$

3

Simplified Models of Missile Control Components

This chapter briefly covers basic missile control components mathematic models.

1) Seeker

The most important control component of the missile is the seeker. Because of its importance, we will discuss it in a separate chapter (see Chapter 5).

2) Actuator

Another important component of the missile is the actuator. There are many kinds of actuators, such as electrical actuator, hydraulic actuator, hot gas actuator and cold gas actuator. The hydraulic actuator has the highest load capacity and the widest frequency bandwidth. However, the technology of the electrical actuator is developing rapidly nowadays, and its load capacity and frequency bandwidth are continually improving. Therefore, today the most often used actuators are electrical. Hot gas servo and cold gas actuators have limited load capacity, but their frequency bandwidths are not low. Currently, they are limited to use in low-end, short-range missiles.

The actuator is the main component that limits the bandwidth of the missile autopilot. Its dynamic model is generally given in first-order or second-order models.

The expression of the first-order model is

$$\frac{\delta(s)}{\delta_c(s)} = \frac{1}{Ts + 1}e^{-\tau s}, \qquad (3-1)$$

where T is the actuator time constant, $\omega = \frac{1}{T}$ is the first-order bandwidth, and τ is the time delay.

The expression of the second-order model is

$$\frac{\delta(s)}{\delta_c(s)} = \frac{\omega_n^2}{s^2 + 2\mu\omega_n s + \omega_n^2}e^{-\tau s}, \qquad (3-2)$$

where ω_n is the undamped natural frequency of the actuator, μ is the damping coefficient and τ is the time delay.

In practical engineering applications, it is often necessary to obtain several commonly used characteristic parameters of the actuator by testing and identifying its transfer function.

(1) Transfer function -3 dB bandwidth ω_{-3dB}.

(2) Transfer function $-90°$ bandwidth $\omega_{-90°}$.

(3) Damping coefficient μ.

(4) Time delay τ.

(5) Angular velocity capability of the actuator with and without load (Fig. 3-1)

$$\dot{\delta} = \frac{\delta^*}{\Delta t}. \qquad (3-3)$$

Generally the actuator frequency band is wide, so it can respond to high-frequency noise signals. This not only exacerbates the actuator's friction effect, but more importantly, when the high-frequency noise is superimposed on a normal actuator response to a command δ_c, it could hit the bidirectional saturation zone of the actuator. After the actuator output δ effects

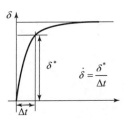

Fig. 3-1 Actuator output δ in response to unit command

have been filtered by the missile body, the effective δ angle will be different from δ_c and an actuator response error occurs.

For this reason, in the normal autopilot design, a notch filter is always added in front of the actuator to eliminate the high frequency noise effect.

3) Angular rate gyro

The transfer function of the angular rate gyro is usually expressed with a second-order transfer function.

$$\frac{\dot{\vartheta}_s(s)}{\dot{\vartheta}(s)} = \frac{\omega_n^2}{s^2 + 2\mu\omega_n s + \omega_n^2}. \qquad (3-4)$$

where $\dot{\vartheta}(s)$ is the angular velocity of the missile and $\dot{\vartheta}_s(s)$ is the output of the angular rate gyro, ω_n is the undamped natural frequency of the angular rate gyro, and μ is its damping ratio. The angular rate gyro currently used generally has a wide frequency bandwidth, often up to 80 Hz. Due to its wide bandwidth, it has little impact on the autopilot design.

4) Accelerometer

Similar to the angular rate gyro, the transfer function of an accelerometer is also generally expressed as a second-order one.

$$\frac{a_s(s)}{a(s)} = \frac{\omega_n^2}{s^2 + 2\mu\omega_n s + \omega_n^2}. \qquad (3-5)$$

where $a(s)$ is the normal acceleration of the missile, $a_s(s)$ is the output of the accelerometer, ω_n is the undamped natural frequency of the accelerometer, and μ is the damping ratio. In practice, the accelerometer frequency bandwidth is usually very wide, often up to 80 Hz, so it has little impact on the autopilot design.

5) Inertial navigation components and integrated inertial navigation module

An inertial navigation module or integrated inertial navigation module can simultaneously output the missile acceleration, velocity, position, angular velocity and orientation at a very high rate (e.g., 100 - 200 Hz). They greatly enrich the useful information for the missile guidance and autopilot design, thus making possible the use of many advanced guidance laws and the use of more advanced autopilot structures.

4

Guidance Radar

§4.1 Introduction

　　Guidance radar is a necessary part in a command guidance system. It is often adopted in the short to medium range ground-to-air missile guidance to accurately track the target and determine the line-of-sight direction of the target. After launch the missile is guided to the narrow tracking radar beam with the help of an initial guidance system. Once within the beam, the guidance radar will use the echo signal of the missile transponder to determine the deviation of the missile from the center of the beam. The error is then uploaded to the missile in the form of a command, which will guide the missile to move along the center of the beam until it meets the target. Since the guidance radar may be installed on a ship or a launch vehicle, the guidance radar should be defined as a precise angle tracking system mounted on a low-speed moving carrier or a stationary station.

§4.2 Motion Characteristic of Line-of-sight

　　The geometric relationship between the target and the guidance radar is shown in Fig. 4.2 – 1.

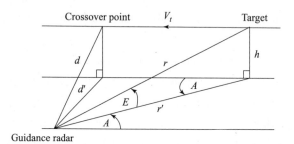

Fig. 4.2 – 1 Geometric relationship between the target and the guidance radar
(**The target moves at a constant speed along a straight line**)

　　Suppose that the target follows along a line parallel to the ground at a constant speed of V_t. The target flight height is h, and the minimum distance from the target to the guidance radar is d' when the target flies over the guidance radar. The distance between the target and the radar is r at the moment of the time t. The tracking angles of the radar are A in the yaw channel direction, and E in the pitch channel direction, respectively.

In the following analysis, we will focus on the analysis of the angular velocities of the guidance radar yaw and pitch tracking channels $\dot{A}$ and $\dot{E}$ as well as their angular accelerations $\ddot{A}$ and $\ddot{E}$. Take V_t, h, d' (or d) as constant parameters in this analysis, while the variables to be studied are A, $\dot{A}$, $\ddot{A}$, E, $\dot{E}$ and $\ddot{E}$. According to the geometric relationship shown in Fig. 4.2 – 1, the expression of the angular velocity for the yaw channel $\dot{A}$ with the variation of A is

$$\dot{A} = \frac{dA}{dt} = \frac{V_t \sin A}{r'} = \frac{V_t \sin^2 A}{d'} = \frac{V_t}{d'} f_1(A), \qquad (4.2-1)$$

where, $f_1(A) = \sin^2 A$. The function $f_1(A)$ is shown in Fig. 4.2 – 2.

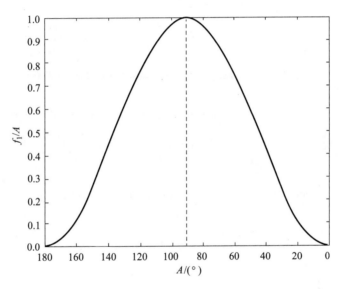

Fig. 4.2 – 2 $f_1(A)$ plot

Taking A in equation (4.2 – 1) as an intermediate variable and taking the derivative of $\dot{A}$ with respect to time t, we can get the expression of $\ddot{A}$ with the variation of A

$$\ddot{A} = \frac{d}{dt}\left(\frac{dA}{dt}\right) = \frac{d}{dA}\left(\frac{dA}{dt}\right) \cdot \frac{dA}{dt}$$

$$= \frac{2V_t \sin A \cos A}{d'} \cdot \frac{V_t \sin^2 A}{d'} = \frac{V_t^2}{(d')^2} \sin 2A \sin^2 A = \frac{V_t^2}{(d')^2} \cdot f_2(A), \qquad (4.2-2)$$

where $f_2(A) = \sin 2A \sin^2 A$.

For a given value of d', $\ddot{A}$ in yaw channel has the maximum value when $A = 60°$ or $120°$ (Fig. 4.2 – 3). Its maximum absolute value is as follows

$$|\ddot{A}_{max}| = \frac{V_t^2}{(d')^2} \sin 120° \cdot \sin^2 60° = \frac{3\sqrt{3}}{8} \cdot \frac{V_t^2}{(d')^2} = 0.65 \frac{V_t^2}{(d')^2}. \qquad (4.2-3)$$

The maximum value of $\ddot{A}$ is positive when $A = 60°$, and negative when $A = 120°$, that is, there is a positive maximum angular acceleration for a coming target $A = 60°$ and a negative maximum angular acceleration for a departing target $A = 120°$.

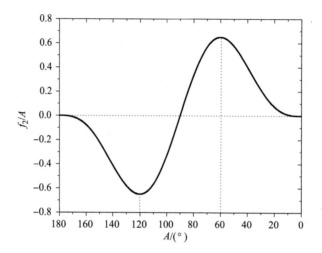

Fig. 4.2-3 Change of $f_2(A)$ with the variation of A

Similarly, for a given d' and h (or $k = h/d'$) in the pitch channel, the expression of E with the variation of A is as follows

$$\tan E = \frac{h}{d'/\sin A} = \frac{h \sin A}{d'} = k \sin A. \qquad (4.2-4)$$

Take d' as a constant, A and k as variables. The expressions of $\dot{E}$ and $\ddot{E}$ with the variation of A are as follows

$$\dot{E} = \frac{V_t}{r}\sin E \cos A = \frac{V_t}{d'}\sin E \cos E \sin A \cos A$$

$$= \frac{V_t \cdot k \cdot \sin^2 A \cos A}{d'(1 + k^2 \sin^2 A)} = \frac{V_t}{d'}f_3(A,k), \qquad (4.2-5)$$

where $f_3(A,k) = \dfrac{k \cdot \sin^2 A \cos A}{(1 + k^2 \sin^2 A)}$, and,

$$\ddot{E} = -\frac{kV_t^2 \sin^3 A(\sin^2 A + k^2 \sin^4 A - 2\cos^2 A)}{(d')^2(1 + k^2 \sin^2 A)^2} = -\frac{V_t^2}{d'^2} \cdot f_4(A,k), \qquad (4.2-6)$$

where $f_4(A,k) = \dfrac{k\sin^3 A(\sin^2 A + k^2 \sin^4 A - 2\cos^2 A)}{(1 + k^2 \sin^2 A)^2}$. Fig. 4.2-4 shows the variation of the function $f_4(A,k)$ with respect to A and k.

It can be proved the $\ddot{E}$ reaches its maximum value when d' and k are given and $A = 90°$ (Fig. 4.2-4), that is,

$$\ddot{E}_{max} = -\frac{kV_t^2}{(d')^2(1 + k^2)}. \qquad (4.2-7)$$

It can be seen from equation (4.2-7) that $\ddot{E}_{max}$ changes with the variation of k for a given d'.

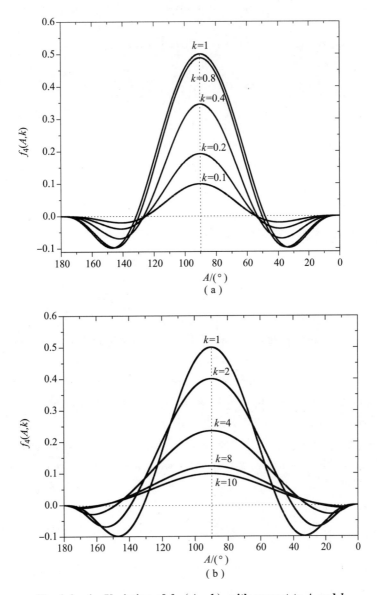

Fig. 4.2-4 Variation of f_4 (A, k) with respect to A and k

Furthermore, $(\ddot{E}_{max})_{max}$ exists as follows when $k = 1$ (that is, $A = 90°$, $E = 45°$)

$$(\ddot{E}_{max})_{max} = -0.5 \frac{V_t^2}{d'^2}(E = 45°, A = 90°).$$

If the target passes directly above the guidance radar (when $d = 0$), then, h can be taken as a constant and E can be regarded as a variable, and the horizontal planes, that is, equation (4.2-1), equation (4.2-2) and equation (4.2-3), can be converted to vertical planes as

$$\dot{E} = \frac{V_t \sin^2 E}{h}, \qquad (4.2-8)$$

$$\ddot{E} = \frac{V_t^2 \sin 2E \sin^2 E}{h^2}, \qquad (4.2-9)$$

$$\ddot{E}_{max} = \frac{3\sqrt{3}}{8} \cdot \frac{V_t^2}{h^2} = 0.65 \frac{V_t^2}{h^2} (E = 45°),$$

$$\ddot{E}_{max} = \frac{3\sqrt{3}}{8} \cdot \frac{V_t^2}{h^2} = 0.65 \frac{V_t^2}{h^2} (E = 45°). \qquad (4.2-10)$$

From the above analysis, it can be seen that even if the target is in a simple straight constant speed flight, the yaw and pitch angular accelerations of the tracking system also exist. This is why the guidance radar tracking system needs to be designed as a type II system with two integrators from the tracking accuracy consideration.

§4.3 Control Loop of the Guidance Radar

According to the analysis in Section 4.2, even if the target is flying at a constant speed, the angular accelerations of both $(\ddot{A})_t$ and $(\ddot{E})_t$ are not constant. Therefore, it is essential for the control loop of the guidance radar to be designed as a type II system at least. Also, its dynamic angle tracking error should meet the system accuracy requirement when tracking the slow varying angular acceleration line of sight.

Fig. 4.3-1 shows the block diagram of a typical guidance radar. The feedback element of the high-gain stabilization loop is generally selected as a rate gyro to ensure that the antenna axis can be stabilized in the inertial space even if the guidance radar may be installed on a slowly moving base. At the same time, the high gain of the stability loop ensures that the angular tracking error remains small under the influence of various disturbance moments. The compensation network of the tracking loop generally adopts a PI compensation scheme to realize a type II system that can stably track the angular acceleration of the line of sight.

Fig. 4.3-2 and Fig. 4.3-3 show the block diagram of the tracking loop of the guidance radar. In Fig. 4.3-3, $k = k_1 k_2 k_3$.

To ensure that the phase shift of the stability loop at the tracking loop crossover frequency is small, the stability loop bandwidth is generally chosen to be no less than five times that of the tracking loop. Fig. 4.3-4 shows the Bode diagram of the PI compensation networks for different T_i values. It can be seen that the compensation position in the frequency domain is different depending on the T_i value. The compensation Bode diagram is shifted toward a high frequency when the T_i value is small and when the T_i value is large, it moves to a low frequency. There are two possible applications for this compensation structure.

1) For a type I system

This structure is used to increase the gain at the loop low frequency to reduce the system steady state error. In this application, the PI compensation network is positioned at low frequencies, and the phase shift induced by it at the system crossover frequency is small (e.g. 5° – 10°). Therefore, its introduction has little effect on the bandwidth of the final design. At this time, the crossover frequency

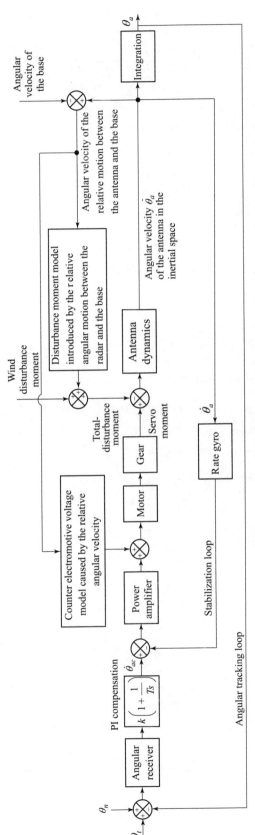

Fig.4.3–1 Control block diagram of a typical guidance radar tracking system

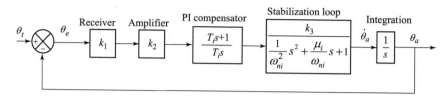

Fig. 4.3-2 Block diagram of the guidance radar

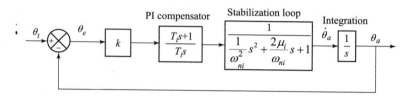

Fig. 4.3-3 Model of the guidance radar control system

of the system can be estimated as $\omega_c \approx k$ rad/s (Fig. 4.3-3).

2) For a type II system

In this application, it is best to modify Fig. 4.3-3 to Fig. 4.3-5 when this scheme is used. The function of the PI compensator $\dfrac{(T_i s + 1)}{T_i s}$ here is to provide an integrator for the type II design requirement and use the numerator $(T_i s + 1)$ term to introduce a lead compensation for system stability consideration because the two integrators will have a $-180°$ phase lag in the tracking loop.

When the PI compensation is used in a type II system, the open loop gain of the system should be $K = \dfrac{k}{T_i}$, and the system crossover frequency is about $\omega_c \approx \sqrt{K}$ rad/s (Fig. 4.3-5).

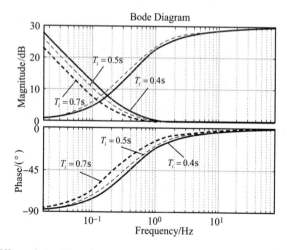

Fig. 4.3-4 Effect of the PI compensation parameter T_i on the compensation function

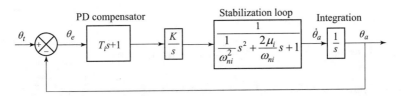

Fig. 4.3 – 5   Another structure of the system radar control model

Fig. 4.3 – 6 shows the Bode diagram of the PD compensation $T_i s + 1$ for different T_i values. The proper time constant T_i should be chosen to give a good phase compensation at the loop crossover frequency at the same time to guarantee a shorter tracking error reduction transient. Therefore, determining the position of the PD compensation in the mid frequency domain demands a distinct process of parameter optimization.

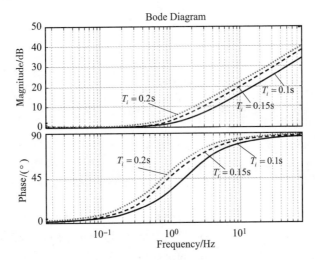

Fig. 4.3 – 6   PD compensation Bode diagram with different T_i

The following is an example to illustrate issues that should be considered when such a tracking system is designed.

It is known that for a certain guidance radar the maximum angular acceleration of the line of sight given is $\ddot{\theta}_t = 28 \text{ mrad/s}^2 = 1.6°/\text{s}^2$. The tracking design requires that the steady state tracking error of the system be less than 0.3 mrad at this steady state angular acceleration. To meet this requirement in the preliminary design, the open loop gain of this radar should be chosen as $K \geq \dfrac{28 \text{ mrad} \cdot \text{s}^{-2}}{0.3 \text{ mrad}} = 93.3 \text{ s}^{-2}$. For example a value $K = 100 \text{ s}^{-2}$. The estimated open loop crossover frequency of this system will be $\omega_c = \sqrt{K} = \sqrt{100} = 10 \text{ rad/s}$. Accordingly, the bandwidth of the second order inner stabilization loop model could be designed as 65 rad/s with a damping ratio of 0.5.

Next, the value of the PD compensation network parameter T_i can be designed according to the following optimal strategy.

Suppose that, for stability consideration, the phase margin should be greater than 40° and the gain margin be greater than 6 dB. At the same time, to achieve a fast and good tracking error reduction transient, an optimization objective function in the form of $J(T_i) = \int_0^1 |\theta_e(t) - \theta_{\text{steady state}}| dt$ can be chosen and the optimization problem will be formulated as:

$$\text{Min } J(T_i) = \min \int_0^T |\theta_e(t) - \theta_{\text{steady state}}| dt, \text{ in which } \theta_{\text{steady state}} = \frac{\ddot{\theta}_{\max}}{K} = \frac{28}{100} = 0.28 \text{ mrad};$$

subject to: $\Delta\phi(T_i) > 40°, \Delta L(T_i) > 6 \text{ dB}$.

Fig. 4.3 –7 shows the variation of the gain margin ΔL and phase margin $\Delta \phi$ with the change of T_i. Fig. 4.3 –8 shows the variation of the object function J with the change of T_i.

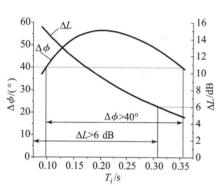 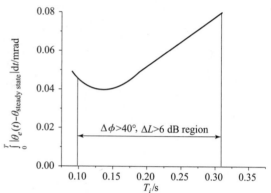

Fig. 4.3 –7 Variation of the system gain and phase margins with the change of T_i

Fig. 4.3 –8 Variation of the objective function $J(T_i) = \int_0^T |\theta_e(t) - \theta_{\text{steady state}}| dt$ with the change of T_i

It can be seen from Fig. 4.3 –7 that the range of T_i, which meets the requirements of both the stability ΔL and $\Delta \phi$ constraints, is $0.10 \text{ s} < T_i < 0.31 \text{ s}$. In this region, the optimal value of T_i can be obtained according to the minimum value of the object function J as $T_i = 0.15$ s. It is known from Fig. 4.3 –8 that this object function J changes smoothly in the vicinity of the optimum solution, that is to say, this optimum solution is also very robust. Fig. 4.3 –9 gives the tracking error transient for the optimum solution $T_i = 0.15$ s together with the transients from the stability boundaries $T_i = 0.10$ s and 0.30 s when the input is a step angular acceleration $\ddot{\theta}_t = 28 \text{mrad/s}^2$. It can be seen that $T_i = 0.10$ s gives a fast response and $T_i = 0.30$ s gives a slow response at the expense of stability margins.

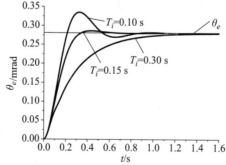

Fig. 4.3 –9 Tracking error transients for different T_i values with a step input of angular acceleration of $\ddot{\theta}_t = 28 \text{mrad/s}^2$

Fig. 4.3 – 10 shows the block diagram of the design results. Fig. 4.3 – 11 shows the open loop Bode diagram of the system. It can be seen that the system crossover frequency is $\omega_c = 2.65$ Hz, the phase margin is $\Delta\phi = 52.9°$ and the gain margin is $\Delta L = 11.8$ dB, in which the PD compensator gives a phase lead compensation of $68.2°$, and the inner stability loop has a phase lag $-15.3°$ at the system crossover frequency. Therefore, the system phase margin is $\Delta\phi = 68.2° - 15.3° = 52.9°$. The closed loop Bode diagram of the system is shown in Fig. 4.3 – 12.

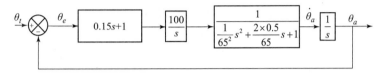

Fig. 4.3 – 10 System block diagram of the design

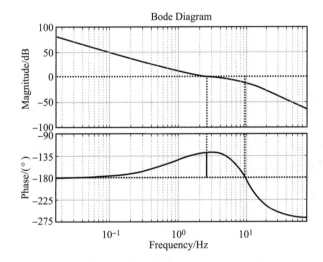

Fig. 4.3 – 11 Open loop Bode diagram of the tracking system

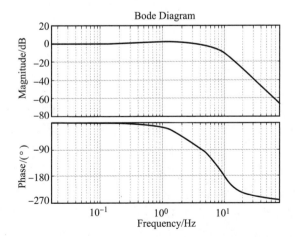

Fig. 4.3 – 12 Closed loop Bode diagram of the tracking system

The system response to a unit step line of sight angle input of $\theta_t = 50$ mrad is given in Fig. 4.3 – 13. Fig. 4.3 – 14 – Fig. 4.3 – 16 are presented to show the tracking error curves of the system when the inputs respectively are $\theta_t = 50$ mrad, $\dot{\theta}_t = 50$ mrad/s and $\ddot{\theta}_t = 28$ mrad/s². It is known that a type-II tracking system has zero steady state error when a constant target angle θ_t and the angular velocity $\dot{\theta}_t$ are tracked, but there is a steady state error $\theta_e = \dfrac{\ddot{\theta}_t}{K}$ when a constant $\ddot{\theta}_t$ input is tracked. In this case, $\ddot{\theta}_t = 28$ mrad/s², $K = 100$ s⁻², and the steady state angular error is $\theta_e = \dfrac{\ddot{\theta}_t}{K} = 0.28$ mrad.

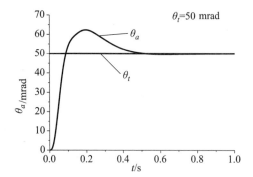

Fig. 4.3 – 13 Response of the radar antenna when a step angle input is tracked

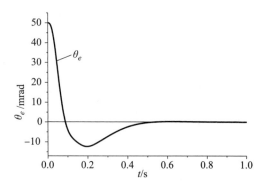

Fig. 4.3 – 14 Tracking angle error with a step input $\theta_t = 50$ mrad

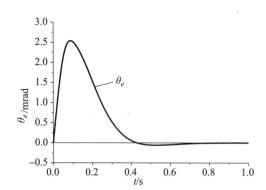

Fig. 4.3 – 15 Tracking angle error with a step input $\dot{\theta}_t = 50$ mrad/s

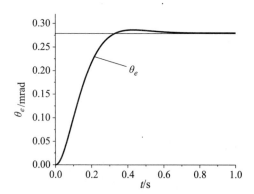

Fig. 4.3 – 16 Tracking angle error with a step input $\ddot{\theta}_t = 28$ mrad/s²

The following example illustrates the rationale of designing a type II angular tracking system using the maximum line of sight angular acceleration $(\ddot{\theta}_t)_{max}$ as its input. Suppose we have the following scenario that the flight speed of the target is $V_t = 250$ m/s, and the minimum slant range is $d' = 1,300$ m. It is known from equation (4.2 – 2) that the maximum line-of-sight angular acceleration will be $(\ddot{\theta}_t)_{max} = 28$ mrad/s² when $A = 60°$. Fig. 4.2 – 3 in Section 4.2 shows the variation curve of $\ddot{A}_t$ with the change of A when the target is flying past and the $\ddot{A}_t$ is not even a constant in a real scenario. Fig. 4. 3 – 17 shows the tracking error curve for the real line of sight variation with a changing $\ddot{\theta}_t$, also given is the estimated tracking error $\theta_{e\ \text{estimated}} = \dfrac{\ddot{\theta}}{K}$. It can be seen that when $\ddot{\theta}_t$ input is changing slowly the estimation formula could predict the tracking error nicely.

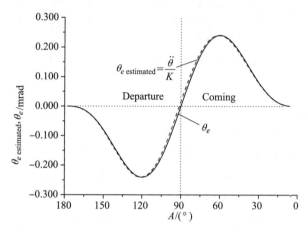

Fig. 4. 3 – 17 Actual tracking error θ_e and estimated tracking error $\ddot{\theta}/K$ curves

It is known from Fig. 4. 3 – 17 that:

(1) In a real tracking scenario, the tracking error can be quite accurately estimated by $\theta_{e\ \text{estimated}} = \dfrac{\ddot{\theta}}{K}$.

(2) The target will have a relatively large third order derivative $\dddot{\theta}_t$ value over the over-head flight region, but its time duration is very short. The target has flown over the guidance radar before the tracking error of the system builds up. This is why we can use a type II, not type III, system structure to design the guidance radar angle tracking system.

§ 4. 4 Effect of the Receiver Thermal Noise on the Guidance Performance

The source of the noise in a radar receiver is mainly thermal noise. Since the noise bandwidth is far greater than that of the tracking system, its effect on the guidance radar is equivalent to a white noise interference input. According to the basic radar equation, the echo signal strength of the target is inversely proportional to the fourth power of the target distance. That is, the signal received by the radar receiver changes drastically as the target distance changes. In order to facilitate the normal operation of the subsequent circuit, the guidance radar receiver is provided with an automatic gain control circuit, and its gain value is inversely proportional to the signal strength, so that the subsequent circuit can always work within the designed signal strength range.

Fig. 4. 4 – 1 shows the schematic of the guidance radar automatic gain control. In the figure, P_s and P_n are the signal strengths of the target and the thermal noise, respectively, before the gain control, while P_{s0} and P_{n0} are the target signal strengths and the thermal noise signal strength after

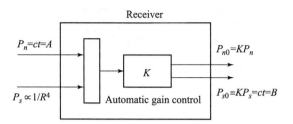

Fig. 4.4-1 Schematic of the guidance radar automatic gain control

the gain control. Generally, the thermal noise strength P_n before the gain adjustment can be considered as a constant. The thermal noise after the gain control $P_{n0} = KP_n$ could be considered as an angular noise by subsequent processing circuits. With this design, as K is proportional to the fourth power of the echoed target signal strength, the angular noise caused by the thermal noise will be proportional to the fourth power of the target distance, that is to say, the thermal noise effect will increase when the signal is weak at a long distance from the target.

Take the white noise (angular noise) power spectrum density $S(\omega)$ caused by the thermal noise as a constant

$$S(\omega) = K_s^2 \ (\text{rad}^2/\text{Hz}).$$

In computer simulation, a white noise could be approximated by a random series of normally distributed digital signal of standard deviation σ and width h (Fig. 4.4-2). The autocorrelation function of the random sequence $R(\tau)$ is

$$R(\tau) = \sigma^2 \quad (\tau = 0 \sim h), \tag{4.4-1}$$

$$R(\tau) = 0 \quad (\tau > h). \tag{4.4-2}$$

Its power spectrum density is

$$S(\omega) = \int_{-\infty}^{\infty} e^{-j\omega\tau} R(\tau) d\tau = \sigma^2 h = K_s^2 \ (\text{rad}^2/\text{Hz}). \tag{4.4-3}$$

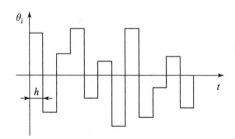

Fig. 4.4-2 The white noise input simulation

Given the thermal noise power spectrum density K_s^2 and the simulation step size h, the variance of the normal distribution random number σ^2 should be taken as

$$\sigma^2 = \frac{K_s^2}{h}. \tag{4.4-4}$$

Suppose that the variation of the power spectrum density of the guidance radar, thermal noise with the change of the target distance is

$$S(\omega) = K_s^2 = 2.51 \times 10^{-14} \left(\frac{R_t^4}{R_0^4}\right) \ (\text{rad}^2/\text{Hz}), \tag{4.4-5}$$

where R_0 is the reference distance ($R_0 = 1\text{km}$). Suppose that the maximum working distance of the radar is $R_t = 32$ km, then the maximum thermal noise power spectrum density will be

$$S(\omega) = 2.51 \times 10^{-14} \times 32^4 = 2.63 \times 10^{-8} \ (\text{rad}^2/\text{Hz}).$$

If the computer simulation step size is taken $h = 0.001$ s, the random number variance will be

$$\sigma^2 = \frac{K_s^2}{h} = 2.63 \times 10^{-5} \ \text{rad}, \ \text{then}$$

$$\sigma = 0.005, 13 \ \text{rad} = 5.13 \ \text{mrad}. \tag{4.4-6}$$

Fig. 4.4-3 shows the thermal noise angle input given in this computer simulation. Moreover, Fig. 4.4-4 gives the response of the radar system to the thermal noise disturbance. The standard deviation of this response σ is 0.53 mrad, and its value is reduced by about an order of magnitude in comparison with the white noise σ. This is because the tracking system has a low frequency bandwidth, which has a strong filtering effect on the white noise input in the high frequency range. It is indicated in Section 4.2 that $\ddot{A}$ and $\ddot{E}$ values will decrease as the target distance R increases. Because the radar tracking error is proportional to $\ddot{A}/K$ and $\ddot{E}/K$, with a given tracking error specification, the system could be designed to automatically reduce its open loop gain K at a longer target distance. This will compensate for the increase of the thermal noise effect at this longer target distance. This self-adaptive system design is a very effective measure to cope with the thermal noise negative effect.

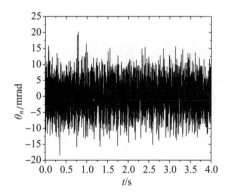

Fig. 4.4-3   Thermal noise input

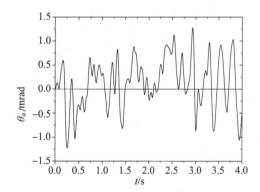

Fig. 4.4-4   Response of the radar angular error due to the thermal noise input

Fig. 4.4-5 shows the implementation of this design scheme.

The output in the low-pass filter in Fig. 4.4-5 gives the low frequency tracking error and the high-pass filter gives the level of the system's response to the noise input. When the relative value of the low-pass filter output is high, the system gain will be automatically increased. Meanwhile the system gain will be reduced when the relative output of the high-pass filter is high. Therefore, this adaptive control strategy can effectively minimize the overall tracking error of the guidance radar at different target distances.

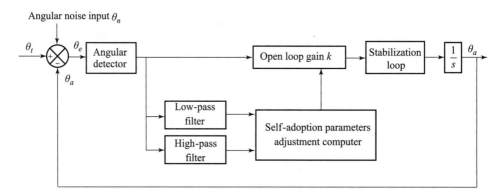

Fig. 4.4-5 Block diagram of the adaptive radar tracking system

§ 4.5 Effect of Target Glint on the Guidance Performance

When the radar illuminates the target, the phase shifts of the echoes from the different parts of the target are different since the target is not an ideal sphere. For this reason, the energy center of the target echo could deviate from the geometric center of the target in pitch and yaw. This phenomenon is called target glint. The glint output can be simulated by a first order colored noise filter output (Fig. 4.5-1).

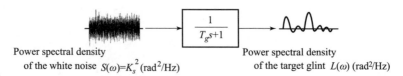

Fig. 4.5-1 Target glint model

In Fig. 4.5-1, the target glint characteristic can be expressed with two parameters, in which K_g^2 is the power spectrum density of the white noise input (m^2/Hz), T_g is the time constant of the colored noise filter, and $L(\omega)$ is the power spectrum density of the target glint. For a certain target, the two parameters of the target glint are known as $K_g^2 = 3.14 \ m^2/Hz$, $T_g = 0.25$ s. Fig. 4.5-2 shows the tracking error when the guidance radar designed in Section 4.3 is tracking a target with

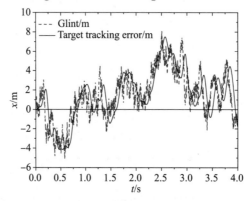

Fig. 4.5-2 Target glint and radar tracking error

glint model as in Fig. 4.5 – 1. Since the glint is a low frequency disturbance, and its frequency bandwidth is in the same order of magnitude as the tracking radar bandwidth, the glint can be followed by the guidance radar. This is why target glint is one of the major sources of guidance radar tracking error.

§ 4.6 Effect of Other Disturbances on the Guidance Performance

4.6.1 Effect of Disturbance Moment

First of all, since the antenna and its base are connected by bearings and many cables, when the two are moving relative to each other, these connections may generate spring and damping disturbance torques. In addition, the tracking antenna has a certain physical size. When there is wind, the wind disturbance moment acted on the antenna can be quite large. It is essential that even with large wind disturbance, the resulting radar tracking error should still satisfy the system specification. Fig. 4.6 – 1 gives a simplified block diagram of the guidance radar tracking system with moment disturbance.

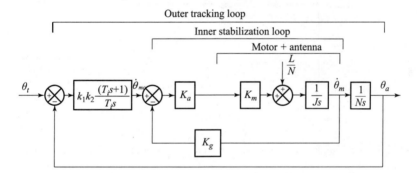

Fig. 4.6 – 1 Block diagram of the guidance radar with disturbance moment input

In the figure, J is the moment of inertia of the antenna, and N is the transmission ratio of the motor gears. The transfer function $\dfrac{\theta_a(s)}{L(s)}$ from the disturbance moment L to the antenna angle error θ_a is given in equation (4.6 – 1). It is known from the equation that the antenna error is inversely proportional to the stabilization open loop gain of K_a, K_m, the tracking loop gain $\dfrac{k_1 k_2}{T_i}$ and the transmission ratio N.

$$\frac{\theta_a(s)}{L(s)} = \frac{\dfrac{T_i}{NK_a K_m k_1 k_2} s}{\dfrac{T_i NJ}{k_a k_m k_1 k_2} s^3 + \dfrac{T_i NK_g}{k_1 k_2} s^2 + T_i s + 1}. \qquad (4.6-1)$$

In other words, the increase of both the tracking loop and the stability loop bandwidth will help reduce the influence of the disturbance moment. In addition, it is known from equation (4.6 – 1) that due to the PI correction in the tracking loop, the steady state error output of the system is zero under the disturbance of the steady state moment.

Suppose the parameters of an example tracking system are as follows: $k_1 k_2 = 1,300$, $T_i = 0.13$ s, $N = 100$, $K_a K_m = 80$ (N·m)/V, $J = 1.2$ kg·m², $K_g = 1$V/(rad·s^{-1}). The Bode diagram of $\theta_a(s)/L(s)$ is shown in Fig. 4.6 – 2 where the responses of the tracking system to the low frequency interference and the high frequency interference can both be neglected, but there may be a large response error for an alternating moment disturbance near the radar system bandwidth due to the resonance effect (such as the effects of wind gust).

Fig. 4.6 – 3 shows the angular error response of the above system under $L/N = 31$ N·m constant moment disturbance.

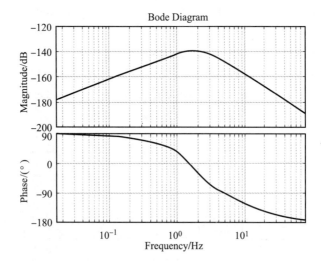

Fig. 4.6 – 2 Bode diagram of the system $\dfrac{\theta_a(s)}{L(s)}$ closed-loop transfer function

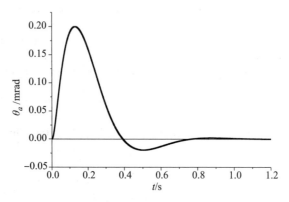

Fig. 4.6 – 3 Angular error response curve with constant moment disturbance ($L/N = 31$ N·m)

Generally, the power spectrum density of the wind disturbance moment model is taken as

$$S(\omega) = \frac{1}{2\pi} \cdot \frac{cV^4}{(1 + \omega^2 T_1^2)(1 + \omega^2 T_2^2)} \quad (\text{N}^2 \cdot \text{m}^2)/\text{Hz}, \qquad (4.6-2)$$

where the value of c depends on the shape and the size of the antenna and the relative direction of the wind. V is the average wind speed; the time constant T_1 is about 8.5 s and T_2 is about 0.5 s.

It is known that for a linear time-invariant system, its input and output power spectrum densities $S_i(\omega)$ and $S_o(\omega)$ have the following relation

$$S_o(\omega) = G(j\omega)G(-j\omega)S_i(\omega), \quad (4.6-3)$$

where $S_i(\omega)$ is the input power spectrum density, $S_o(\omega)$ is the output power spectrum density and $G(s)$ is the system transfer function.

When $S_i(\omega)$ is taken as a white noise of equation (4.6-3) and its power spectrum density is $K_s^2 = \frac{1}{2\pi}cV^4$, it can be seen from equation (4.6-2) that the expression of the filter $G(s)$ in the equation (4.6-3) is $G(s) = \frac{1}{T_1 s + 1} \cdot \frac{1}{T_2 s + 1}$. Therefore, the method in Fig. 4.6-4 can be used to simulate the time domain output characteristics of the wind disturbance.

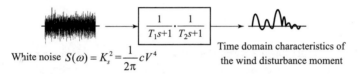

Fig. 4.6-4 The wind disturbance simulation model

Set that $c = 1$, $V = 5$ m/s, $T_1 = 8.5$ s, $T_2 = 0.5$ s, and the power spectrum density of the white noise is $K_s^2 = \frac{1}{2\pi}cV^4 = 99.5$ m²/Hz, the integration step $h = 0.001$ s, and the random number variance of the white noise $\sigma^2 = \frac{K_s^2}{h} = 9.95 \times 10^4$ rad. Fig. 4.6-5 shows the time domain response of the radar system under the influence of the wind disturbance moment. It can be seen that the gust is described by a low frequency disturbance moment model.

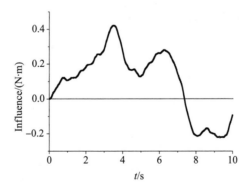

Fig. 4.6-5 Time domain response of the system under the influence of wind disturbance moment

4.6.2 Effect of Target Maneuvers

If the target has a maneuvering acceleration a_t (m/s²) perpendicular to the direction of the missile-target line of sight (Fig. 4.6-6), it corresponds to an angular acceleration $\ddot{\theta} = \dfrac{a_t}{R}$ of the line of sight, and its effect can be handled in accordance with the model for tracking the angular acceleration target in Section 4.3. For example, if the target is making a 4 g maneuver ($a_t = 39.2$ m/s^2), and at the maneuvering time the target distance is $R = 4$ km, then the corresponding angular acceleration will be $\ddot{\theta} = \dfrac{39.2}{4,000} \times 1,000 = 9.8$ (mrad/s^2), which has a value of about 35% of the maximum angular acceleration 28 mrad/s^2 in the example of Section 4.3. Obviously, if such a situation exists, its influence cannot be ignored.

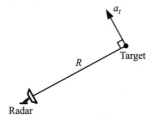

Fig. 4.6-6　Target linear acceleration motion effect on the guidance radar tracking

5

Seekers

§ 5.1 Definitions

Currently, most missile guidance adopts proportional guidance law or its improved guidance law. The main information required for this kind of guidance law is the inertial LOS angular velocity $\dot{q}$. Fig. 5.1 – 1 gives the definitions of the angles related to the seeker and the missile during guidance, where,

ϑ—Angle of the x-axis of the missile with respect to the inertial space;

θ—Angle of the missile velocity vector with respect to the inertial space;

q—LOS angle of the missile and the target relative to the inertial space;

q_d—Angle of the seeker antenna axis with respect to the inertial space;

ε—Angular error of the missile-target line with respect to the seeker axis measured by the seeker detector (optical, radio, laser, etc.).

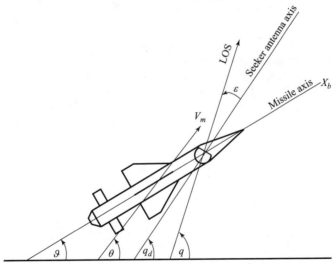

Fig. 5.1 –1 Angle definition diagram

The seeker itself is an angular tracking device which can track the missile-target line in the inertial space. As the target is tracked by the seeker, the $\dot{q}$ signal required for guidance can be obtained by measuring the seeker angular velocity in the inertial space.

Since the seeker is mounted on the moving base of the missile, it is essential for the seeker to isolate the influence of the missile's angular motion on the $\dot{q}$ measurement of the seeker.

§5.2 Different Kinds of Seekers

Essentially, the seekers used on current missiles can be classified as dynamic gyro stabilized seekers, stabilized platform based seekers, semi-strapdown seekers, strapdown seekers and roll-pitch seekers.

5.2.1 Dynamic Gyro Seeker

This type of seeker has two angular degrees of freedom relative to the missile body with the help of the seeker outer gimbal and inner gimbal. The seeker detector is fixed to the inner gimbal, and a dynamic gyro with high moment of momentum or angular momentum can rotate around the detector load through a bearing at high speeds (Fig. 5.2-1).

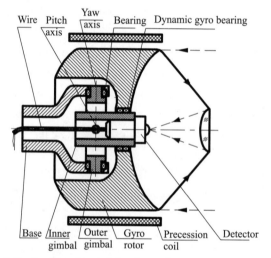

Fig. 5.2-1 Structure of the dynamic gyro stabilized seeker

The functions of the seeker of measuring the LOS angular velocity $\dot{q}$ and isolating the missile's motion can be illustrated by the block diagram (Fig. 5.2-2).

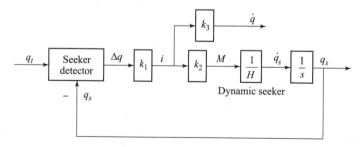

Fig. 5.2-2 Block diagram of the dynamic gyro stabilized seeker system

The angular tracking error signal Δq output by the detector will produce a corresponding precession current i, a procession moment M and a procession angular velocity of the dynamic gyro stabilized seeker $\dot{q}$. Therefore, the seeker precession current can be taken as the rotational angular velocity $\dot{q}$ of the seeker in the inertial space. The open loop gain of the dynamic gyro stabilized seeker tracking loop (usually referred to as the quality factor of the seeker) is denoted as $D = k_1 k_2 / H$ ((rad·s^{-1})/rad), and its physical meaning is the seeker steady state angular velocity D (°/s) per unit seeker detector error 1°. The transfer function of the seeker closed-loop tracking system is

$$\frac{q_s(s)}{q_t(s)} = \frac{1}{T_s s + 1} \left(T_s = \frac{1}{D} \right). \qquad (5.2-1)$$

The momentum moment of the dynamic gyro stabilized seeker $H = J\omega$ is usually designed to be very large, so when the angular motion of the missile body creates a disturbance moment M_d, the corresponding error of the seeker precession angular velocity $\Delta \dot{q} = M_d / H$ will be very small. Since the dynamic gyro has an excellent disturbance rejection capability, the seeker decoupling performance testing procedure for this type of seeker is often omitted in the production process.

Here are some examples of commonly seen dynamic gyro stabilized seekers.

1) Infrared seeker (IR seeker)

The infrared seeker is a passive system seeker and the target infrared signal to be measured is a very weak signal. For this reason, a smaller field of view and a longer focal length are required to reduce the background clutter disturbance. Therefore, most of such seekers use a Cassegrain optical structure design.

(1) Early infrared dynamic gyro seeker (such as the American AIM-9B Sidewinder and Russian Arrow-2M).

Most of the early time infrared seekers adopted uncooled infrared detectors. Since a cooling device is not necessary, the infrared detector device could be made small and it can be installed in the seeker's inner gimbal. In this case, the detector does not need to be placed in the center of the inner and outer gimbal axes (Fig. 5.2-3).

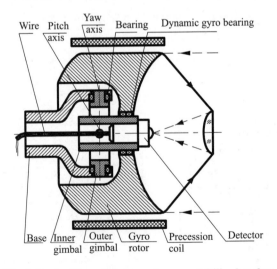

Fig. 5.2-3 **Uncooled dynamic gyro stabilized seeker**

(2) Strapdown cooled infrared dynamic gyro seeker (such as the American Sidewinder and Israeli Python 3 missile).

Since the cooled infrared detector assembly is equipped with a cooling device, its volume is large. The infrared detector is usually fixed on the missile body, and it is required to be installed in the orthogonal center of the inner and outer seeker gimbal axes (Fig. 5.2 – 4).

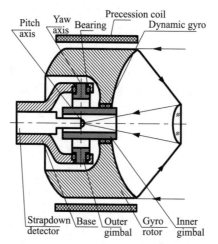

Fig. 5.2 – 4 **Strapdown infrared detector dynamic gyro of the seeker**

2) Semi-active laser seeker

Since the strength of the laser signal reflected by the illuminated target is much stronger than the background clutter in this system, a large field of view design could be adopted. For example, the seeker field of view can often be set as large as 15° to 30°. This eliminates the need for a Cassegrain structure.

(1) The laser semi-active dynamic gyro seeker of Hellfire missiles.

The American Hellfire is a helicopter-mounted laser semi-active air-to-surface guided missile. The four-quadrant laser detector used is mounted in front of the inner gimbal of the seeker's dynamic gyro (Fig. 5.2 – 5).

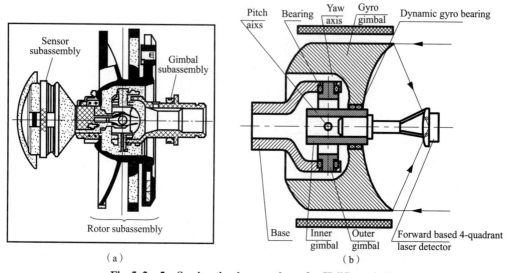

Fig. 5.2 – 5 **Semi-active laser seeker of a Hellfire missile**

(2) The semi-active dynamic gyro laser seeker of the Russian guided projectile (Fig. 5.2–6).

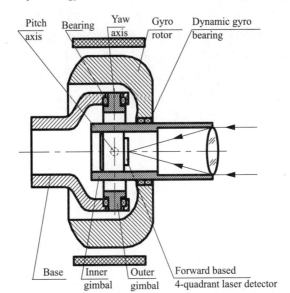

Fig. 5.2–6 Semi-active dynamic gyro laser seeker of the Russian guided projectile

3) Dynamic gyro seeker driven by the motor

In this structure, the precession moment of the dynamic gyro is provided by two electrical motors which are connected to the two gimbals of the dynamic gyro. Since the precession direction of the dynamic gyro is 90° from the precession moment, the actual implementation block diagram of the dynamic gyro seeker is shown in Fig. 5.2–7.

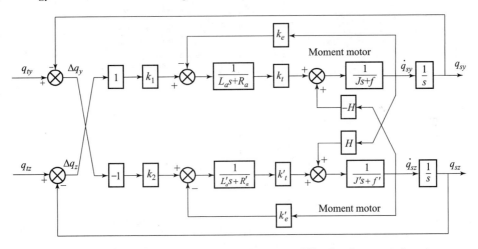

Fig. 5.2–7 Block diagram of the dynamic gyro stabilized seeker control system

4) Dynamic gyro stabilized seeker with additional tracking gimbal

One of the disadvantages of the dynamic gyro seeker is that when the seeker has a gimbal angle ϕ, the effective precession moment generated by the precession coil is only $\cos\theta$ times the precession moment when $\phi = 0$, so such a seeker cannot work under a large gimbal angle (usually the upper limit of the gimbal angle is 40°). In order to solve this problem, there is a type of dynamic gyro seeker designed with an additional tracking gimbal which tracks the central axis of the seeker, and the precession coil is mounted on the tracking gimbal. Since in this way there is only a small angle between the precession moment and the moment of momentum axis of the dynamic gyro, the precession moment loss caused by a large gimbal angle can be avoided, so that the seeker can work under a quite large gimbal angle (the gimbal angle can be increased to 60°). At the same time, since the angle between the seeker load and the tracking gimbal is small, the influence of the gimbal angle disturbance moment on the seeker will be greatly reduced.

Fig. 5.2 – 8 shows the schematic diagram of the seeker, its control block diagram and definitions of its related angles.

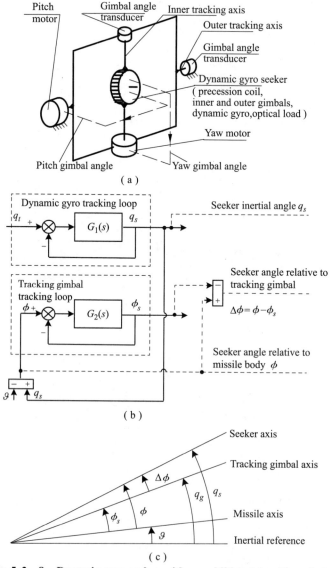

Fig. 5.2 –8 Dynamic gyro seeker with an additional tracking gimbal
(a) Schematic diagram of the dynamic gyro seeker with an additional tracking gimbal, (b) Block diagram of the dynamic gyro seeker control system with an additional tracking gimbal,
(c) Definitions of angles relative to the dynamic gyro seeker with an additional tracking gimbal

In Fig. 5.2 – 8, ϑ——Missile body attitude angle;

q_s——Seeker inertial space angle;

q_g——Tracking gimbal inertial space angle;

ϕ——Angle between the optic axis of the seeker and the missile-axis;

ϕ_s——Angle between the tracking gimbal and the missile-axis;

$\Delta\phi$——Gimbal angle of the optic axis of the seeker with respect to the tracking gimbal.

5.2.2 Stabilized Platform Seeker

1) Two-gimbal stabilized platform based seeker

The platform based seeker is composed of two loops. The inner high-gain inertial space angular velocity stabilization loop provides the ability to reject the angular motion disturbance of the missile body motion and to provide the angular velocity output of the missile-target line of sight in the inertial space. The outer tracking loop ensures the tracking of the missile-target line (Fig. 5.2 – 9). In comparison with the dynamic gyro stabilized seeker, the stabilized platform based seeker has faster response, and it also allows a large gimbal angle and bigger seeker load. The American GBU – 15 guided missile, the AGM – 130 air-to-ground missile, and the Polyphem fiber optic guided missile jointly developed by France, Germany and Italy all adopt image guidance seekers with such a structure. The US Sparrow air-to-air missile uses a two-gimbal stabilized platform radar seeker.

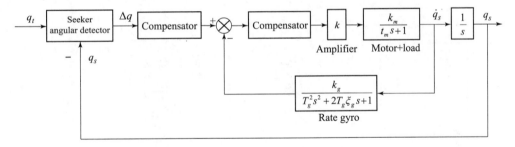

Fig. 5.2 – 9 Block diagram of a stabilized platform seeker

Fig. 5.2 – 10 shows a two-gimbal stabilized platform based radar seeker designed by Marconi Space Defense System (MSDS) Co., Ltd. This structure uses a conventional permanent magnet DC motor and a 20:1 gear reduction unit.

2) Three-gimbal stabilized platform based seeker

In order to eliminate the impact of the missile roll motion on the seeker image stabilization, an additional roll gimbal can be added on the basis of the two-gimbal platform for roll stabilization (Fig. 5.2 – 11). The literature suggests that some Russian air-to-surface image guided missiles and bombs have adopted this structure.

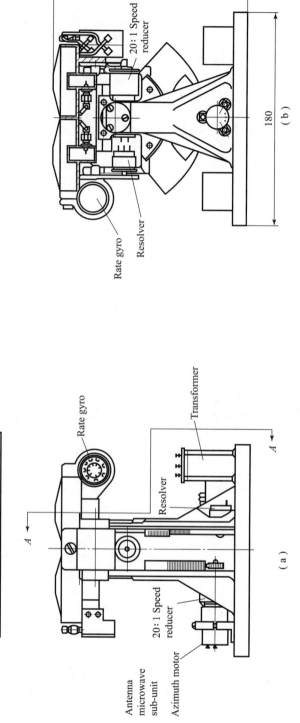

Fig. 5.2-10 Seeker designed by Marconi Space Defense System (MSDS) Co., Ltd (DC motor and angular rate gyro)

(a) View with gyro choke removed; (b) Section A—A with transformer removed

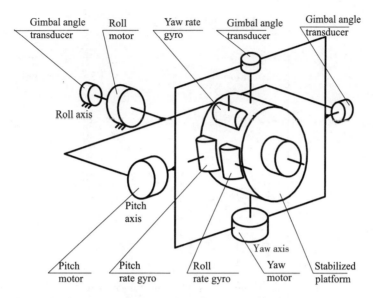

Fig. 5.2 – 11 Schematic of the three-gimbal stabilized platform based seeker

3) Attitude gyro stabilized platform based seeker

In this structure, the error angle generated by the radar detector is used to drive the attitude gyro located on the platform. The inner and outer gimbals of the seeker track the motion of the attitude gyro axis, that is, the gimbal angle of the attitude gyro is the control command of the inner and outer gimbals of the seeker. Fig. 5.2 – 12 is the attitude gyro stabilized radar seeker.

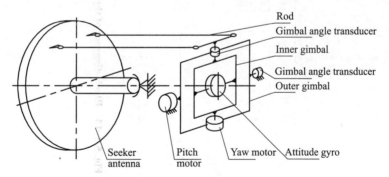

Fig. 5.2 – 12 Attitude gyro stabilized platform based seeker (radar guidance system)

Fig. 5.2 – 13 is the block diagram for the attitude gyro stabilized platform based seeker. Suppose that x represents the variable relative to the inertial space; $\tilde{x}$ represents the variable relative to the missile coordinate system, q_g represents the seeker angle relative to the inertial space; $\tilde{q}_g$ represents the seeker angle relative to the missile coordinate system.

Unlike the standard platform based seeker, which forms a stabilization loop with a high gain angular velocity loop, the stabilization loop of this scheme is an attitude loop. The seeker tracking loop cannot have a very wide frequency bandwidth due to the low sampling frequency and delay of the target detector. However, the sampling frequency of the attitude gyro gimbal angle sensor and the seeker detector platform gimbal angle sensor in this angular stabilization loop could be high, so that

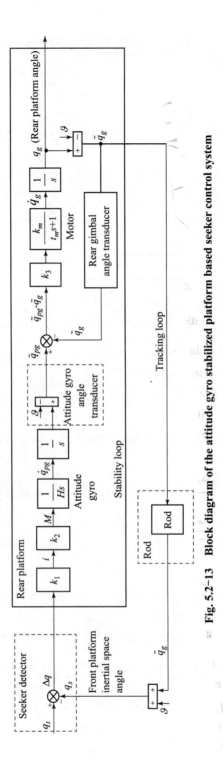

Fig. 5.2-13　Block diagram of the attitude gyro stabilized platform based seeker control system

they can form an angular tracking loop with wide bandwidth to eliminate the angular disturbance of the missile angular motion and stabilize the seeker optic axis.

5.2.3 Detector Strapdown Stabilized Optic Seeker

Compared with the traditional method of placing the detector on the platform for stabilization, there is also a scheme in which the seeker detector is strapped down on the missile body, and the optical axis of the missile body installed detector is stabilized by rotating a moving mirror. The schematic diagram of such a seeker is given in Fig. 5.2 – 14. The angle definitions of this design are shown in Fig. 5.2 – 15.

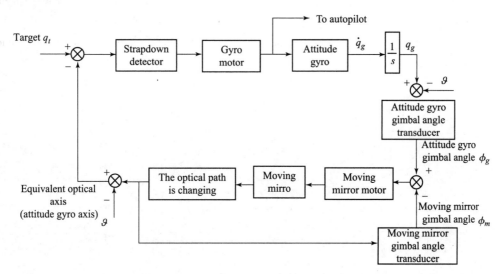

Fig. 5.2 – 14 Schematic diagram of the seeker

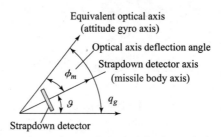

Moving mirror rotation produces optical axis deflection angle ϕ_m to make the optical axis coincide with the detector axis

Fig. 5.2 – 15 Angle definitions of the seeker

In this scheme, the orientation of the attitude gyro is taken as the reference axis. The attitude gyro axis is followed by the mirror rotation to compensate for the optical axis change due to missile angle variation, and also to keep the optical axis coincident with the detector axis.

The stabilization loop of such a seeker is also an angular stabilization loop, which provides the decoupling of the missile body motion disturbance, and the principle of optic axis stabilization is the same as the attitude stabilization loop of the attitude gyro stabilized platform based seeker in Section 5.2.2.

5.2.4 Semi-strapdown Platform Seeker

The angular velocity gyro of the platform based seeker is mounted on the seeker inner gimbal. In order to reduce the load on the seeker platform and lower the cost, the angular velocity gyro on the seeker platform can be removed. In this design, the angular velocity $\dot{q}_s$ of the seeker platform in the inertial space is obtained by adding the angular velocity $\dot{\vartheta}$ of the missile measured by the onboard inertial navigation system and the derivative $\dot{\phi}$ of the seeker gimbal angle ϕ. This scheme can greatly reduce the load and the size of the seeker platform. As a result, the missile can have a more sharpened head, which could greatly reduce the drag of the missile and increase its flight speed.

Since in this scheme only the gyro is strapped down but the seeker detector is still on the platform, it is generally referred to as the so-called semi-strapdown platform based seeker scheme (Fig. 5.2 – 16).

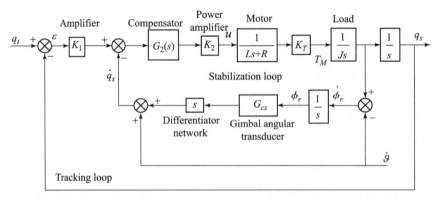

Fig. 5.2 – 16 Block diagram of the semi-strapdown seeker control system

5.2.5 Strapdown Seeker

1) Phased array strapdown seeker

In this design, this phased array antenna is fixed to the missile body (Fig. 5.2 – 17). Here $x_b y_b z_b$ is the missile coordinate system, $x_p y_p z_p$ is the beam coordinate system and $x_i y_i z_i$ is the inertial coordinate system. Taking the pitch channel as an example, the relationship between the angles involved are shown in Fig. 5.2 – 18 where θ_B is the beam angle formed by the beam direction x_p and the missile axis x_b, ϑ is the pitch attitude angle of the missile, and ε_p is the phase array antenna measured target LOS error.

The scheme that is used for the phased array seeker to measure the line-of-sight angular velocity $\dot{q}$ is shown in Fig. 5.2 – 19. Similar to the ordinary platform based seeker, the seeker output $\dot{q}_s$ is still proportional to the angular error ε_p between the beam direction and the target direction. However, since the beam pointing control is in relation to the missile coordinate system, the angular velocity $\dot{\vartheta}$ of the missile should be subtracted from the $\dot{q}_s$ output and the integration is made to obtain the required phased array beam command θ_{BC} relative to the missile body. The expected beam angle θ_B is obtained by the phase array beam control design. As a result, a closed-loop tracking of the target can be achieved.

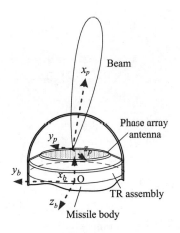

Fig. 5.2 – 17 Structure of the strapdown phased array seeker

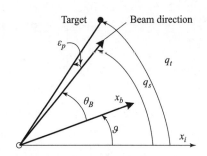

Fig. 5.2 – 18 Relationship of the strapdown phased array antenna angles

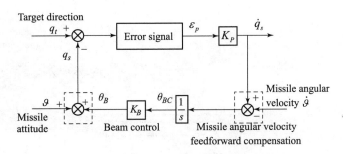

Fig. 5.2 – 19 Control system of the strapdown phased array seeker

It should be noted that the control from the phased array beam command θ_{BC} to the actual beam direction θ_B is an open loop control, and its accuracy completely depends on the calibration process in the phased array antenna production. Due to the strict requirements for the seeker decoupling performance, this calibration requires very high precision and represents an enormous amount of work. At present, it is known that it is required during production to carry out two-dimensional angular calibration of the relationship between the beam command θ_{BC} and the beam deflection angle θ_B for every phase array operational frequency.

2) Image strapdown seeker

The image detector is a highly accurate linear detector. However, after the detector is strapped to the missile body, it requires the detector field of view to cover the gimbal angle range of the conventional seeker, that is, it requires the image detector to have a large enough field of view. If it can be ensured that the target always remains in the image detector field of view during the whole guidance phase, then the angular velocity $\dot{q}$ of the missile-target line of sight can be acquired by adding the derivative of the angular error signal ε measured by the detector onboard to the rate gyro output $\dot{\vartheta}$ of the missile (Fig. 5.2 – 20). That is $\dot{q} = \dot{\vartheta} + \dot{\varepsilon}$.

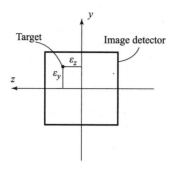

Fig. 5.2 – 20 Target deviation error acquired by the strapdown image detector

Fig. 5.2 – 21 shows the appearance of the US extra atmospheric interceptor. The last stage of the missile uses an infrared image strapdown seeker, the direct force attitude control and trajectory control schemes. It is also known that the Israel short range guided missile Spike-SR also used the same infrared strapdown image seeker solution.

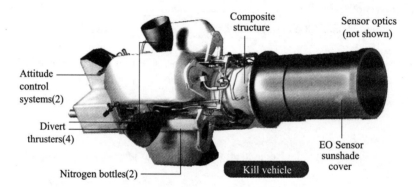

Fig. 5.2 – 21 EKV atmospheric interceptors used by KEI intercept missiles

3) Laser strapdown seeker

Generally, the laser seeker detector uses a 4-quadrant scheme to measure the target angular error. In order to ensure the accuracy of the angle measurement, the size of the diffused laser image is usually very small, that is, the linear range of the measurement for a laser detector is small. However, since the reflected energy of the laser signal is very strong in comparison with the background clutter signal, the nonlinear range of the detector can be very large, that is, the laser seeker detector can choose a large field of view design. Theoretically, a large linear region can be obtained by increasing the size of the diffused laser image, so that the strapdown seeker scheme can be used in the missile guidance (Fig. 5.2 – 22). However, it should be noted that the accuracy of the angle measurement in this solution is low. That is, the output accuracy of the seeker angular velocity $\dot{q} = \dot{\vartheta} + \dot{\varepsilon}$ of the missile-target line depends entirely on the angular error measurement calibration accuracy of each specific seeker detector and its level of sensitivity to environmental change. However, it is known that this scheme indeed has been applied to some guided bombs and guided rockets that do not require very high guidance accuracy. Obviously, the accuracy of angle measurement can be improved by increasing the number of detector pixels. For example, the US APKWS 70mm guided rocket used a seven-pixel laser detector scheme.

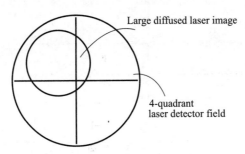

Fig. 5.2-22 **Full strapdown laser seeker solution**

5.2.6 Roll-pitch Seeker

The United States AIM – 9X and the European "Meteor" require a 90° gimbal angle to increase their interception boundary. At the same time, in order to reduce the flight drag and increase the operation range, it is required that the size of the seeker be as small as possible. The following measures have been taken for these purposes:

(1) Adopting the roll-pitch two-gimbal seeker scheme so that a gimbal angle of 90° can be achieved.

(2) Using strapdown infrared image detectors, missile body fixed stirling refrigerator and Kurdish flexible optical path deflection to achieve the detector strapdown and remove the detector and the J-T refrigerator from the seeker platform, which served as the seeker load in the previous seeker design.

(3) Adopting the semi-strapdown stabilization scheme and removing the angular velocity gyro from the seeker platform.

If the seeker has a roll gimbal angle ϕ_R and a pitch gimbal angle ϕ_P when using the Kurd optical path, it can be known that if ε_y and ε_z are the pitch and yaw angle errors on the seeker optical lens and ε_y^* and ε_z^* are the pitch and yaw angle errors on the strapdown detector, the following relation exists

$$\begin{bmatrix} \varepsilon_y^* \\ \varepsilon_z^* \end{bmatrix} = \begin{bmatrix} \cos(\phi_R + \phi_P) & -\sin(\phi_R + \phi_P) \\ \sin(\phi_R + \phi_P) & \cos(\phi_R + \phi_P) \end{bmatrix} \begin{bmatrix} \varepsilon_y \\ \varepsilon_z \end{bmatrix}. \qquad (5.2-2)$$

That is to say, the image of the target on the strapdown detector will generate an angular rotation around the optical axis ($\phi_R + \phi_P$).

Fig. 5.2-23 shows the image of the target seen from the seeker optical lens and the same image on the strapdown detector when $\phi_R = -45°$ and $\phi_R = -25°$.

The function of the traditional seeker stabilization loop is to stabilize the optical axis and reject the disturbance of the missile body motion. For a normal two-gimbal seeker this decoupling function is achieved by the seeker pitch and yaw gimbal bearing freedom to reject the missile pitch and yaw

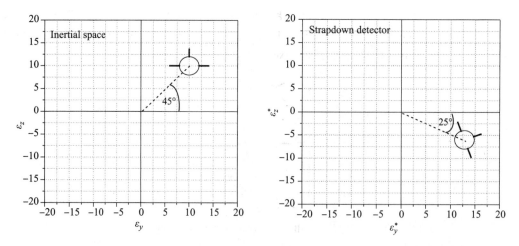

Fig. 5.2 – 23 Image variations of the target on seeker optical lens and the strapdown detector

motion. But in the pitch-roll seeker situation, to do this the seeker has to make a roll rotation to let the pitch gimbal orientate in the composite missile angular motion direction and use pitch bearing alone to isolate the missile's two axis motions (Fig. 5.2 – 24).

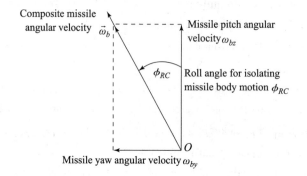

Fig. 5.2 – 24 Principle of the missile body perturbation rejection by pitch gimbal alone

When the missile has a random pitch and yaw angular motion, the roll gimbal will need to rotate back and forth at a high speed. So even if this method is adopted, the seeker decoupling performance must be poor compared with the traditional two-gimbal stabilized seeker and it can be difficult to meet the design specifications.

The real difficulty here is that the main task for a seeker is to track the target. Normally the roll gimbal angles required for target tracking and missile body motion decoupling are not the same. For this reason in reality it is impossible for the pitch-roll seeker to work in the same way as the traditional seeker.

As the pitch-roll seeker cannot track the target accurately, a different approach has to be taken to obtain the guidance required LOS angular velocity $\dot{q}$ output. Suppose with a proper balance of the target tracking and base decoupling requirements, the seeker optical axis can be controlled to maintain the target always in the seeker field view. Then from the navigation system supplied missile angular velocities $\dot{\vartheta}$ and $\dot{\phi}$, seeker gimbal angles ϕ_P and ϕ_R and seeker tracking errors ε_P and ε_y, the required LOS angular velocity could be derived as below.

As a simple example, in one plane only, the $\dot{q}$ expression can be given as equation (5.2-3)
$$\dot{q} = \omega_i + \dot{\phi}_i + \dot{\varepsilon}_i. \qquad (5.2-3)$$

As this scheme is adopted, the target image could still constantly translate and rotate on the strapdown detector, which will greatly increase the image processing difficulty.

§5.3　Anti-disturbance Moment of the Seeker

The platform based seeker consists of two loops, the angular tracking loop and the stabilization loop (Fig. 5.3-1). When the seeker has a disturbance moment input M_d, a signal $\dot{q}_s$ will appear in the stabilization loop feedback, and a signal $\dot{q}_c$ will appear in the tracking loop. The difference of the two ($\dot{q}_c - \dot{q}_s$) will produce the required anti-disturbance moment $k_2(\dot{q}_c - \dot{q}_s)$ to balance the disturbance of M_d. With the disturbance M_d as the input, the output transfer functions of $\dot{q}_s$ and $\dot{q}_c$ are respectively expressed as

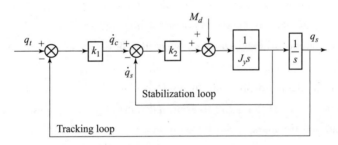

Fig. 5.3-1　The platform based seeker tracking loop

$$\frac{\dot{q}_c(s)}{M_d(s)} = -\left(\frac{1}{k_2}\right)\frac{1}{\dfrac{J_y s^2}{k_2 k_1} + \dfrac{s}{k_1} + 1}, \qquad (5.3-1)$$

$$\frac{\dot{q}_s(s)}{M_d(s)} = \left(\frac{1}{k_1 k_2}\right) \times \left(\frac{s}{\dfrac{J_y s^2}{k_2 k_1} + \dfrac{s}{k_1} + 1}\right). \qquad (5.3-2)$$

It can be seen that the transmission coefficient for $\dot{q}_c$ is $1/k_2$, and for $\dot{q}_s$ is $1/k_1 k_2$. Therefore, a stabilization loop with high open loop gain k_2 can simultaneously reduce the negative effect of the disturbance moment on the guidance command $\dot{q}_s$ or $\dot{q}_c$.

For convenience, the disturbance moment can be converted into a disturbance angular velocity $\dot{q}_d$, $\dot{q}_d = M_d/k_2$. At the same time, take $K_2 = k_2/J_y$, $K_1 = k_1$, and Fig. 5.3-1 can now be simplified as Fig. 5.3-2.

Under the action of disturbance $\dot{q}_d$, the output transfer functions of $\dot{q}_s$ and $\dot{q}_c$ can be expressed as

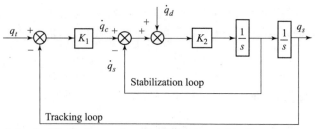

Fig. 5.3-2 Equivalent seeker block diagram with disturbance input

$$\frac{\dot{q}_c(s)}{\dot{q}_d(s)} = -\frac{1}{\frac{s^2}{K_2 K_1} + \frac{s}{K_1} + 1}. \quad (5.3-3)$$

$$\frac{\dot{q}_s(s)}{\dot{q}_d(s)} = \left(\frac{1}{K_1}\right) \times \left(\frac{s}{\frac{s^2}{K_2 K_1} + \frac{s}{K_1} + 1}\right). \quad (5.3-4)$$

Since the missile guidance signal $\dot{q}$ can be taken as $\dot{q}_c$ or $\dot{q}_s$, it is clear that the disturbance effect for different guidance signal extraction points is quite different.

It can be seen from the two transfer functions that there is always the following relationship between the stabilization loop command $\dot{q}_c(s)$ and the stabilization loop feedback $\dot{q}_s(s)$.

$$\dot{q}_c(s) = -\frac{K_1}{s}\dot{q}_s(s), \quad (5.3-5)$$

where $\dot{q}_c(s)$ is the anti-disturbance signal from the outer loop, and $-\dot{q}_s(s)$ is the anti-disturbance signal from the stabilization loop. From the relation $\dot{q}_c(s) = (K_1/s)(-\dot{q}_s(s))$ it is clear that the components $\dot{q}_c(s)$ and $-\dot{q}_s(s)$ are always perpendicular to each other, and $\dot{q}_c(s)$ lags behind $-\dot{q}_s(s)$ by 90°. Fig. 5.3-3 gives the vector diagram of $(\dot{q}_c - \dot{q}_s)$ and the $\dot{q}_d$ disturbance.

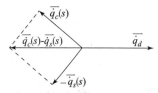

Fig. 5.3-3 Vector synthesis of the control moment

Take the parameters of a typical platform seeker, $K_1 = 10$ and $K_2 = 120$ for example. Fig. 5.3-4 shows the Bode diagram of its $\dot{q}_c(s)/\dot{q}_d(s)$ as well as $-\dot{q}_s(s)/\dot{q}_d(s)$ transfer functions.

It is known from $\dot{q}_c(s) = -K_1\dot{q}_s(s)/s$ that when $\omega < K_1$ rad/s (for low frequency disturbance), the decoupling level of $\dot{q}_s(s)$ is superior to $\dot{q}_c(s)$, when $\omega \approx K_1$ rad/s, $\dot{q}_s(s)$ and $\dot{q}_c(s)$ will have equivalent decoupling level; when $\omega > K_1$ rad/s (for high frequency disturbance), the decoupling level of $\dot{q}_c(s)$ is better than that of $\dot{q}_s(s)$. The above conclusions can be clearly seen from the decoupling level of the two for three frequencies of low frequency ($\omega = K_1/2$ rad/s), medium frequency ($\omega = K_1$ rad/s), high frequency ($\omega = 2K_1$ rad/s) (Table 5.3-1) and the anti-disturbance vector diagram of $\dot{q}_c(s) - \dot{q}_s(s)$ (Fig. 5.3-5).

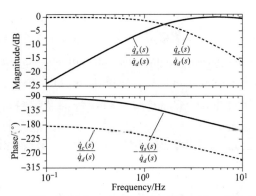

Fig. 5.3 – 4 Bode diagram of the disturbance effect for different guidance signal extraction points

Table 5.3 – 1 Coupling level for different disturbance frequencies

Frequency/(rad · s^{-1})	Coupling level	
	$|\dot{q}_c/\dot{q}_d|$	$|\dot{q}_s/\dot{q}_d|$
Low frequency $\omega = K_1/2 = 5$	0.909	0.455
Medium frequency $\omega = K_1 = 10$	0.737	0.737
High frequency $\omega = 2K_1 = 20$	0.474	0.949

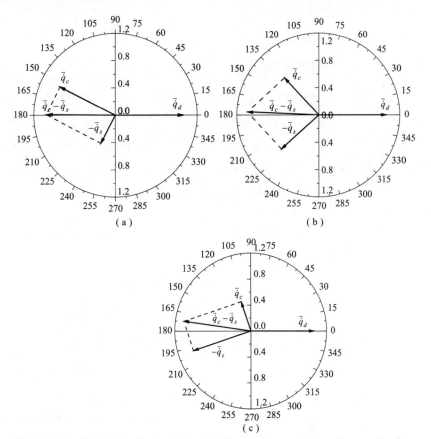

Fig. 5.3 – 5 Vector synthesis of the disturbance moment decoupling mechanism

(a) $\omega = 5$ rad/s; (b) $\omega = 10$ rad/s; (c) $\omega = 20$ rad/s

When there is no disturbance moment input and the seeker is tracking a target $\dot{q}_t$, because the bandwidth of the stabilization loop is very high, we will have $\dot{q}_c \approx \dot{q}_s$. That is to say, there is no difference in taking $\dot{q}_s$ or $\dot{q}_c$ as the guidance command, but if disturbance exists, taking $\dot{q}_s$ or $\dot{q}_c$ as the guidance signal will be totally different and so its effects on the missile guidance performance are different. This will be one of the main topics to be discussed in the next section.

§5.4 Body Motion Coupling and the Parasitic Loop

5.4.1 Body Motion Coupling Dynamics Model

The seeker block diagram under the influence of the missile body motion coupling disturbance moment is shown in Fig. 5.4 – 1.

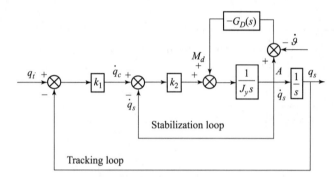

Fig. 5.4 – 1 Seeker block diagram under the influence of the body motion coupling disturbance moment

In Fig. 5.4 – 1, $G_D(s)$ is the transfer function of the disturbance moment M_d caused by the seeker-missile body relative motion. Usually the decoupling level of the platform based seeker is very high, and the output signal $\dot{q}_s$ of the stabilization loop caused by missile angular velocity is small in comparison with missile angular velocity $\dot{\vartheta}$, usually not more than 5%. Therefore, the influence of the change of $\dot{q}_s$ at point A in Fig. 5.4 – 1 can be neglected, and a basic model for studying the effect of the missile body disturbance is shown in Fig. 5.4 – 2.

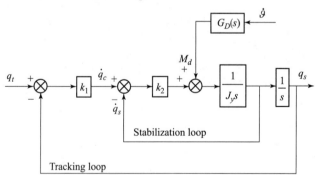

Fig. 5.4 – 2 Equivalent seeker block diagram

Here, it should be noted that the seeker block diagram model in early Russian publications as shown in Fig. 5.4 – 3 has been adopted in most published guidance control books and reference papers.

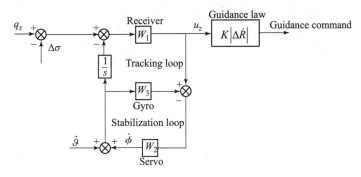

Fig. 5.4 – 3 Model of taking the motor output angular
velocity as the relative angular velocity

In Fig. 5.4 – 3, $\Delta\sigma$ is the radome refraction error, $\dot{\vartheta}$ is the missile angular velocity, q_s is the line-of-sight angle of the target, u_z is the seeker line-of-sight angular velocity output and ϕ is the antenna gimbal angle.

There is a serious mistake when this model is applied in a motor driven platform. Here the angular acceleration and angular velocity generated by the motor driven platform are respectively mistakenly regarded as the seeker gimbal angular acceleration $\ddot{\phi}$ and the angular velocity $\dot{\phi}$ and it is considered that the angular velocity $\dot{q}_s$ of the platform in the inertial space is the sum of $\dot{\phi}$ and $\dot{\vartheta}$. Actually, according to Newton's Law, the motor moment M generated seeker platform angular acceleration and angular velocity are relative to the inertial space rather than to the missile body coordinate system. It should be noted that the missile body motion can affect the seeker only through the wire moment, bearing moment and other moments that occur when the seeker has a relative motion with respect to the missile. When there is no moment disturbance, the missile motion has no effect on the seeker due to the gimbal bearing decoupling function.

From the above analysis, it is known that the seeker block diagram shown in Fig. 5.4 – 2 can truly reflect the missile body motion coupling characteristics. For simplification, the influence of the disturbance moment M_d can be changed to a disturbance angular acceleration input $\ddot{q}_d$, i. e., $\ddot{q}_d = M_d/J_y$, $\bar{G}_D(s) = G_D(s)/J_y$, $K_2 = k_2/J_y$ and $K_1 = k_1$. The equivalent diagram is shown in Fig. 5.4 – 4.

According to this model, the transfer function from $\dot{\vartheta}$ to $\Delta \dot{q}_s$ is

$$-\frac{\Delta\dot{q}_s(s)}{\dot{\vartheta}(s)} = \left(\frac{\bar{G}_D(s)}{K_1 K_2}\right) \times \left(\frac{s}{\frac{s^2}{K_2 K_1} + \frac{s}{K_1} + 1}\right), \quad (5.4-1)$$

and the transfer function from $\dot{\vartheta}$ to $\Delta \dot{q}_c$ is

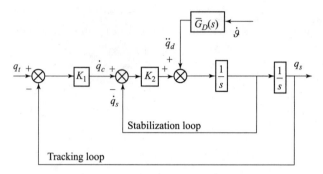

Fig. 5.4-4 Equivalent seeker block diagram

$$\frac{\Delta \dot{q}_c(s)}{\dot{\vartheta}(s)} = -\left(\frac{\bar{G}_D(s)}{K_2}\right) \times \left(\frac{1}{\frac{s^2}{K_2 K_1} + \frac{s}{K_1} + 1}\right). \qquad (5.4-2)$$

From the difference between the two transfer functions, it is indicated that regardless of the disturbance moment model $\bar{G}_D(s)$, there is always the following relationship between $\dfrac{\Delta \dot{q}_c(s)}{\dot{\vartheta}(s)}$ and $\dfrac{\Delta \dot{q}_s(s)}{\dot{\vartheta}(s)}$,

$$\frac{\Delta \dot{q}_c(s)}{\dot{\vartheta}(s)} = -\frac{K_1}{s} \frac{\Delta \dot{q}_s(s)}{\dot{\vartheta}(s)},$$

that is, the two have an equivalent order of magnitude decoupling level when $\omega \approx K_1$ rad/s, and the decoupling level for the stabilization loop command $\Delta \dot{q}_c$ is better than that of the stabilization loop feedback $\Delta \dot{q}_s$ at a high frequency range (when $\omega > K_1$ rad/s). While the conclusion is the opposite in the low frequency range (when $\omega < K_1$ rad/s), and the $\Delta \dot{q}_s$ always lags behind the $\Delta \dot{q}_c$ by 90°.

For an actual seeker, the disturbance moment model varies greatly depending on the different seeker structure designs, and it can be known only by specific design identification. In the following, the damping moment will be taken as the disturbance model to analyze the decoupling problem and its influence on the missile guidance parasitic loop when the seeker takes the stabilization loop command $\dot{q}_c$ or stabilization loop feedback $\dot{q}_s$ as its output. The damping disturbance is related to the damping moment generated by the seeker bearing and the connecting wire when the missile is moving relative to the seeker. In general, this disturbance moment model can be given as

$$\ddot{q}_d(s) = K_\omega \dot{\vartheta}(s). \qquad (5.4-3)$$

In equation (5.4-3), $K_\omega = k_\omega / J_y$, and the equivalent diagram is given in Fig. 5.4-5.

Therefore, the transfer function from $\dot{\vartheta}(s)$ to $\Delta \dot{q}_s$ becomes

5 Seekers

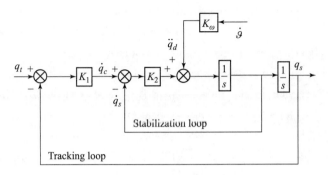

Fig. 5.4-5 Equivalent diagram of the seeker control system

$$\frac{\Delta \dot{q}_s(s)}{\dot{\vartheta}(s)} = \left(\frac{K_\omega}{K_1 K_2}\right) \times \left(\frac{s}{\frac{s^2}{K_2 K_1} + \frac{s}{K_1} + 1}\right). \quad (5.4-4)$$

The transfer function from $\dot{\vartheta}(s)$ to $\Delta \dot{q}_c$ is

$$\frac{\Delta \dot{q}_c(s)}{\dot{\vartheta}(s)} = -\left(\frac{K_\omega}{K_2}\right) \times \left(\frac{1}{\frac{s^2}{K_2 K_1} + \frac{s}{K_1} + 1}\right). \quad (5.4-5)$$

Take the tracking loop parameter $K_1 = 10$, and the stabilization loop parameter $K_2 = 169$. Suppose $K_\omega = 1$ and the Bode diagram of the above second order transfer function can be obtained (Fig. 5.4-6). It can be seen that when the disturbance moment is the damping moment, the relationship between the amplitude and the phase of the two is still consistent with our conclusions about the general disturbance moment $\overline{G}_D(s)$.

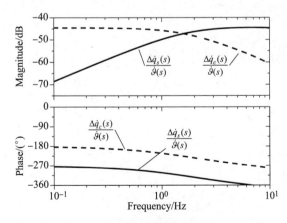

Fig. 5.4-6 Bode diagram of the coupling level transfer function at different output points for a seeker under the damping moment disturbance

5.4.2 Guidance Parasitic Loop Introduced by Seeker-missile Coupling

It is known that in guidance loop operation the seeker output $\dot{q}$ is used to generate an acceleration command a_c to the missile autopilot through the guidance law $a_c = NV_c\dot{q}$. For any seeker without disturbance moments, its output is independent of the missile's angular motion. However, when there is a disturbance moment produced by the missile motion $\dot{\vartheta}$, it will generate a seeker disturbance output $\Delta \dot{q}$ through its coupling model. This seeker-missile coupling output $\Delta \dot{q}$ will produce an inner parasitic loop in the guidance loop. The presence of this parasitic loop can seriously affect the performance of the missile guidance system. Fig. 5. 4 – 7 shows the block diagram of this parasitic loop. In order to facilitate the analysis of control system stability by using the negative feedback theory, a negative feedback is introduced in the loop, so that the coupling transfer function in the parasitic loop takes the form $- \Delta \dot{q}(s)/\dot{\vartheta}(s)$.

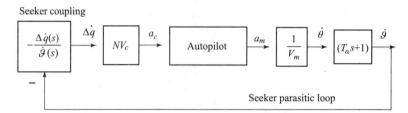

Fig. 5. 4 –7   Parasitic loop of the seeker coupling

In Fig. 5. 4 –7, T_α is the angle of attack time constant, N is the effective navigation ratio, V_c is the missile-target relative velocity, V_m is the missile flight velocity, a_c is the acceleration command of the autopilot, and a_m is the acceleration response of the missile. According to the previous analysis, the coupling transfer function has different characteristics for different extraction points of the seeker guidance signal. Fig. 5. 4 – 8 and Fig. 5. 4 – 9 show the seeker parasitic loop composed of different coupling transfer functions for different output points of the seeker.

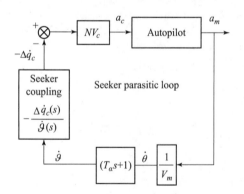
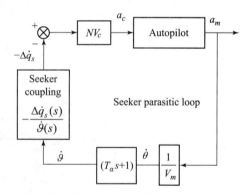

Fig. 5. 4 –8   Parasitic loop model when extracting the seeker guidance signal at point C

Fig. 5. 4 –9   Parasitic loop model when extracting the seeker guidance signal at point S

In the figures above, $\Delta \dot{q}_c$ is the coupling line-of-sight angular velocity generated when the guidance signal is extracted from point C (the seeker stabilization loop command), and $\Delta \dot{q}_s$ is the coupling line-of-sight angular velocity generated when the guidance signal is extracted from point S (the seeker stabilization loop feedback).

From the previous analysis, it is known that as $\dot{q}_c(s) = -(K_1/s)\dot{q}_s(s)$, point C and point S

have equivalent decoupling level under the disturbance only when $\omega = K_1$ rad/s.

For the coupling model (Fig. 5.4 – 10), as an example, take the loop parameters as $K_1 = 10$ and $K_2 = 169$. When $\omega = K_1 = 10$ (1.6 Hz), the coupling level or magnitude is the same no matter whether the guidance signal is extracted from point C or point S. As the damping moment coefficient $K_\omega = 7.86$, at the 1.6Hz both of the coupling levels are 3%. The coupling level transfer function for different extraction points C, S and different frequency ω are shown in Fig. 5.4 – 11.

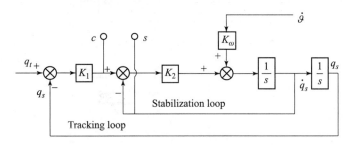

Fig. 5.4 – 10 Seeker coupling block diagram

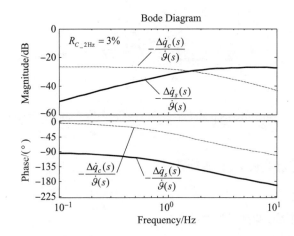

Fig. 5.4 – 11 Coupling transfer function for the given example

Comparing the Bode diagrams of $-\Delta\dot{q}_c(s)/\dot{\vartheta}(s)$ and $-\Delta\dot{q}_s(s)/\dot{\vartheta}(s)$, we can observe that, in terms of the transfer function magnitude, the decoupling level for the case of a signal from point S in the low frequency region performs better while the point C case in the high frequency region performs better. Meanwhile, in terms of phase, the point S case has a 90° phase lag in comparison with the point C case. This will have an adverse effect on the stability of the parasitic loop.

Based on the above coupling transfer function model, suppose the seeker disturbance moment parameter is R_ω ($R_\omega = K_\omega/K_2$), and the parasitic loop non-dimensional parameters are taken as $\overline{T}_\alpha$ and $NV_c R_\omega/V_m$ (Fig. 5.4 – 12). Fig. 5.4 – 13 shows the stable region and unstable region of the parasitic loop taking non-dimensional $\overline{T}_\alpha$ and $NV_c R_\omega/V_m$ as parameters. Here $\overline{s} = T_g s$, $\overline{T}_\alpha = \dfrac{T_\alpha}{T_g}$ and T_g is the system guidance time constant.

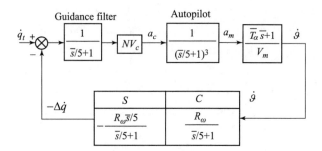

Fig. 5.4 – 12 Seeker parasitic loop with damping moment coupling

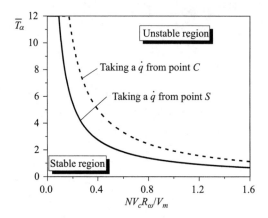

Fig. 5.4 – 13 Parasitic loop stabilization region with damping moment coupling

§ 5.5 A Real Seeker Model

5.5.1 A Real Seeker Model

The effect of the seeker parasitic loop on the missile guidance performance can be clearly described with a real world seeker model with parasitic loops. When there is no parasitic loop in the seeker, the guidance block diagram from the target $\dot{q}_t$ to the autopilot output a_m is shown in Fig. 5.5 – 1. When a parasitic loop exists, the modified block diagram is shown in Fig. 5.5 – 2.

Fig. 5.5 – 1 $\dot{q}_t$ to a_m block diagram with no parasitic loop

The real seeker model $\dot{q}_s(s)/\dot{q}_t(s)$ or $\dot{q}_c(s)/\dot{q}_t(s)$ can be obtained as shown in Fig. 5.5 – 3 by equivalent transformation of Fig. 5.5 – 2.

It can be observed from Fig. 5.5 – 4 that when the real seeker model is used to replace the seeker model with no coupling disturbance, the block diagram from $\dot{q}_t$ to a_m will be the same as before.

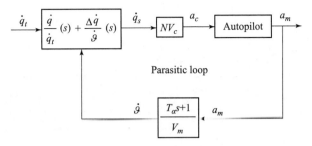

Fig. 5.5-2 $\dot{q}_t$ to a_m block diagram with parasitic loop

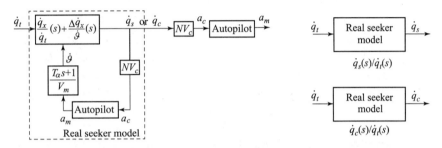

Fig. 5.5-3 Real seeker model

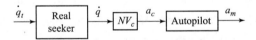

Fig. 5.5-4 Guidance model after introducing a real seeker model

Next, the impact of the parasitic loop on the real seeker model is analyzed with the damping disturbance as an example. The parameters of the seeker, guidance law and autopilot are given in Table 5.5-1.

Table 5.5-1 Typical seeker model parameters

Parameters	Values
Tracking loop K_1	10 rad/s (2 Hz)
Stabilization loop K_2	169 rad/s (27 Hz)
Relative velocity V_c	1,200 m/s
Missile velocity V_m	800 m/s
Angle of attack time constant T_α	1 s
Navigation ratio N	4
Autopilot damping ratio ζ_b	0.65
Autopilot frequency ω_b	2 Hz

The damping moment parameter K_ω is adjusted so that the coupling levels $\Delta \dot{q}/\dot{\vartheta}$ of the two signal output points C and S are 1%, 2%, 3%, and 4% when $\omega = 2$ Hz. The Bode diagram of

$\dot q_c(s)/\dot q_t(s)$ and $\dot q_s(s)/\dot q_t(s)$ transfer functions are given in Fig. 5.5 – 5.

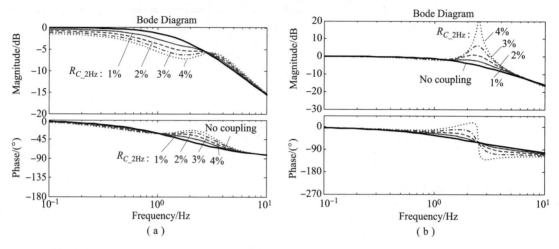

Fig. 5.5 – 5 Real seeker models with point C and point S signal as the outputs respectively

(a) $\dot q_c(s)/\dot q_t(s)$; (b) $\dot q_s(s)/\dot q_t(s)$

Fig. 5.5 – 6 shows the time domain response of the two real seeker models to a step input signal $\dot q_t$. From the frequency domain and time domain characteristics of the two, it is known that the seeker achieves good stability when the signal is taken from point C, but the low frequency gain will decrease when there is a parasitic loop. This phenomenon does not occur when the $\dot q$ signal is taken from point S, but the seeker will have less stability.

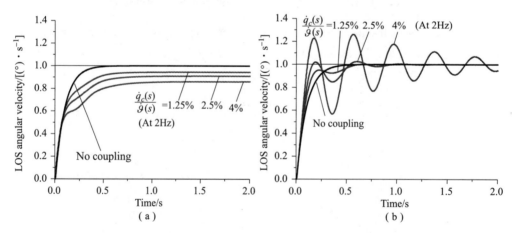

Fig. 5.5 – 6 Time domain characteristics of the real seeker model under the influence of the damping moment

(a) Extraction point C; (b) Extraction point S

Based on the above analysis, it can be seen that the parasitic loop of the seeker at different output extraction points will have different effects on the time domain and frequency domain responses of the seeker, and will affect the guidance performance differently.

For practical seeker design, the real coupling moment model could be very complicated depending on the specific platform design. It is almost impossible to use a simple model to simulate its effect. For this reason, in real guidance system design, hardware-in-the-loop simulation is always used to evaluate the missile-seeker coupling effect on the guidance system performance.

Owing to the restricted resources, only a limited number of hardware-in-the-loop simulations can be performed. But if a real seeker model $\dfrac{\dot{q}_s(s)}{\dot{q}_t(s)}$ or $\dfrac{\dot{q}_c(s)}{\dot{q}_t(s)}$ can be obtained via test identification, the guidance system performance evaluation can then be performed by way of mathematical simulation. Furthermore, the mathematical simulation results can help to determine which seeker output $\dot{q}_s(s)$ or $\dot{q}_c(s)$ should be chosen in the final guidance system implementation.

5.5.2 Testing Methods

Based on the above analysis, two testing methods can be used to obtain the real seeker's transfer function that includes the missile-seeker coupling parasitic loop effect.

1) Direct testing method

The direct testing method arrangement is shown in Fig. 5.5 −7 and Fig. 5.5 −8.

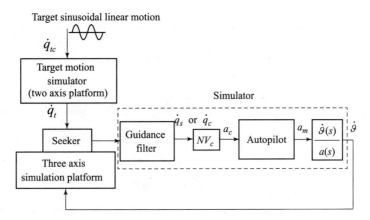

Fig. 5.5 −7 **Schematic of the real seeker transfer function testing arrangement**

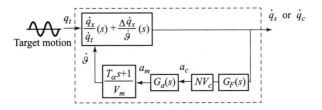

Fig. 5.5 −8 **Block diagram for the direct testing**

The input of the test system is the target line-of-sight angular velocity $\dot{q}_t$, which is generated by a two-axis target simulator. The output of the test system is the seeker stabilization loop command $\dot{q}_c$ and the rate gyro feedback signal $\dot{q}_s$. In this testing, the test system should have the guidance filter model, the guidance law model and the autopilot model installed in the simulation computer and the simulation computer output (the missile attitude angular velocity $\dot{\vartheta}$ and angle ϑ) should be used to

2) Indirect testing method

This method tests both the seeker model with no coupling $\dfrac{\dot{q}_x(s)}{\dot{q}_t(s)}$ and the coupling model $\dfrac{\Delta\dot{q}_c(s)}{\dot{\vartheta}(s)}$ and $\dfrac{\Delta\dot{q}_s(s)}{\dot{\vartheta}(s)}$ separately, and combining the two will give the required result $\dfrac{\Delta\dot{q}_c(s)}{\dot{q}_t(s)}$ or $\dfrac{\Delta\dot{q}_s(s)}{\dot{q}_t(s)}$ (Fig. 5.5–9).

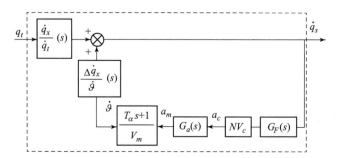

Fig. 5.5–9　Schematic of the indirect testing method

Through the seeker transfer function testing, the seeker transfer function without coupling can be obtained as

$$G_s(s) = \dfrac{\dot{q}(s)}{\dot{q}_t(s)}. \qquad (5.5-1)$$

Through the missile seeker coupling testing, the coupling transfer function can be obtained as

$$R_{DRE}(s) = \dfrac{\Delta\dot{q}_c(s)}{\dot{\vartheta}(s)} \text{ or } \dfrac{\Delta\dot{q}_s(s)}{\dot{\vartheta}(s)}. \qquad (5.5-2)$$

Finally, the real seeker transfer function with the coupling parasitic loop included can be obtained as

$$R_{seeker}(s) = \dfrac{\dfrac{\dot{q}_x(s)}{\dot{q}_t(s)}}{1 + \dfrac{\Delta\dot{q}(s)}{\dot{\vartheta}(s)} G_a(s) G_F(s)(T_\alpha s + 1)\dfrac{NV_c}{V_m}}. \qquad (5.5-3)$$

where $G_F(s)$ is the guidance filter model and $G_a(s)$ is the autopilot model.

§5.6　Other Parasitic Loop Models

5.6.1　Parasitic Loop Model for a Phase Array Strapdown Seeker

Using the phased array strapdown seeker block diagram shown in Fig. 5.6–1, the seeker output can be given by equation (5.6–1):

5 Seekers

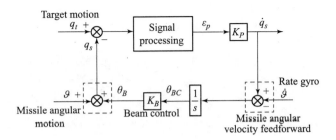

Fig. 5.6-1 Phase array strapdown seeker

$$\dot{q}_s(s) = \frac{\dfrac{1}{K_B}}{\dfrac{1}{K_B K_P}s + 1}\dot{q}_t + \frac{1 - \dfrac{1}{K_B}}{\dfrac{1}{K_P K_B}s + 1}\dot{\vartheta}. \qquad (5.6-1)$$

According to equation (5.6-1), when the beam control gain is $K_B = 1$, the missile angular velocity $\dot{\vartheta}$ disturbance to the seeker can be completely eliminated from the seeker output $\dot{q}_s$. At this time, the transfer function of the phase array strapdown seeker will be simply as follows:

$$\frac{\dot{q}_s(s)}{\dot{q}_t(s)} = \frac{\dfrac{1}{K_B}}{\dfrac{1}{K_B K_P}s + 1} \approx \frac{1}{T_s s + 1} \quad \left(T_s = \frac{1}{K_P}\right), \qquad (5.6-2)$$

where K_B is the beam control gain. Since the beam control from its command θ_{BC} to the beam reflection angle θ_B is an open loop control, the gain K_B can only be obtained through the strict calibration for different beam angles and operation frequencies. K_B can be adjusted to 1 in theory by computer compensation after calibration. When the gain K_B cannot be perfectly compensated to 1, a parasitic loop will be generated and its effect on the seeker output is given by

$$\frac{\Delta\dot{q}_s(s)}{\dot{\vartheta}(s)} = \frac{1 - \dfrac{1}{K_B}}{T_s s + 1}. \qquad (5.6-3)$$

When an error exists in the gain K_B, $K_B = 1 + \Delta K_B$, in which ΔK_B is the calibration error, that is $\Delta K_B = \dfrac{\theta_B - \theta_{BC}}{\theta_{BC}}$. Therefore

$$1 - \frac{1}{K_B} = 1 - \frac{1}{1 + \Delta K_B} \approx 1 - (1 - \Delta K_B) = \Delta K_B. \qquad (5.6-4)$$

So the coupling transfer function will be

$$\frac{\Delta\dot{q}_s(s)}{\dot{\vartheta}(s)} = \frac{\Delta K_B}{T_s s + 1} \qquad (5.6-5)$$

Since this decoupling model is the same as the radome slope error model, the analysis of its influence can be found in the following section.

5.6.2 Parasitic Loop Model Due to Radome Slope Error

When the target echo passes through the missile radome, the random dielectric constant and its conical shape could deviate its path from a straight line. As this happens, the seeker will track a false target with a tracking error Δq as shown in Fig. 5.6-2.

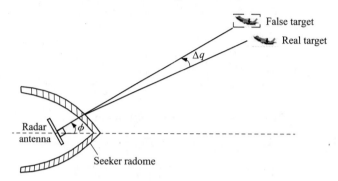

Fig. 5.6-2 Mechanism diagram of radome beam pointing error

Usually, the error of the Δq model is set to be proportional to the seeker gimbal angle ϕ, that is $\Delta q = R_{\text{dom}} \phi$, where R_{dom} is the random error coefficient.

When the seeker tracks the target in the inertial space, and the missile has certain angular velocity $\dot{\vartheta}$, the seeker gimbal angular will be $\phi = -\vartheta$ and $\dot{\phi} = -\dot{\vartheta}$.

Therefore, taking into account the beam pointing error caused by the radome, the seeker model is shown in Fig. 5.6-3 or a simplified model in Fig. 5.6-4.

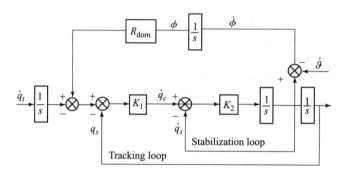

Fig. 5.6-3 Seeker model with radome slope error effect

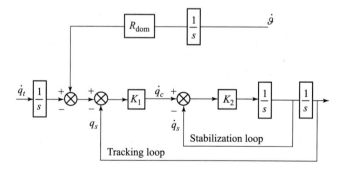

Fig. 5.6-4 Simplified seeker model with radome slope error effect

Therefore, the coupling transfer function caused by the radome error R_{dom} will be

$$\frac{\Delta \dot{q}_c(s)}{\dot{\vartheta}(s)} \approx \frac{\Delta \dot{q}_s(s)}{\dot{\vartheta}(s)} = \frac{-R_{\text{dom}}}{\frac{1}{K_1}s + 1} = \frac{-R_{\text{dom}}}{T_s s + 1}. \tag{5.6-6}$$

5.6.3 Beam Control Gain Error ΔK_B of the Phased Array Seeker and the Radome Slope Error R_{dom} Effect on the Seeker's Performance

It is seen from the above two subsections that the seeker coupling caused by ΔK_B of the phase array and R_{dom} of the radome slope error are respectively

$$\frac{\Delta \dot{q}_c(s)}{\dot{\vartheta}(s)} = \frac{\Delta K_B}{T_s s + 1}. \tag{5.6-7}$$

$$\frac{\Delta \dot{q}_c(s)}{\dot{\vartheta}(s)} = \frac{-R_{\text{dom}}}{T_s s + 1}. \tag{5.6-8}$$

We can see that the two transfer functions have the same form. Thus their influence on the parasitic loop can be expressed with a unified transfer function

$$\frac{\Delta \dot{q}_c(s)}{\dot{\vartheta}(s)} = \frac{-R}{T_s s + 1}. \tag{5.6-9}$$

In the above unified transfer function, $R = R_{\text{dom}} = -\Delta K_B$.

After introducing this coupling model, the parasitic loop model is shown in Fig. 5.6-5.

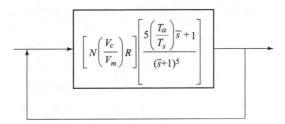

Fig. 5.6-5 A unified parasitic loop model

From Fig. 5.6-5, it can be seen that when the coupling coefficient R is negative, the parasitic loop is a negative feedback, and when R is positive, the parasitic loop is a positive feedback. Due to the lead compensation of the missile transfer function $(T_\alpha s + 1)$, the positive feedback case may still be stable when the parasitic open loop gain is low. However, for the same absolute value $|R|$, the parasitic loop is definitely less stable as the feedback is positive. When R is taken both as positive and negative, its different effects on the parasitic loop stability can be clearly seen from the stabilization region diagram in Fig. 5.6-6, which is given with non-dimensional parameters $N\left(\dfrac{V_c}{V_m}\right)R$ and $\dfrac{T_\alpha}{T_g}$ as variables.

Fig. 5.6-6 shows that the parasitic loop stability is poor when the parameter $\dfrac{T_\alpha}{T_g}$ is high. At high altitudes, the value of the angle of attack time constant T_α is large. In order to avoid the

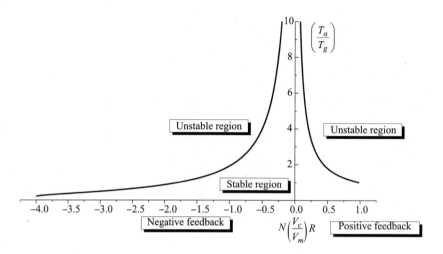

Fig. 5.6−6 Stabilization region of the parasitic loop with $N\left(\dfrac{V_c}{V_m}\right)R$ and $\dfrac{T_\alpha}{T_g}$ as variables

instability problem of the parasitic loop, some missiles adopt an approach of reducing the autopilot bandwidth at high altitudes to increase the value of T_g, so that the value of $\dfrac{T_\alpha}{T_g}$ is not too large when the missile enters high altitudes. Another parameter $N\left(\dfrac{V_c}{V_m}\right)R$ shows that the increase of the proportional navigation constant N, the relative velocity V_c to the missile velocity V_m ratio $\dfrac{V_c}{V_m}$ as well as the coupling coefficient R will all have a negative effect on the seeker parasitic loop stability.

The above analysis of the parasitic loop is based on the assumption that $t_{go} = \infty$ and the gain of the guidance loop is zero, i.e. (Fig. 5.6−7), there is no influence of the guidance loop on the stability of the parasitic loop. However, when the missile is approaching the target and t_{go} is not large, the presence of the guidance loop can indeed reduce the stability of the parasitic loop. Fig. 5.6−7 shows the block diagram of the stability analysis of the parasitic loop with presence of a guidance loop.

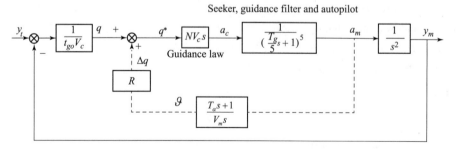

Fig. 5.6−7 Block diagram of the stabilization region analysis of the parasitic loop with a guidance loop

Take the same non-dimensional parameters $N\left(\dfrac{V_c}{V_m}\right)R$ and $\dfrac{T_\alpha}{T_g}$ as variables and the non-dimensional time-to-go $\frac{t_{go}}{T_g}$ as parameter, the new stabilization region diagram is given as Fig. 5.6 – 8 $\left(\text{hereby taking } \frac{t_{go}}{T_g} = \infty, 10, 5\right)$. From the figure, we can observe the parasitic loop stability will deteriorate as the missile approaches the target $\left(\text{i. e. , as } \frac{t_{go}}{T_g} \text{ gets smaller}\right)$.

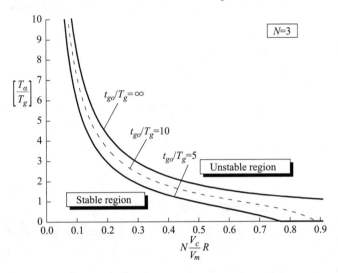

Fig. 5.6 – 8 Influence of guidance loops on the stabilization region of the parasitic loop

It should be noted that as t_{go} changes, the guidance loop becomes a time-varying system, and it is wrong to analyze the stability of the parasitic loop with a time-invariant system analysis method. However, for qualitative analysis, the conclusion that small t_{go} will reduce the parasitic loop stability is still qualitatively correct.

5.6.4　A Novel Online Estimation and Compensation Method for Strapdown Phased Array Seeker Disturbance Rejection Effect Using Extended State Kalman Filter

5.6.4.1　Mathematical model

The basic missile-target geometry is shown in Fig. 5.6 – 9.

The apparent LOS angle q^* can be expressed as

$$q^* = q_t + (q_s - \vartheta)R. \tag{5.6-10}$$

In the stable tracking situation, the tracking error angle ε can be ignored and the radome error slope R is far less than 1 in practice. Hence, equation (5.6 – 10) can be substituted into

$$q^* \cong q_t + (q_t - \vartheta)R = q_t(1 + R) - \vartheta R = q_t - \vartheta R. \tag{5.6-11}$$

5.6.4.2　Design the online estimation and compensation method for disturbance rejection effect (DRE) parasitic loop

1) ESKF design (extended-state Kalman filter)

Consider the following class of nonlinear time-varying uncertain systems,

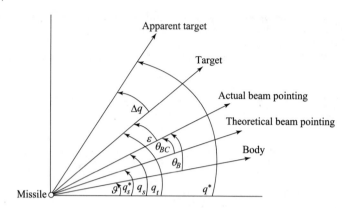

Fig. 5.6–9 The basic missile-target geometry model of strapdown phased array radar seeker

q_t—The true LOS angle; q_s—The LOS angular rate; q^*—The apparent LOS angle; θ_B—The beam pointing angle; θ_{BC}—The beam pointing angle instruction; ε—The tracking error angle; ϑ—The attitude angle.

$$\begin{cases} X_{k+1} = \bar{A}_k X_k + \bar{B}_k F(X_k, k) + w_k \\ Z_k = \bar{C}_k X_k + n_k, k = 0, 1, \ldots, \end{cases} \quad (5.6-12)$$

where $X_k \in \mathbb{R}^n$ is the state, $\bar{A}_k$, $\bar{B}_k$ and $\bar{C}_k$ are known time-varying matrixes with $\bar{A}_k \in \mathbb{R}^{n \times n}$, $\bar{B}_k \in \mathbb{R}^{n \times l}$, $\bar{C}_k \in \mathbb{R}^{m \times n}$. $F(\bar{X}_k, k) \in \mathbb{R}^l$ is the nonlinear uncertain dynamics in the system, and its nominal model is the known function $\bar{F}(X_k, k)$. $w_k \in \mathbb{R}^n$ and $n_k \in \mathbb{R}^m$ are the process noise and measurement noise respectively. $Z_k \in \mathbb{R}^m$ is the measurement output.

In the model, the uncertain dynamics are divided into three parts: the known linear part $\bar{A}_k X_k$, the time-varying nonlinear uncertain dynamics $F(X_k, k)$, and the noise (w_k, n_k). The ESKF filter method suggested using different methods to deal with different kinds of uncertainties. $F(X_k, k)$ is treated as an extended state to be estimated as well as compensated for, and (w_k, n_k) is attenuated by the optimization technique of the Kalman filter. Therefore, the system can be equivalently transformed to

$$\begin{cases} \begin{bmatrix} X_{k+1} \\ F_{k+1} \end{bmatrix} = A_k \begin{bmatrix} X_k \\ F_k \end{bmatrix} + B_k G_k + \begin{bmatrix} w_k \\ 0 \end{bmatrix} \\ Z_k = C_k \begin{bmatrix} X_k \\ F_k \end{bmatrix} + n_k \end{cases}, \quad (5.6-13)$$

where

$$F_k \triangleq F(X_k, k), G_k = F_{k+1} - F_k, A_k = \begin{bmatrix} \bar{A}_k & \bar{B}_k \\ 0 & I \end{bmatrix}, B_k = \begin{bmatrix} 0 \\ I \end{bmatrix}, C_k = [\bar{C}_k \quad 0]$$

Design of the extended-state observer (ESO) is based on the extended model in equation (5.6–14):

$$\begin{bmatrix} \hat{X}_{k+1} \\ \hat{F}_{k+1} \end{bmatrix} = A_k \begin{bmatrix} \hat{X}_k \\ \hat{F}_k \end{bmatrix} + B_k \hat{G}_k - K_k \left(Z_k - C_k \begin{bmatrix} \hat{X}_k \\ \hat{F}_k \end{bmatrix} \right), \quad (5.6-14)$$

where, $\hat{G}_k$, the estimate of G_k, is used to correct the estimation error of the state and the uncertainty by making full use of the model information. Thus, use the nominal model of G_k, $\bar{G}_k = \bar{F}(X_{x+1}, k+1) - \bar{F}(X_k, k)$. Then the estimate is denoted as $\hat{\bar{G}}_k = \bar{F}(\bar{A}_k \hat{X}_k + \bar{B}_k \hat{F}_k, k+1) - \bar{F}(X_k, k)$. According to the estimate of the nominal model $\bar{G}_k$, $\hat{G}_k$ is designed as $\hat{G}_{k,i} = \text{sat}(\hat{\bar{G}}_{k,i}, \sqrt{\bar{q}_{k,i}})$, $i = 1, 2, \ldots$, where $\text{sat}(\cdot)$ is the saturation function defined by $\text{sat}(f, b) = \max\{\min\{f, b\}, -b\}, b > 0$. The saturation function $\text{sat}(\cdot)$ is used to ensure the boundedness of $\hat{G}_{k,i}$. Then

$$Q_{1,k} = \begin{bmatrix} \mathbf{0}_{n\times n} & \mathbf{0}_{n\times l} \\ \mathbf{0}_{l\times n} & 4\bar{\mathbf{Q}}_k \end{bmatrix}. \tag{5.6-15}$$

where $\bar{\mathbf{Q}}_k \triangleq l \cdot \text{diag}([\bar{q}_{k,1} \ \bar{q}_{k,2} \ \cdots \ \bar{q}_{k,l}])$.

Design θ to decouple the cross terms of estimation error and the uncertainties,

$$\theta = \sqrt{\frac{\text{tr}(Q_{1,0})}{\text{tr}(P_0)}} \tag{5.6-16}$$

and

$$Q_{2,k} = \begin{bmatrix} S_k & \mathbf{0}_{n\times l} \\ \mathbf{0}_{l\times n} & \mathbf{0}_{l\times l} \end{bmatrix}. \tag{5.6-17}$$

Let P_k satisfy the iteration equation,

$$P_{k+1} = (1+\theta)(A_k + K_k C_k) P_k (A_k + K_k C_k)^T + K_k R_k K_k^T + \left(1 + \frac{1}{\theta}\right) Q_{1,k} + Q_{2,k}, \tag{5.6-18}$$

with the initial value P_0.

Define $K_k^* = \arg\min_{K_k} \{(1+\theta)(A_k + K_k C_k) P_k (A_k + K_k C_k)^T + K_k R_k K_k^T\}$, since P_k is a positive semi-definite matrix and R_k is a positive definite matrix, $C_k P_k C_k^T + \frac{1}{1+\theta} R_k$ is the positive definite matrix. It is straightforward to obtain

$$K_k^* = -A_k P_k C_k^T \left(C_k P_k C_k^T + \frac{1}{1+\theta} R_k\right)^{-1}. \tag{5.6-19}$$

As a consequence, the ESKF can be designed as follows

$$\begin{bmatrix} \hat{X}_{k+1} \\ \hat{F}_{k+1} \end{bmatrix} = A_k \begin{bmatrix} \hat{X}_k \\ \hat{F}_k \end{bmatrix} + B_k \hat{G}_k - K_k \left(Z_k - C_k \begin{bmatrix} \hat{X}_k \\ \hat{F}_k \end{bmatrix}\right). \tag{5.6-20}$$

$$K_k = -A_k P_k C_k^T \left(C_k P_k C_k^T + \frac{1}{1+\theta} R_k\right)^{-1}. \tag{5.6-21}$$

$$P_{k+1} = (1+\theta)(A_k + K_k C_k) P_k (A_k + K_k C_k)^T + K_k R_k K_k^T + \left(1 + \frac{1}{\theta}\right) Q_{1,k} + Q_{2,k}, \tag{5.6-22}$$

where

$$Q_{1,k} = \begin{bmatrix} \mathbf{0}_{n\times n} & \mathbf{0}_{n\times l} \\ \mathbf{0}_{l\times n} & 4\bar{\mathbf{Q}}_k \end{bmatrix}, Q_{2,k} = \begin{bmatrix} S_k & \mathbf{0}_{n\times l} \\ \mathbf{0}_{l\times n} & \mathbf{0}_{l\times l} \end{bmatrix}, \bar{\mathbf{Q}}_k \triangleq l \cdot \text{diag}([\bar{q}_{k,1} \ \bar{q}_{k,2} \ \cdots \ \bar{q}_{k,l}]),$$

$$\theta = \sqrt{\frac{\text{tr}(Q_{l,0})}{\text{tr}(P_0)}}, \hat{G}_{k,i} \triangleq \text{sat}(\bar{G}_{k,i}, \sqrt{\bar{q}_{k,i}}), \bar{G}_k \triangleq \bar{G}(\hat{X}_k, k),$$

$$\text{sat}(f,b) = \max\{\min\{f,b\}, -b\}, b > 0.$$

For the system with uncertainty and noise, the tuning of $\boldsymbol{K}_k$ becomes a tradeoff between the disturbance rejection and the noise sensitivity. This is because higher $\boldsymbol{K}_k$ leads to faster tracking, but also results in higher levels of noise. Thus, different from ESO, $\boldsymbol{K}_k$ is no longer statically and manually tuned here. Instead, $\boldsymbol{K}_k$ is optimized to ensure the mean square estimation error is minimal at each step. For more details about ESKF, please refer to reference [1].

2) Online estimation and compensation method for DRE parasitic loop based on ESKF filter

The beam pointing error slope R^* and the radome error slope R (which affect the DRE parasitic loop of the strapdown phased array radar seeker) are set as the state variables. The beam pointing error slope estimate value $\hat{R}^*$, the radome error slope estimate value $\hat{R}$, and the true LOS angular rate estimate value $\hat{\dot{q}}_t$ are directly obtained with the ESKF filter. Thereby, the online compensation for the DRE parasitic loop is completed.

Setting the system state variables of the ESKF filter to
$$\boldsymbol{X} = [X_1, X_2, X_3, X_4, X_5]^T = [q_t, \dot{q}_t, \theta_B, R, R^*]^T, \qquad (5.6-23)$$

The LOS angular rate $\dot{q}_s$ is a measured output and expressed as,
$$Z = \dot{q}_s. \qquad (5.6-24)$$

The attitude angle ϑ_m and the attitude angular rate $\dot{\vartheta}_m$ are measured by the inertial navigation system on board the missiles. They are the input quantities for the ESKF, the corresponding noise standard deviations are σ_{u1} and σ_{u2}, respectively. Thus, the ESKF model for the DRE parasitic loop can be obtained in Fig. 5.6 – 10.

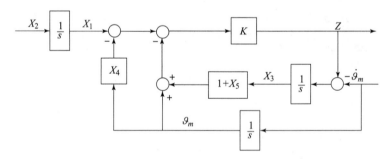

Fig. 5.6 – 10 The block diagram of the ESKF model for DRE parasitic loop

According to Fig. 5.6 – 10, the state equation and the measurement equation of the ESKF filter are obtained as follows,
$$\boldsymbol{X}(t) = f[\boldsymbol{X}(t)] + \boldsymbol{m}w(t), \qquad (5.6-25)$$
$$\boldsymbol{Z}(t) = h[\boldsymbol{X}(t)] + v(t), \qquad (5.6-26)$$

where,
$$f[\boldsymbol{X}(t)] = \begin{bmatrix} X_2 \\ 0 \\ KX_1 - K\vartheta_m - KX_5\vartheta_m - KX_3 - KX_3X_4 - \dot{\vartheta}_m \\ 0 \\ 0 \end{bmatrix}$$

$$h[X(t)] = KX_1 - KX_5\vartheta_m - KX_3 - KX_3X_4 - K\vartheta_m - \dot{\vartheta}_m,$$
$$w(t) = [0 \quad w_2(t) \quad 0 \quad w_4(t) \quad w_5(t)]^T,$$

where $f[X(t)]$ represents the state matrix and $h[X(t)]$ represents the measurement matrix. $w(t)$ and $v(t)$ are the precess noise and measurement noise, respectively. Design S_{w2}, S_{w4}, and S_{w5} are the process noise power spectral densities. S_k is the system noise covariance matrix. $v(t)$ is the zero-mean Gauss white-noise and its standard deviation is represented as σ_v. R_k is the measurement noise covariance matrix.

The nonlinear uncertain dynamics F_k can be set as

$$F_k = \begin{bmatrix} \dot{R} \\ R^* \end{bmatrix}, \tag{5.6-27}$$

then the state equation and measurement equation can be substituted as shown in equation (5.6-28),

$$\begin{cases} X_{k+1} = \begin{bmatrix} 1 & \Delta t & 0 & 0 & 0 \\ 0 & 1 & 0 & 0 & 0 \\ K\Delta t & 0 & 1 - K\Delta t - K\hat{X}_4\Delta t & -K\hat{X}_3\Delta t & -K\vartheta_m\Delta t \\ 0 & 0 & 0 & 1 & 0 \\ 0 & 0 & 0 & 0 & 1 \end{bmatrix} X_k + \begin{bmatrix} 0 & 0 \\ 0 & 0 \\ 0 & 0 \\ \Delta t & 0 \\ 0 & \Delta t \end{bmatrix} \begin{bmatrix} \dot{R} \\ R^* \end{bmatrix} + mw_k, \\ Z_k = [K \quad 0 \quad -K - K\hat{X}_4 \quad -K\hat{X}_3 \quad -K\vartheta_m]\hat{X}_k + n_k, k = 0,1,\ldots, \end{cases}$$
$$(5.6-28)$$

where Δt is sample time. Then, the discrete fundamental matrix $\bar{A}[\hat{X}(t)]$ is shown in equation (5.6-29),

$$\bar{A}[\hat{X}(t)] = \begin{bmatrix} 1 & \Delta t & 0 & 0 & 0 & 0 & 0 \\ 0 & 1 & 0 & 0 & 0 & 0 & 0 \\ K\Delta t & 0 & 1 - K\Delta t - K\hat{X}_{4,k-1}\Delta t & -K\hat{X}_{3,k-1}\Delta t & -K\vartheta_m\Delta t & 0 & 0 \\ 0 & 0 & 0 & 1 & 0 & \Delta t & 0 \\ 0 & 0 & 0 & 0 & 1 & 0 & \Delta t \\ 0 & 0 & 0 & 0 & 0 & 1 & 0 \\ 0 & 0 & 0 & 0 & 0 & 0 & 1 \end{bmatrix}, \tag{5.6-29}$$

the disturbance matrix $\bar{B}[\hat{X}(t)]$ and the discrete measurement matrix $\bar{C}[\hat{X}(t)]$ are obtained in equation (5.6-30) and equation (5.6-31), respectively, that is,

$$\bar{B}[\hat{X}(t)] = \begin{bmatrix} 0 & 0 \\ 0 & 0 \\ 0 & 0 \\ 0 & 0 \\ 0 & 0 \\ 1 & 0 \\ 0 & 1 \end{bmatrix}, \tag{5.6-30}$$

and

$$\bar{C}[\hat{X}(t)] = [K \quad 0 \quad -K - KX_{4,k/k-1} \quad -KX_{3,k/k-1} \quad -K\vartheta \quad 0 \quad 0]. \quad (5.6-31)$$

The unknown disturbance nominal matrix G_k can be written as

$$G_k = \begin{bmatrix} \dot{R} + w_4(k) \\ \dot{R}^* + w_5(k) \end{bmatrix}. \quad (5.6-32)$$

The extended-state observer (ESO) method is applied to estimate and compensate the unknown disturbance. Hence, the nominal model of G_k is used as

$$\bar{G}_k = \begin{bmatrix} X_{4,k} - \hat{X}_{4,k-1} \\ X_{5,k} - \hat{X}_{5,k-1} \end{bmatrix}. \quad (5.6-33)$$

The extended model disturbance estimation covariance matrix Q_{1k} is

$$Q_{1k} = \begin{bmatrix} \mathbf{0}_{5\times5} & \mathbf{0}_{5\times2} \\ \mathbf{0}_{2\times5} & \bar{Q}_k \end{bmatrix}, \quad (5.6-34)$$

where

$$\bar{Q}_k = 4 \times 2 \times \begin{bmatrix} R_{k+1} + w_{4,k+1} - R_k - w_{4,k} & 0 \\ 0 & R^*_{k+1} + w_{5,k+1} - R^*_k - w_{5,k} \end{bmatrix},$$

and the design of the extended model noise covariance matrix Q_{2k} is

$$Q_{2k} = \begin{bmatrix} 0 & 0 & 0 & 0 & 0 & 0 & 0 \\ 0 & \dfrac{S_{w2,k-1}}{\Delta t} & 0 & 0 & 0 & 0 & 0 \\ 0 & 0 & 0 & 0 & 0 & 0 & 0 \\ 0 & 0 & 0 & \dfrac{S_{w4,k-1}}{\Delta t} & 0 & 0 & 0 \\ 0 & 0 & 0 & 0 & \dfrac{S_{w5,k-1}}{\Delta t} & 0 & 0 \\ 0 & 0 & 0 & 0 & 0 & 0 & 0 \\ 0 & 0 & 0 & 0 & 0 & 0 & 0 \end{bmatrix}. \quad (5.6-35)$$

Then the design of the discrete system noise driven covariance matrix of the extended model is

$$\Gamma_k \approx m \cdot \Delta t = \begin{bmatrix} \Delta t & 0 & 0 & 0 & 0 \\ 0 & \Delta t & 0 & 0 & 0 \\ 0 & 0 & \Delta t & 0 & 0 \\ 0 & 0 & 0 & \Delta t & 0 \\ 0 & 0 & 0 & 0 & \Delta t \end{bmatrix}. \quad (5.6-36)$$

Furthermore, the design measurement noise covariance matrix of the extended model is

$$R_k = \frac{S_v}{\Delta t} = (\sigma_v)^2. \quad (5.6-37)$$

Then the online estimation and compensation method for the DRE parasitic loop using ESKF (Equation (5.6-38) – equation (5.6-39)) is designed as

$$K_k = -\bar{A}_k P_k \bar{C}_k^T \left(\bar{C}_k P_k \bar{C}_k^T + \frac{1}{\Delta t + \theta} R_k \right)^{-1}, \quad (5.6-38)$$

$$P_{k+1} = (\Delta t + \theta)(\bar{A}_k + K_k\bar{C}_k)P_k(\bar{A}_k + K_k\bar{C}_k)^{\mathrm{T}} + K_kR_kK_k^{\mathrm{T}} + \left(1 + \frac{1}{\theta}\right)Q_{1,k} + \Gamma_kQ_{2,k}\Gamma_k^{\mathrm{T}}.$$

(5.6-39)

5.6.4.3 Mathematical simulation and result analysis

In this section, the ESKF method is applied to achieve online estimation and compensation for the DRE parasitic loop by estimating the radome error slope R, beam pointing error slope R^*, and the LOS angular rate $\dot{q}_t$. The simulation parameters are designed in Table 5.6-1.

Table 5.6-1 The simulation parameters of DRE model

Parameters	σ_v	σ_{u1}	σ_{u2}	S_{w2}	S_{w4}	S_{w5}
Values	0.2°/s	0.1°/s	0.1°/s	0.000,01	0.001	0.001

The strapdown phased array radar seeker forward loop gain was set to $K = 10$. The actual LOS angle was set as $q_t = 1t^*$. Consequently, the LOS angular rate is $\dot{q}_t = 1^*/s$. Also, it is assumed that the missile body motion varies sinusoidally with amplitude $3°$ and frequency of 2 Hz. The initial value of the state estimation was defined as $\hat{X}_0 = [0, 0, 0, 0, 0]$, $\hat{F}_0 = [0, 0]$, and the initial value of the extended model estimation error covariance matrix was set as $P_0 = I_{7 \times 7}$. For simplicity, "ESKF" is used to represent the proposed method in the following simulations.

The method introduced in this paper has the same inputs as the traditional EKF. To reflect the robustness and higher estimation precision of the proposed method, select the EKF, which has been widely used in engineering practice, and the STUKF as comparative items for the following simulation.

Considering the radome error slope R is a random value within the range of $(-0.03, 0.03)$ and the beam pointing error slope R^* is a random value in the range of $(-0.05, 0.05)$. This part of simulation is only looking for the EKF, STUKF and ESKF performance differences of the estimation of these two error slopes. The simulation results are shown in Fig. 5.6-11.

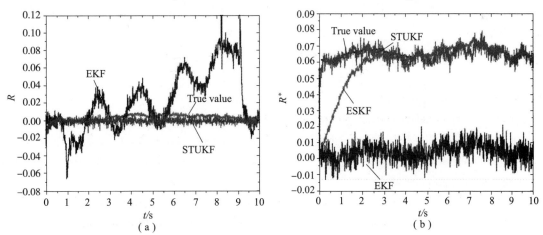

Fig. 5.6-11 The simulation results of ESKF, EKF and STUKF for the DRE parasitic loop

(a) The estimation of radome error slope R; (b) The estimation of beam pointing error slope R^*

As shown in Fig. 5.6 – 11 (a) the performance of EKF is the worst. The ESKF can be estimated more accurately and more stably than STUKF under the small initial error condition. There is an initial estimation error using STUKF. Fig. 5.6 – 11 reveals that the ESKF costs more time to achieve a stable and precise estimation than low nonlinear condition. Although the STUKF can be estimated fast, there is still an estimation error which cannot be ignored.

§ 5.7 Platform Based Seeker Design

5.7.1 Stabilization Loop Design

It is known from the previous sections that to reduce the influence of the missile angular motion interference on the seeker performance, it is necessary to increase the open loop gain of the stabilization loop as much as possible. Under the condition that the seeker has certain load and the driving motor bandwidth is limited, the stabilization loop low frequency gain can be increased by designing a PI compensator (or a lag compensator) (because the missile interference motion lies in the low frequency range) to increase the seeker's ability to reject the missile seeker coupling.

The following example illustrates how to implement this approach. With a stabilization loop structure as shown in Fig. 5.7 – 1, the rate gyro has the undamped natural frequency of $\omega_{gn} = 80$ Hz, the damping coefficient of $\zeta_g = 0.7$, and the motor time constant of $T = 0.000,4$ s. The missile angular motion interference frequency is about 2 Hz. It is required to select the optimal parameters of the PI compensator K_p and K_i so that under the design constraints that the gain margin is greater than 6 dB and the phase margin greater than 40°, the stabilization loop is optimized to reject the disturbance of the missile body angular motion. The compensator parameter K_p and K_i should be chosen to maximize the loop gain at the interference frequency 2 Hz. This optimal design problem can be described as follows.

Objective function:
$$\text{Max } M_{2\text{ Hz}}(K_p, K_i),$$

Subject to:
$$\text{Gain margin } L(K_p, K_i) \geq 6 \text{ dB},$$
$$\text{Phase margin } \Delta\Phi(K_p, K_i) \geq 40°.$$

In the above equation, $M_{2\text{ Hz}}$ is the stabilization loop gain at 2 Hz.

With a commercial non-linear programming software, the solution of this optimization problem can be obtained as $K_p = 181$, $K_i = 1.0 \times 10^4$ and the open loop gain at 2 Hz is 65.4. Fig. 5.7 – 2 gives the contour of the objective function. It can be seen that the maximum gain at 2 Hz is restricted by the phase margin constraint.

With this design, the stabilization open loop gain at 2 Hz is 65.4. When the stabilization loop is designed with the same gain and phase-margin constraints and with no PI compensator added, the open loop gain at 2 Hz is only 22.2. So the open loop gain at 2 Hz is increased by a factor of 3 under the same phase and gain margin constraints after the PI compensation optimization design is adopted,

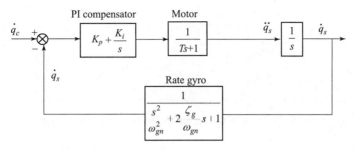

Fig. 5.7 – 1 Block diagram of the stabilization loop with PI compensation

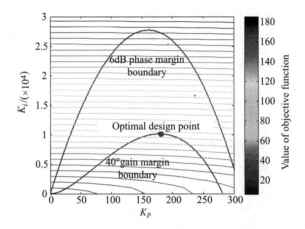

Fig. 5.7 – 2 Contour of the objective function with gain margin and phase margin constraint

that is, the decoupling level under the 2 Hz missile motion disturbance is improved by a factor of 3.

5.7.2 Tracking Loop Design

The design of the platform based seeker is very different from type II ground tracking radars. In order to reduce the phase shift of the inner loop at the open loop crossover frequency of the tracking radar outer loop, the inner loop is usually designed to have an inner and outer loop bandwidth ratio of around 5. However, for the missile seeker, the main function of the inner loop is to ensure the decoupling of the missile motion, so its frequency bandwidth is generally designed to be very wide (usually at 15 – 30 Hz). The seeker angle tracking loop detector usually has a low sampling frequency (laser detector 20 Hz, image and radar head 50 – 100 Hz). If the sampling period of the detector is τ_1, the sample holding time will cause an equivalent pure delay of $\tau_1/2$. Suppose that the pure delay of the detector is τ_2, then the transfer function of the seeker detector becomes $e^{-\tau s}$ ($\tau = \tau_1/2 + \tau_2$). The presence of this pure time delay transfer function causes the tracking loop bandwidth to be much lower than that of the stabilization loop, even though the phase shift of the stabilization loop at the tracking loop crossover frequency is very small. In order to speed up the dynamic response of the seeker, a PD compensation network can be added to the tracking loop. An example will be used in the following to illustrate the effectiveness of this design.

Suppose that the block diagram of the seeker tracking loop is as shown in Fig. 5.7 – 3.

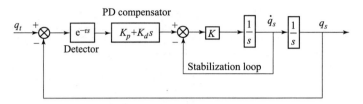

Fig. 5.7-3 Block diagram of the seeker tracking loop

It is known that the open loop gain of the stabilization loop $K = 188$ (bandwidth 30 Hz), and the seeker time delay $\tau = 40$ ms. It is required to select the optimal parameters of the PD compensation network K_p and K_d to achieve the fastest step response of the tracking loop. Suppose $e(t)$ is the tracking error for a unit step input, then the design objective function can be taken as $J = \int_0^T t|e(t)|\mathrm{d}t$ together with the stability constraint as gain margin 6 dB and phase margin 40°.

The mathematical model for this design is:

Objective function:
$$\min J = \int_0^T t|e(t)|\mathrm{d}t;$$

Subject to:

The tracking loop gain margin: $L(K_p, K_d) \geq 6$ dB;

The tracking loop phase margin: $\Delta\Phi(K_p, K_d) \geq 40$.

Note that the integral time is taken as the tracking loop settling time $T = 0.6$ s. The objective function contour and the gain and phase margins are shown in Fig. 5.7-4 by using a nonlinear programming software for searching the optimum solution.

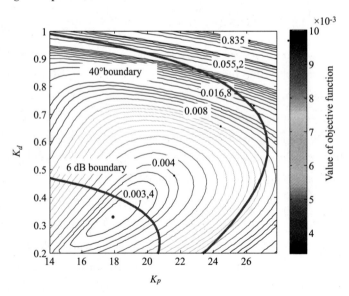

Fig. 5.7-4 Contour lines of the time domain objective function

It can be seen that the result of this optimal design is $K_p = 17.8$ and $K_d = 0.33$ (that is, the open loop gain of the seeker is 17.8 and its corresponding bandwidth is 2.83 Hz). Fig. 5.7 – 5 shows the step response of the seeker design. It can be clearly seen from the figure that the design with PD compensation can significantly improve the response of the seeker.

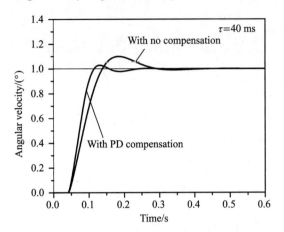

Fig. 5.7 –5   Time domain characteristics of the designed seeker

6

Missile Autopilot Design

In previous chapters, we gave an overview of the missile guidance control system, its fully nonlinear equations of motion, linearized simplified equations of motion, and a brief description of basic missile control component mathematics models. In this chapter, we shall focus on autopilot design.

§ 6.1 Acceleration Autopilot

The purpose of guidance is to change the missile's flight path. Since $a = V\dot{\theta}$, controlling the missile normal acceleration can lead to the change of the missile's velocity vector direction and its flight path. An acceleration autopilot could not only achieve accurate tracking of the guidance normal acceleration command, but also help to speed up the autopilot response speed through feedback.

6.1.1 Two-loop Acceleration Autopilot

The structure of a typical two-loop autopilot is shown in Fig. 6.1 – 1.

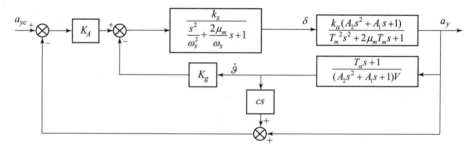

Fig. 6.1 – 1 Two-loop acceleration autopilot

The autopilot has a damping loop, its feedback coefficient is given as K_g. Assume that the position of the feedback accelerometer is placed at c (m) in front of the center of gravity, which causes the missile angular acceleration $\ddot{\vartheta}$ introduced normal acceleration component $c\ddot{\vartheta}$ to be added to the output of the accelerometer.

Fig. 6.1 – 1 could also be given in the form of Fig. 6.1 – 2.

Omitting the small terms A_1 and A_2, the transfer function from the missile normal acceleration to the signal output at position A of Fig. 6.1 – 2 will be

6 Missile Autopilot Design

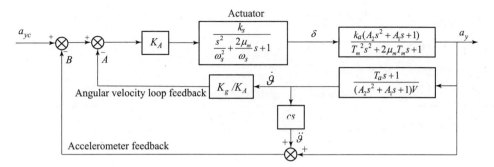

Fig. 6.1-2 Alternative two-loop acceleration autopilot block diagram

$$K_g\left(\frac{T_\alpha}{K_A V}s + \frac{1}{K_A V}\right)a_y, \tag{6.1-1}$$

and the signal output at position B will be

$$\left(\frac{cT_\alpha}{V}s^2 + \frac{c}{V}s + 1\right)a_y. \tag{6.1-2}$$

Adding these two signal outputs will give the total feedback of the missile normal acceleration as

$$\left[\frac{cT_\alpha}{V}s^2 + \left(\frac{K_t T_\alpha}{K_A V} + \frac{1}{V}c\right)s + \left(1 + \frac{K_g}{K_A V}\right)\right]a_y. \tag{6.1-3}$$

That is, taking into account all the feedbacks, the feedback structure is equivalent to a second-order lead compensation network (Fig. 6.1-3), where the angular velocity loop is a first-order compensation and the accelerometer preposition constitutes a second-order compensation of the s^2 term. This ensures that the autopilot's bandwidth can be increased by phase lead compensation when the actuator lag is determined.

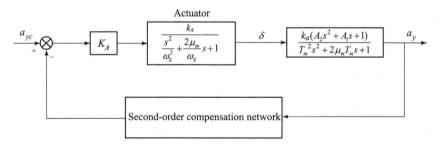

Fig. 6.1-3 Two-loop acceleration autopilot feedback structure

This phase lead compensation network is likely to provide a lead phase angle of 70° at the autopilot crossover frequency. The phase shift of the uncontrolled missile body is about −180° at the crossover frequency as it has very low aerodynamic damping. When the actuator is allowed to have a phase lag of 20° to 25° at the crossover frequency, this configuration could give the autopilot design within a phase margin of 45° to 50°.

In the following example of the two-loop autopilot structure, the missile dynamic coefficients (Table 6.1-1) given in reference [5] are used. In addition, the first-order model of the actuator is given as $\dfrac{-0.017,5}{0.013,3s+1}$. It has time constant of $0.013,3s$, the bandwidth of about 12 Hz, and the other autopilot parameters are $K_A = 0.000,65$, $K_g = 0.072,8$. The corresponding control diagram of the autopilot is shown in Fig. 6.1 – 4.

Table 6.1 – 1 Missile dynamic coefficients

a_α/s^2	a_δ/s^{-2}	a_ω/s^{-1}	b_α/s^{-1}	b_δ/s^{-1}	c/m	$V/(\text{m}\cdot\text{s}^{-1})$
72.4	471	1.5	1.27	0.477	0.66	1,140

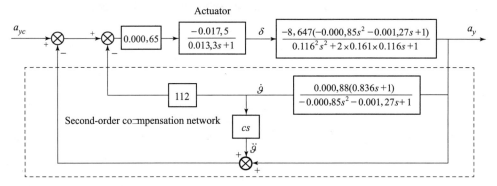

Fig. 6.1 – 4 Two-loop acceleration autopilot example

The second-order compensation network Bode diagram corresponding to Fig. 6.1 – 4 in this example is shown in Fig. 6.1 – 5. The advantages of this second-order lead compensation for the normal acceleration a_y output are clear from the graph. Without forward positioning of the accelerometer (as $c = 0$ m), the lead compensation becomes a first order lead compensation network.

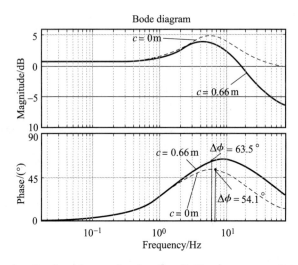

Fig. 6.1 – 5 First-order ($c = 0$ m) and second-order ($c = 0.66$ m) compensation network Bode diagrams

Since the second-order lead compensation network provides a leading phase angle of 63.5° at the crossover frequency of 5.96 Hz, where the lag angle of the actuator is $-26.5°$ and the lag phase angle of the missile body is $-176.7°$. Thus the phase margin of this autopilot design is $40.3°$. The open loop Bode diagram of this example is shown in Fig. 6.1-6.

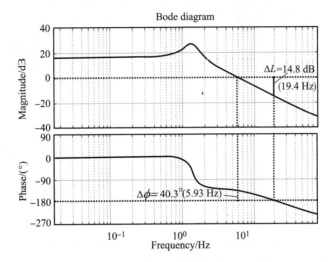

Fig. 6.1-6 Open loop Bode diagram of this two-loop acceleration autopilot system

The closed-loop Bode diagram for this example is shown in Fig. 6.1-7. The time-domain response for the unit step autopilot acceleration command is shown in Fig. 6.1-8.

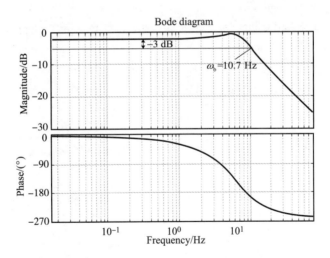

Fig. 6.1-7 Closed-loop Bode diagram of this two-loop acceleration autopilot

The disadvantage of this type of autopilot structure is that, to stabilize the gain from the command a_c to the autopilot output a, a high autopilot open loop gain K has to be taken. However, very often the actuator bandwidth will limit the autopilot open loop gain to a lower than expected value. For example, the open loop gain of this autopilot is only 3.68, and its closed-loop gain is only 0.787. Due to the poor robustness of the closed-loop gain of autopilots of this structure, this autopilot structure is rarely used nowadays in practical applications.

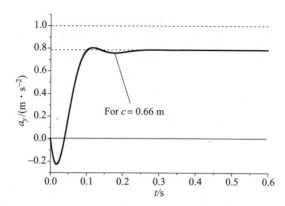

Fig. 6.1-8 The autopilot time-domain response for the unit step acceleration command

6.1.2 Two-loop Acceleration Autopilot with PI Compensation

To improve the robustness of the two-loop autopilot mentioned in the previous section, a PI compensator or a lag compensator can be introduced in the autopilot forward-loop. This will greatly increase the autopilot open loop gain at low frequencies and make the autopilot steady state closed-loop gain more robust.

Normally, the PI compensator transfer function is given as $1 + \dfrac{\omega^*}{s}$. Here ω^* is the corner frequency of the PI compensator. A lower designed value of ω^* will lead to a slow, steady, stable error elimination transient and a higher value of ω^* will lead to lower autopilot stability. A properly designed value of ω^* should make the steady-state error eliminated roughly at the end of the autopilot transient. For this reason, $\omega^* = 0.3$ rad/s is chosen for the above example. The block diagram of the autopilot design is shown in Fig. 6.1-9.

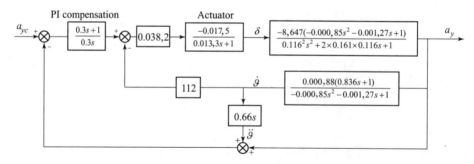

Fig. 6.1-9 Block diagram of the two-loop autopilot with PI compensation

The open loop and closed loop Bode diagrams of the autopilot with PI compensation and without PI compensation are respectively shown in Fig. 6.1-10 and Fig. 6.1-11.

Table 6.1-2 shows the difference of the stability margin of the autopilot with and without PI compensation.

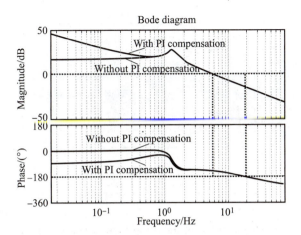

Fig. 6.1 – 10 Open loop Bode diagrams of the autopilot with PI compensation and without PI compensation

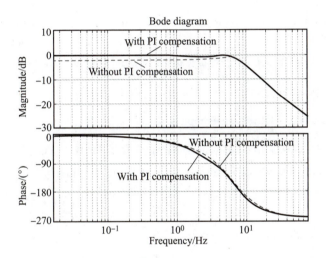

Fig. 6.1 – 11 Closed loop Bode diagrams of the autopilot with PI compensation and without PI compensation

Table 6.1 – 2 The difference of the stability margin of the autopilot with and without PI compensation

Items	Gain margin $\Delta L/\mathrm{dB}$	Phase margin $\Delta\phi/(°)$
Without PI compensation	14.8	40.3
With PI compensation	14.8	39.7

Fig. 6.1 – 12 shows the improvement of the steady state value of the autopilot time domain response with PI compensation.

The following example illustrates that it is possible to replace the PI compensator with a lag compensator and still get the same autopilot closed loop robustness. It is known that the standard

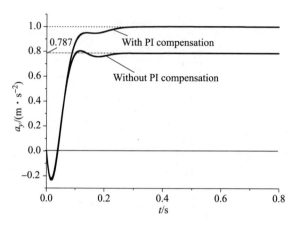

Fig. 6.1-12 Unit step response of the autopilot with and without PI compensation

structure of a PI compensation is $1 + \dfrac{\omega^*}{s}$, and that improvement of a lag compensation is $\beta \left(\dfrac{T_b s + 1}{\beta T_b s + 1} \right)$, where the value of β will increase the autopilot low frequency gain by a factor of β. When the value of T_b is chosen as $\dfrac{1}{T_b} \approx \omega^*$ and β is large enough, the two compensator structures will have similar compensation effects for reducing autopilot steady static error. The Bode diagrams for a PI compensator $1 + \dfrac{1}{0.3s}$ and lag compensator $10 \times \left(\dfrac{0.3s + 1}{10 \times 0.3s + 1} \right)$ are shown in Fig. 6.1-13. Both designs have a corner frequency of 3.33 rad/s (0.531 Hz). The low frequency gain of the autopilot with lag compensation has increased by a factor of ten. Fig. 6.1-14 shows the unit step response of the autopilot under the above PI compensation, lag compensation and with no compensation network system. The PI compensation and lag compensation effectively reduce the system static error, and the effects are similar.

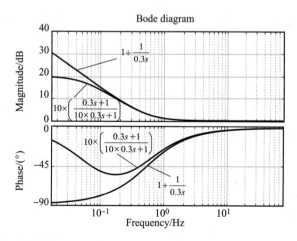

Fig. 6.1-13 Bode diagram of the designed PI compensator and lag compensator

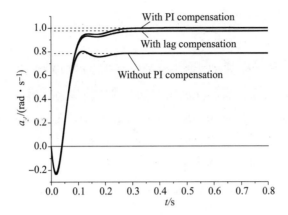

Fig. 6.1 – 14 Unit step response for autopilots with no compensation, PI compensation and lag compensation

6.1.3 Three-loop Autopilot with Pseudo Angle of Attack Feedback

Fig. 6.1 – 15 shows the structural diagram of the missile body transfer function $\dfrac{\alpha(s)}{\delta(s)}$ as the object being controlled for an acceleration autopilot.

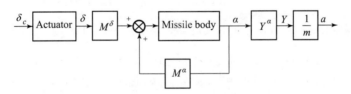

Fig. 6.1 – 15 Diagram of transfer function $\dfrac{\alpha(s)}{\delta(s)}$

Generally, after the autopilot design is completed, the gain of the missile body transfer function $\dfrac{\alpha(s)}{\delta(s)}$ has to be changed from its nominal value by a certain amount to check its effect on autopilot robustness. The range of deviation could be, for example, $\left(\dfrac{1}{1.6}\right) \sim 1.6$, that is from 0.625 to 1.6. It can be seen from Fig. 6.1 – 15 that the value of the parameters Y^α, M^δ and m generally do not change much at the chosen set points, but the value $M^\alpha = -x^* Y^\alpha$ (here, x^* is the distance between the center of pressure and center of gravity of the missile) could change a great deal, when the static stability x^* is low. This is because the relative value of x^* could vary greatly with the uncertainty of the positions of the center of gravity and the center of pressure. It is known the missile body transfer function gain $\dfrac{\alpha}{\delta} = \left(\dfrac{M^\delta}{M^\alpha}\right) \approx \dfrac{a_\delta}{a_\alpha}$. Since $a_\alpha \propto x^*$, $\dfrac{\alpha}{\delta} \propto \dfrac{1}{x^*}$. Suppose ε is the possible absolute variation of x^*, and its effect on x^* is shown in Fig. 6.1 – 16.

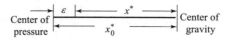

Fig. 6.1–16 The distance change between the center of gravity and the center of pressure

As an example, the dynamic coefficient values of a typical air-to-air missile given in the reference [7] (Table 6.1–3) are taken as the parameters of the missile body and a second-order actuator $\omega_n = 150$ rad/s, $\mu = 0.7$ are chosen for the autopilot. Suppose that the nominal distance between the missile center of gravity and center of pressure is $x^* = 50$ mm, and the nominal control gain is set $\dfrac{\alpha}{\delta} = \dfrac{a_\delta}{a_\alpha} = 1$. If the x^* uncertainty for this example is $\varepsilon = \pm 5$ mm, the range of x^* variation will be 45~55 mm and the range of a_α 216–264 s^{-2}. With this x^* variation, the control gain $\dfrac{\alpha}{\delta}$ may vary by $-10\% \sim +11\%$.

Table 6.1–3 H_I model parameters of the missile body

Height/m	a_α/s^{-2}	a_δ/s^{-2}	a_ω/s^{-1}	b_α/s^{-1}	b_δ/s^{-1}	$V/(\mathrm{m}\cdot\mathrm{s}^{-1})$
9.14×10^3	240	240	3	1.17	0.239	914

Let a damping feedback loop be added to the missile body H_I in Table 6.1–3 to form a new missile body H_II (Fig. 6.1–17). In this instance, the transfer function of the new missile body becomes $\dfrac{\alpha(s)}{\delta^*(s)}$ (here, δ^* is the input of the new missile body formed after the damping loop is added, and it can be called the pseudo actuator deflection angle). The angle of attack response for unit step pseudo actuator input δ^* for different ε values and static stability are shown in Fig. 6.1–18. Obviously, the missile body with sufficient static stability has a good robustness.

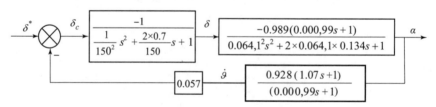

Fig. 6.1–17 Structural diagram of the new missile body H_II (missile body H_I with a damping loop) ($x^* = 50$ mm, $a_\alpha = 240$ s^{-2})

However, if a missile body L_I is of low static stability, for example, its nominal value of x^* is only 6 mm, then the nominal value of a_α will be reduced from 240 to $\left(\dfrac{6}{50}\right) \times 240 = 28.8$ s^{-2}, and the nominal control ratio will increase to approximately $\dfrac{\alpha}{\delta} = \dfrac{240}{28.8} = 8.3$. That is, a missile body with low static stability can greatly improve its control ratio, but we must also pay close attention to the problem of robustness reduction for a missile body with low static stability.

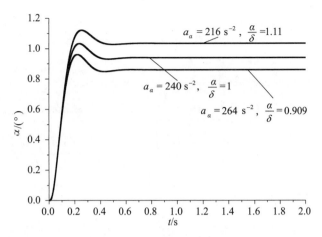

Fig. 6.1–18 Angle of attack response for unit step pseudo actuator input of the new missile body H_Ⅱ

If the uncertainty between center of gravity and center of pressure x^* is still taken as ± 5 mm, its variation will range from 1 mm to 11 mm. This makes the missile control ratio fluctuate from -44% to $+600\%$. The response of the angle of attack for the unit step actuator δ input with this different static stability change is shown in Fig. 6.1–19. Obviously, the uncertainty of x^* has a great influence on the performance of a weakly-stabilized missile body. Since the missile body is simply a component of the autopilot loop, a large variation of the missile body gain will make the autopilot design unacceptable when final robustness evaluation is performed.

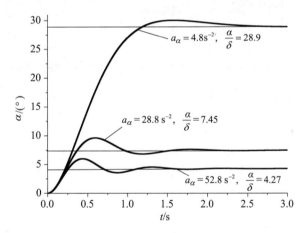

Fig. 6.1–19 Angle of attack for the unit step actuator deflection angle of missile body L_Ⅰ

The solution to the above problems is to construct an angle of attack feedback loop around the missile body to form an artificial restoring moment and achieve stability augmentation for the missile body with low static stability, and to stabilize the control ratio $\dfrac{\alpha}{\delta^*}$ of the new missile body. Currently, the angle of attack feedback signal is generally obtained by using the angular rate gyro output, and the pseudo attack angle signal is generated through the transfer function from the angular velocity to the angle of attack. The block diagram of the new missile body with pseudo angle of attack loop is shown in Fig. 6.1 – 20, where, K_g, K_s are respectively the design values of the autopilot damping loop and the stability augmentation loop of the pseudo angle of attack; K_s is the gain of the actuator; T_s, μ_s are the time constant and damping of the second-order actuator. The definitions of the remaining parameters of the missile body are given in Chapter 2.

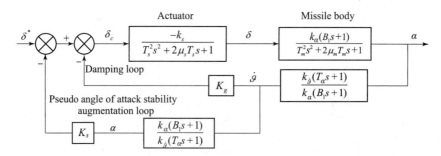

Fig. 6.1 – 20 Block diagram of the new missile body L_Ⅱ with pseudo angle of attack feedback

The block diagram and its chosen design parameters for the above-mentioned weak-static stability missile body are shown in Fig. 6.1 – 21.

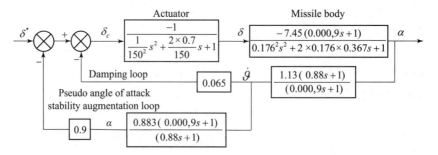

Fig. 6.1 – 21 Block diagram of the new missile body L_Ⅱ with pseudo angle of attack feedback

A missile body L_Ⅱ is formed after the pseudo angle of attack feedback is added. The angle of attack response of the new missile body L_Ⅱ for unit step pseudo actuator input δ^* is shown in Fig. 6.1 – 22.

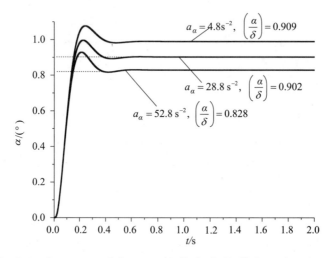

Fig. 6.1 – 22 Angle of attack response of the new missile body L_Ⅱ for unit step pseudo actuator input

It can be seen that the use of a pseudo angle of attack feedback loop allows a new missile with low static stability to have a stable gain from δ^* to α.

The possibility of simplification of the pseudo angle of attack feedback transfer function $\dfrac{\alpha(s)}{\dot{\vartheta}(s)}$ is discussed below. As we know from Chapter 2, the value of B_1 in equation (6.1-4) is very small and can be omitted. Therefore

$$\frac{\alpha(s)}{\dot{\vartheta}(s)} = \frac{k_\alpha(B_1 s + 1)}{k_{\dot{\vartheta}}^* (T_\alpha s + 1)} \approx \frac{k_\alpha}{k_{\dot{\vartheta}}^*}\left(\frac{1}{T_\alpha s + 1}\right) = \frac{k}{(T_\alpha s + 1)}, \qquad (6.1-4)$$

where

$$T_\alpha = \frac{a_\delta}{a_\delta b_\alpha - a_\alpha b_\delta} = \frac{1}{b_\alpha\left(1 - \dfrac{a_\alpha b_\delta}{a_\delta b_\alpha}\right)} = \frac{1}{b_\alpha(1 - x^*/\ell_\delta)} \approx \frac{1}{b_\alpha} \ (\text{as } x^*/\ell_\delta \text{ is a small item}).$$

So the value of k could be simplified as

$$k = \frac{a_\delta}{a_\delta b_\alpha - a_\alpha b_\delta} \approx \frac{1}{b_\alpha}.$$

The angle of attack response for unit input δ^* of the simplified pseudo angle of attack feedback structure in Fig. 6.1-21 is given in Fig. 6.1-23.

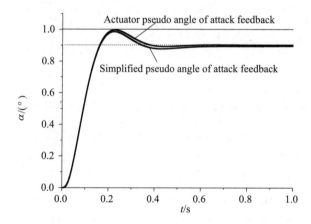

Fig. 6.1-23 Angle of attack response of the new missile body L_Ⅱ (missile body L_Ⅰ + stability augmentation loop + damping loop) for unit step input

It can be seen from the simulation results that it is completely feasible to construct a pseudo angle of attack loop with a simplified pseudo angle of attack transfer function. At present, autopilots with pseudo-attack angle feedback often adopt this simplified transfer function $\dfrac{\alpha(s)}{\dot{\vartheta}(s)}$ in implementation.

The following describes why a low static stability missile body design approach is often adopted. With the ever increasing maneuverability requirements for air-to-air missiles, high maneuvering ability can only be achieved by relying on high angles of attack for the given missile flight speed and aerodynamic configuration. For example, nowadays the allowable flight angle of attack for some missiles can reach 50°. To obtain a large angle of attack requires a large actuator deflection angle, but the allowable actuator deflection angle is usually no more than 30°. In order to obtain a large angle of attack with a small actuator deflection, the static stability of the missile body must be reduced. The introduction of a pseudo angle of attack feedback can simultaneously ensure a high control ratio and acceptable robustness of the new missile body. For example, if the missile of the above example is required to have an angle of attack output of 50°, the new missile design will require a pseudo actuator command $\delta^* = -55.5°$, but the actual required steady state actuator deflection angle is only $\delta_c(\infty) = -6.72°$. The response of an angle of attack with a steady state value of 50° and the related actuator angle δ and pseudo actuator angle δ^* inputs are given in Fig. 6.1 − 24.

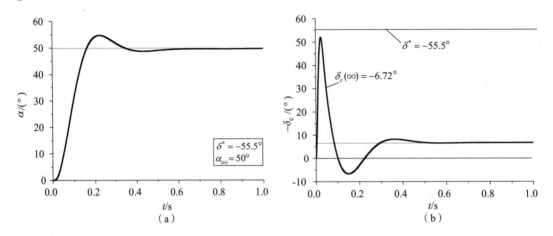

Fig. 6.1 − 24 The response of angle of attack and related δ and δ^* inputs

It should be noted that introducing a pseudo angle of attack feedback can only reduce the steady state actuator angle required. For fast angle of attack response a large transient actuator angle δ is still demanded. This problem can be solved by introducing saturation limit for the actuator angle δ, with a slightly reduced angle of attack response. Fig. 6.1 − 25 shows the effect of an actuator angle limit of 25° on the angle of attack response and related δ input.

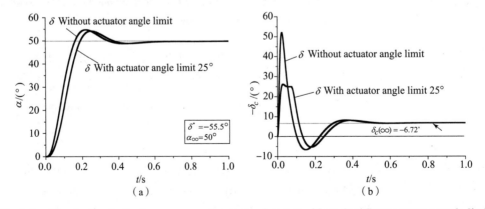

Fig. 6.1 − 25 Angle of attack response and actuator input δ with and without actuator angle limit

In the simulation, the actual actuator is set to have amplitude limit of 25°. This is because the low static stability missile can only reduce the steady state actuator deflection angle required to generate the angle of attack, and it cannot reduce the dynamic actuator deflection angle needed for the dynamic response of the missile body.

Fig. 6.1 – 26 shows the structure of a standard pseudo angle of attack feedback three-loop acceleration autopilot. The design methods for its acceleration loop and PI or lag compensators are the same as for two-loop autopilot with PI or lag compensation. The autopilot bandwidth obtained from this autopilot structure is close to the maximum allowed by the given actuator bandwidth constraint, while the use of a pseudo angle of attack feedback could reduce the steady state actuator angle for a given angle of attack output demand and it has a good missile body control ratio robustness. The use of PI or lag compensators will reduce the static error of the autopilot response. Due to these advantages, this autopilot structure is the most frequently used acceleration autopilot structure today.

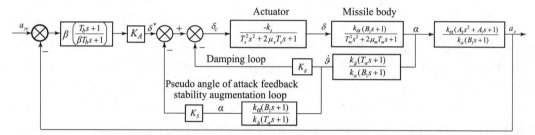

Fig. 6.1 – 26 Standard pseudo angle of attack feedback three-loop acceleration autopilot

6.1.4 Classic Three-loop Autopilot

The three-loop autopilot discussed here is the earliest classic three-loop autopilot structure used by the US Sparrow Air-to-Air Missile (Fig. 6.1 – 27). This structure was later called a three-loop autopilot in a large number of documents, so without additional explanation, the term of three-loop autopilot is given to this control structure.

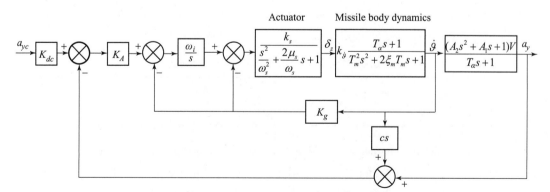

Fig. 6.1 – 27 Three-loop autopilot structure

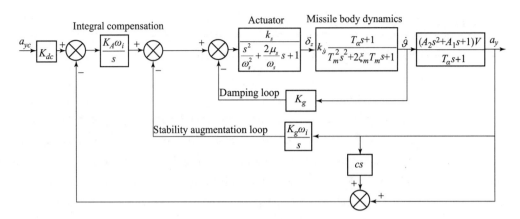

Fig. 6.1-27 Three-loop autopilot structure (Continued)

The inner loop of this structure is a damping loop, and the middle loop is a simplified stability augmentation loop. Here, instead of taking angle of attack as the feedback, the integration of the angular velocity $\dot{\vartheta}$, that is ϑ, is used for stability augmentation feedback. It is known that the attitude angle ϑ is equal to the sum of the angle of attack α and the flight path angle θ, but the angle of attack changes much more than the flight path angle in the missile short-period transient motion due to the large inertia of the flight path angle. Therefore, the attitude feedback loop can be approximately used as a stability augmentation loop. For analyzing the structural characteristics of the three-loop autopilot, the dynamic coefficients (Table 6.1-4) of an air-to-air missile given in reference [7] are used as an example.

Table 6.1-4 Missile dynamic coefficients

Height/m	a_α/s^{-2}	a_δ/s^{-2}	a_ω/s^{-1}	b_α/s^{-1}	b_δ/s^{-1}	$V/(\mathrm{m}\cdot\mathrm{s}^{-1})$
9.14×10^3	240	204	0	1.17	0.239	914

Table 6.1-5 shows the design values K_g, ω_i, K_A of the feedback gains of the three autopilot loops and gain K_{dc} is introduced to adjust the autopilot closed-loop gain as 1. When the actuator dynamics is omitted, the block diagram of the autopilot will be shown as Fig. 6.1-28. Its unit step response shown in Fig. 6.1-29 is very close to a first-order system response.

Table 6.1-5 Design values of the autopilot

K_g	ω_i	K_A	K_{dc}
0.264	5.10	0.000,929	1.31

Conclusions about the characteristics of this autopilot structure can be obtained by analyzing the characteristic roots of each loop for the selected parameters (Table 6.1-6). It can be seen that, the natural frequency of the missile body itself is 2.46 Hz and its damping ratio is as low as 0.04. After the damping loop is introduced, its damping ratio increases to 1.62, but the second-order root frequency only increases slightly to 2.70 Hz, which is the characteristic of a general

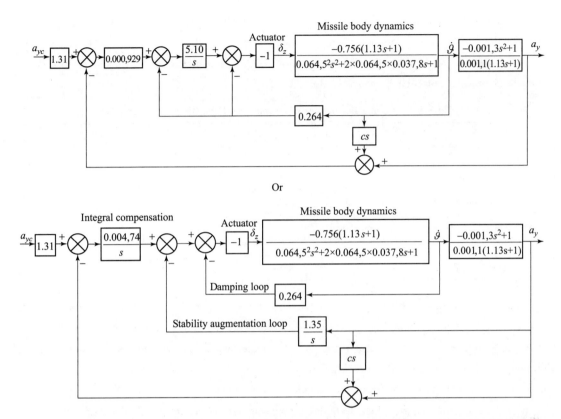

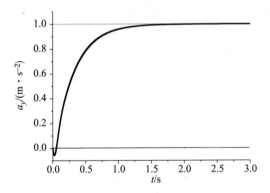

Fig. 6.1-28 Design diagram of the Sparrow Air-to-Air three-loop autopilot

Fig. 6.1-29 a_y unit step response

damping loop. After the introduction of the stability augmentation loop, the new missile body frequency increases to 5.71 Hz with a damping of 0.76. This is the feature of the stability augmentation loop. Its function is to increase the static stability of the new missile body, and to increase its control ratio robustness. The outermost feedback loop of this structure adopts an integral compensation and introduces the main first-order root. The final design gives a second-order root frequency of 5.44 Hz, but the first-order root frequency is only 0.53 Hz. Hence, it is concluded that the closed loop characteristic of this autopilot structure is basically determined by the integral compensation of the outer loop. The frequency bandwidth of the stability augmentation loop is very high, so the missile body parameter variation will have little effect on the characteristics of the autopilot. That is to say, this autopilot structure is highly robust. The disadvantage of this design approach is that the autopilot response is slow.

Table 6.1 – 6 Some performance parameters of the three-loop autopilot

Items	Closed loop second order frequency/Hz	Damping	First order time constant	First order root bandwidth frequency/Hz
Missile body itself	2.46	0.04	—	—
New missile body with damping loop	2.70	1.62	—	—
New missile body with stability augmentation loop	5.71	0.76	—	—
Three-loop autopilot	5.44	0.70	0.3	0.53

Fig. 6.1 – 30 shows the transient components for the first mode and second mode in the autopilot step response. It can be seen that the steady state value of the first order low frequency model component accounts for approximately 114% of the closed loop response steady state value of the autopilot. While the second order high-frequency model component only accounts for − 13%. Therefore, the first order low-frequency mode is the main mode for this three-loop autopilot structure. And the second order high-frequency mode has less effect on the time domain transient of the autopilot. That is to say, the variation in the missile parameter will have little effect on the characteristics of this autopilot structure. Of course, this advantage is obtained by reduced autopilot response speeds. With the current need for autopilot fast response, the designer should properly weigh the advantage and disadvantage of this early autopilot design approach.

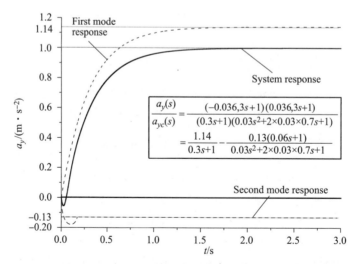

Fig. 6.1 – 30 Curves of unit step response mode decomposition

One of the disadvantages of this autopilot structure is that its closed loop gain is different from the standard integral compensation control loop. It can be derived from Fig. 6.1 – 17 that its closed loop gain is not 1 but

$$\frac{a_y(\infty)}{a_{yc}(\infty)} = \frac{k_A \cdot V}{k_g + k_A \cdot V}. \qquad (6.1-5)$$

The gain of this autopilot structure will change with the change of missile flight speed. Therefore, there is a need to introduce a parameter K_{dc} to adjust the autopilot gain as 1.

$$K_{dc} = \frac{k_g + k_A \cdot V}{k_A \cdot V}. \qquad (6.1-6)$$

For the given example, the unadjusted autopilot gain is $\dfrac{a_y(\infty)}{a_{yc}(\infty)} = \dfrac{k_A \cdot V}{k_g + k_A \cdot V} = \dfrac{0.000\,929 \times 914}{0.264 + 0.000\,929 \times 914} = 0.763$, so the parameter K_{dc} value for nominal missile speed of 914 m/s should be taken as 1.31.

6.1.5 Discussion of Variable Acceleration Autopilot Structures

The above two-loop autopilot, two-loop autopilot with PI compensation, three-loop autopilot with pseudo angle of attack feedback and classic three-loop autopilot can be unified to a basic structure as shown in Fig. 6.1 – 31. The characteristics of the compensated new missile body and its compensation network for each structure are shown in Table 6.1 – 7.

Table 6.1 – 7 **Comparison of the structural characteristics of different acceleration autopilots**

Items	Feedback compensation formed new missile body	Compensation network and its characteristics
Two-loop autopilot	Damping increases with the angular velocity feedback.	No compensator in forward loop. With static error.
Two-loop autopilot with PI compensation	Damping increases with the angular velocity feedback.	PI compensation (No static error. The PI compensator phase shift should not be large at the autopilot crossover frequency. The robustness of the new missile body has a great effect on the autopilot performance, but it has a fast response.)
Three-loop autopilot with pseudo angle of attack feedback	Stability augmentation of pseudo angle of attack feedback. Damping increases with the angular velocity feedback.	PI compensation (No static error. The PI compensation phase shift should not be large at the autopilot crossover frequency. The robustness of the new missile body has great effect on the autopilot performance, but it has a fast response.)
Classic three-loop autopilot	Stability augmentation of missile attitude feedback. Damping increases with the angular velocity feedback.	Integral compensation (The integral compensator has a 90° phase shift at the autopilot crossover frequency. The robustness of the new missile body has less effect on the autopilot performance, and the robustness of the autopilot is high. The static error is related to the flight speed of the missile.)

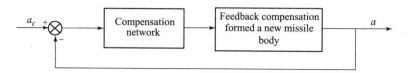

Fig. 6.1 – 31 Basic structure of the acceleration autopilot

6.1.6 Hinge Moment Autopilot

One of the simplest ways to generate the normal acceleration a_y required for guidance is to use the following actual angle δ and steady state missile normal acceleration relation.

$$a_y = \frac{\frac{1}{2}\rho V^2 S c_y^a}{m}\alpha = \frac{\frac{1}{2}\rho V^2 S c_y^\alpha}{m}\left(\frac{a_\delta}{a_\alpha}\right)\delta = K\delta. \tag{6.1-7}$$

It is possible to use the actuator command $\delta_c = \dfrac{a_{yc}}{K}$ to replace the acceleration command a_{yc} to form a very simple aerodynamic feedback autopilot. When the dynamic pressure of the missile $q = \dfrac{1}{2}\rho V^2$ does not change much and its static stability is relatively high (i.e., the value of a_δ/a_α changes little with the change of the positions of the center of gravity and the center of pressure). Even without an artificial acceleration feedback, the missile body itself constructs a very simple aerodynamic autopilot (Fig. 6.1 – 32).

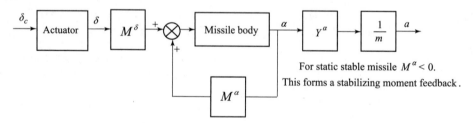

Fig. 6.1 – 32 Schematic diagram of an aerodynamic autopilot

Usually, anti-tank missiles will fly at a constant speed above the ground under the action of the sustainer rocket. Their dynamic pressure will not change much, and a high static stability aerodynamic design will provide enough stabilizing moment to stabilize their control ratio α/δ. For the above reasons, many anti-tank missiles have not adopted the artificial acceleration autopilot approach in their realization of control to simplify the system and reduce its cost.

However, the flight speed and flight height of ground-to-air missiles and air-to-air missiles could vary greatly (i.e., their dynamic pressure q changes greatly). A simple method to solve the impact of dynamic pressure changes is to make the actuator deflection angle inversely proportional to the missile dynamic pressure change. In this way, the missile acceleration output will remain unchanged even when the missile dynamic pressure variates. The early American "Sidewinder," Israel's "Python 3" and many other countries' early air-to-air missiles all adopted this simple autopilot structure.

To implement this idea, this type of autopilot uses the actuator driving moment command as the autopilot command (Fig. 6.1 – 33). Since the actuator's hinge moment produced by the actuator incident angle of attack $\delta + \alpha$ is proportional to missile flight dynamic pressure, the actuator hinge moment is balanced by the actuator driving moment. The generated corresponding normal acceleration will not be sensitive to the dynamic pressure change.

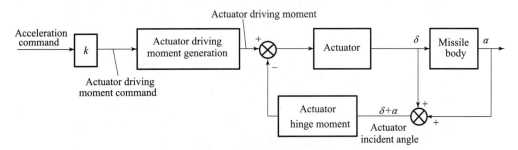

Fig. 6.1 – 33 Block diagram of the hinge moment autopilot

The actuator hinge moment H consists of two parts. One part is related to the actuator deflection angle δ and the other part is related to the angle of attack α of the missile. The hinge moment transfer function expression is

$$H(s) = S \cdot q \cdot d(C_{H_\delta}(s) \cdot \delta(s) + C_{H_\alpha}(s) \cdot \alpha(s)), \tag{6.1-8}$$

$$\frac{H(s)}{\delta(s)} = S \cdot q \cdot d\left(C_{H_\delta}(s) + C_{H_\alpha}(s) \frac{\alpha(s)}{\delta(s)}\right), \tag{6.1-9}$$

where S and d are respectively the reference area and reference length of the actuator center of pressure position to the actuator axis, and q is the dynamic pressure.

The following describes the characteristics of this type of autopilot, using the example given in reference [2] "*Automatic Control of Aircraft and Missiles.*" This type of typical hinge moment autopilot has two configurations. One of the more complex structures is shown in Fig. 6.1 – 34. It uses rate gyro feedback as the damping loop, and constructs an actuator deflection angle feedback loop to speed up the actuator response. To improve the actuator mechanism stability, a phase lead compensator is introduced. It should be noted that in this design, the main feedback is the hinge moment feedback, and if not, the nature of the actuator will be completely changed.

The early hinge moment autopilot was simple in structure with no rate gyro and actuator deflection angle feedback loop included. Its structure is shown in Fig. 6.1 – 35.

For the above two examples, the related transfer functions are

$$\frac{H(s)}{\delta(s)} = \frac{-12.73(s^2 + 1.56s + 811.72)}{s^2 + 1.652s + 424.63} \left(\frac{\text{N} \cdot \text{m}}{\text{rad}}\right), \tag{6.1-10}$$

$$\frac{\dot{\vartheta}(s)}{\delta(s)} = \frac{303.6(s + 1.75)}{s^2 + 1.652s + 424.63} \left(\frac{\text{rad/s}}{\text{rad}}\right), \tag{6.1-11}$$

$$\frac{a_y(s)}{\dot{\vartheta}(s)} = \frac{-0.55(s^2 + 8.7s + 8,050)}{s + 1.75} \left(\frac{\text{m/s}^2}{\text{rad/s}}\right). \tag{6.1-12}$$

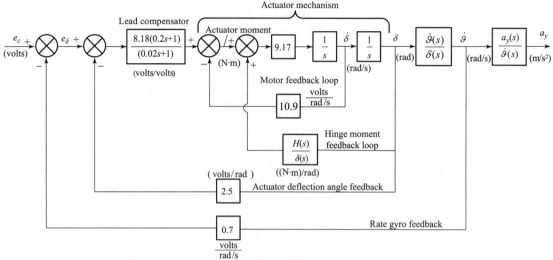

Fig. 6.1 – 34 Structure of the hinge moment autopilot (I)

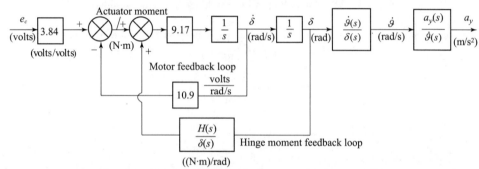

Fig. 6.1 – 35 Structure of hinge moment autopilot (II)

The normal acceleration time responses for unit step actuator driving moment input e_c (volts) of these two autopilots are shown in Fig. 6.1 – 36. It can be seen from this figure that the hinge moment autopilot can provide a steady state normal acceleration that does not change with the dynamic pressure variation, and the hinge moment autopilot with angular rate gyro and actuator deflection angle feedback responds faster and more smoothly.

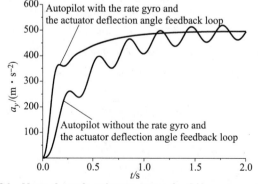

Fig. 6.1 – 36 Normal acceleration response for hinge moment autopilots under the action of the unit step actuator driving moment command

6.1.7 Questions Concerning Acceleration Autopilot Design

1) Missile elastic vibration mode effect on autopilot design

When considering missile body elasticity, the actuator deflection will not only cause a rigid rotation of the missile body, but also cause a missile body elastic angular motion. This elastic angular motion is theoretically composed of the many vibration modes, but only its first order mode or sometimes the second order mode needs to be considered in autopilot design. The least elastic vibration-sensitive positions for the autopilot rate gyro and accelerometer are shown in Fig. 6.1 - 37.

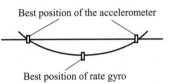

Fig. 6.1 - 37 **Elastic response of the missile body**

When the missile body elastic motion is taken into account, the block diagram of the autopilot angular velocity loop will be changed (Fig. 6.1 - 38). Here ω_R is the missile's elastic variation angular velocity at the rate gyro position.

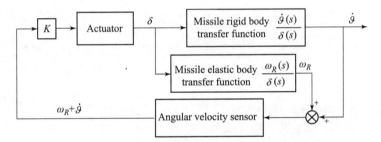

Fig. 6.1 - 38 **Block diagram of the autopilot angular velocity loop with elastic response of the missile body included**

Since the damping of the missile elastic vibration motion is very low, the magnitude value of its transfer function Bode diagram could be very high at its main mode frequency. This could make the autopilot open loop gain magnitude above 0 dB at this high elastic mode frequency and cause the autopilot to be unstable. At the same time, the actuator friction and saturation introduced by this high frequency are also unacceptable.

For this reason, in the design of the autopilot, a notch filter is always added in front of the actuator input to avoid control system instability and its negative impact on the actuator operation. The form of the standard transfer function of the notch filter commonly used is

$$G_F(s) = \frac{\frac{1}{\omega_i^2}s^2 + \frac{2\xi_1}{\omega_i}s + 1}{\frac{1}{\omega_i^2}s^2 + \frac{2\xi_2}{\omega_i}s + 1}, \quad (6.1-13)$$

where ω_i is the center frequency of the notch filter and is usually taken as the estimated value of the main elastic frequency of the missile body, and the denominator damping coefficient ξ_2 generally takes the value of 1. The notch depth and width are dependent on the value of numerator damping coefficient ξ_1.

Fig. 6.1-39 shows the Bode diagram of a notch filter when $\omega_i = 40$ Hz, $\xi_2 = 1$, $\xi_1 = 0.1$, 0.15, 0.2. The corresponding notch depths are -20 db, -16.5 db, and -14 db respectively.

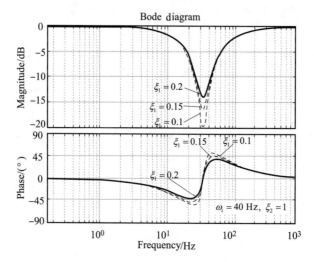

Fig. 6.1-39 Bode diagram of the example notch filter

In practical engineering design, it should be ensured that the notch filter can still provide sufficient signal attenuation when the real elastic frequency has small deviation from its nominal value.

2) Canard control and rear control

According to Section 6.1.1, the transfer function of the missile body as an object being controlled by the autopilot is

$$\frac{a_y(s)}{\delta_z(s)} = \frac{k_\alpha(A_2 s^2 + A_1 s + 1)}{T_m^2 s^2 + 2\mu_m T_m s + 1}. \tag{6.1-14}$$

For rear controlled missile $A_1 < 0, A_2 < 0$. Therefore, its transfer function can also be given as

$$\frac{a_y(s)}{\delta_z(s)} = \frac{k_\alpha(-|A_2| s^2 - |A_1| s + 1)}{T_m^2 s^2 + 2\mu_m T_m s + 1}. \tag{6.1-15}$$

Since this transfer function has a zero in the right half plane, a rear-controlled missile is a non-minimum phase system.

For canard control $A_1 > 0$, $A_2 > 0$, so a canard controlled missile has a transfer function as

$$\frac{a_y(s)}{\delta_z(s)} = \frac{k_\alpha(A_2 s^2 + A_1 s + 1)}{T_m^2 s^2 + 2\mu_m T_m s + 1}. \tag{6.1-16}$$

The zeros and the poles of this transfer function are all in the left half plane, so a canard controlled missile is a minimum phase system.

Fig. 6.1-40 shows the typical acceleration autopilot response of a rear controlled missile to a $a_{yc} = 50$ m/s^2 step acceleration command. In the figure, the related actuator deflection angle δ and angle of attack α are also given. It should be noted that the acceleration autopilot output at the beginning of the rear control is opposite to the expected acceleration response. The expected acceleration direction appears only when the lift generated by the angle of attack is greater than the negative actuator force after the angle of attack is large enough. This is the unique disadvantage of non-minimum phase systems. The positive actuator deflection produced in the intermediate process is to suppress the overshoot of the response. The ratio of the angle of attack and the actuator deflection angle in the steady state reflects the control ratio of the missile.

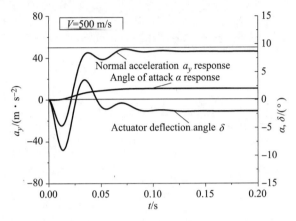

Fig. 6.1 – 40 The rear controlled missile autopilot response to a step acceleration command $a_{yc} = 50$ m/s^2

Fig. 6.1 – 41 shows the response of a canard controlled acceleration autopilot to a $a_{yc} = 50$ m/s^2 step command. Also shown are the actuator deflection angle δ and the angle of attack α response. In this case, the actuator initially generated forces are in the same direction as lift produced by the angle of attack, so its control is more efficient. The actuator deflection angle is reduced in the intermediate phase of control to suppress the overshoot of the response. The ratio of the angle of attack and the actuator deflection angle in its steady state reflects the control ratio of this missile. The canard controlled autopilot shows the typical transient response for a minimum phase control system.

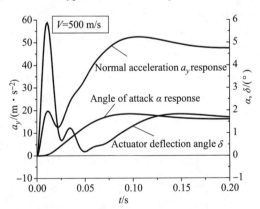

Fig. 6.1 – 41 The canard controlled autopilot response for an $a_{yc} = 50$ m/s^2 step acceleration command

From the above analysis, it can be known that the steady state acceleration response for a canard controlled missile is generated by adding the same direction lift and the actuator force, while the steady state acceleration response for a rear controlled missile is caused by the difference between the lift force and the actuator force. Suppose that the actuator angle generated force is $b_\delta \delta$. Then the corresponding steady state angle of attack α produced force will be $b_\alpha \alpha = b_\alpha \left(\dfrac{a_\delta}{a_\alpha}\right)\delta$. Therefore, the ratio of the two forces will be $K = b_\alpha \cdot \dfrac{a_\delta}{a_\alpha} \cdot \delta \cdot \dfrac{1}{b_\delta \delta} = \dfrac{b_\alpha}{b_\delta} \cdot \dfrac{a_\delta}{a_\alpha}$. That is to say, if the lift value is 1, the actuator force will be $\dfrac{1}{K}$. So the gain ratio of the canard controlled missile and the rear controlled missile will be

$$\frac{\text{Canard controlled missile gain}}{\text{Rear controlled missile gain}} = \frac{1 + \dfrac{1}{K}}{1 - \dfrac{1}{K}}.$$

Suppose that $K = \dfrac{a_\delta b_\alpha}{a_\alpha b_\delta} = 6$, then the gain ratio of the canard control and the rear control will be

$$\frac{\left(1 + \dfrac{1}{6}\right)}{\left(1 - \dfrac{1}{6}\right)} = \frac{1.17}{0.83} = 1.4.$$

Although canard control has the above advantages, it has a drawback. That is, its incident angle of the actuator is the sum of the angle of attack and the actuator deflection angle. While in rear control, the actuator incident angle is the difference of the angle of attack and the actuator angle. For this reason, the usable actuator deflection angle for a canard controlled missile is smaller than that of a rear controlled missile due to the saturation limit for the actuator incident angle. Therefore, most current tactical missile designs adopt the rear controlled aerodynamic configuration.

3) Static unstable missile control

The problem of static unstable missile control may occur for a short period of time in the initial phase of the powered flight for some tactical missiles. A static unstable missile body itself is a divergent system without an autopilot. Fig. 6.1 – 42 shows the response of the actuator deflection angle, the angle of attack, and acceleration under an $a_{yc} = 50$ m/s^2 step acceleration command input for an acceleration autopilot of a rear controlled static unstable missile.

Similarly, a negative actuator deflection angle at the beginning of the control is still needed to rotate the missile body to produce a positive angle of attack. However, the difference from a static stable missile body is that the actuator static state deflection angle in this case should be positive to balance the destabilizing moment generated by the acceleration output required angle of attack (Fig. 6.1 – 43).

Also, a pseudo angle of attack feedback described above can also be used to make the compensated new missile body stable.

4) Discussion of the angle of attack autopilot

Some scholars attempted to use the advanced robust control theory for the design of an

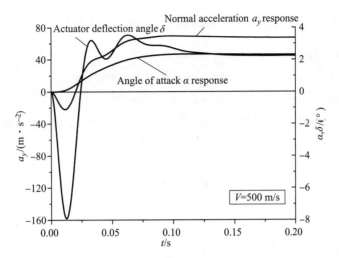

Fig. 6.1 – 42 Response of the step command of the static unstable missile body acceleration autopilot

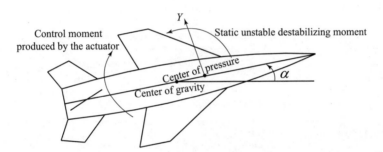

Fig. 6.1 – 43 Moment equilibrium relationship when a static unstable missile body produces a steady angle of attack

acceleration autopilot. However, a rear controlled missile body $\dfrac{a_y(s)}{a_{yc}(s)}$ is a non-minimum phase system. But the transfer function from the actuator angle to the angle of attack is a minimum phase system, since B_1 is greater than 0.

$$\frac{\alpha(s)}{\delta(s)} = k_\alpha \frac{B_1 s + 1}{T_m^2 s^2 + 2\mu_m T_m s + 1}. \qquad (6.1-17)$$

Theoretically, an angle of attack autopilot can be used indirectly to achieve acceleration control (Fig. 6.1 – 44).

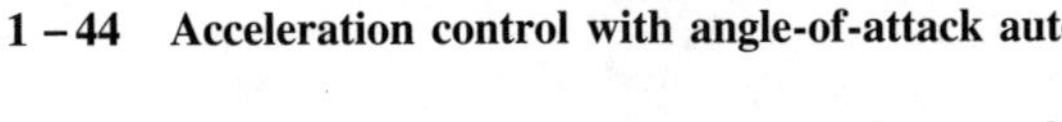

$$a_c \rightarrow \boxed{\text{Angle of attack autopilot}} \xrightarrow{\alpha} \boxed{\dfrac{Y(s)}{\alpha(s)}} \xrightarrow{Y} \boxed{\dfrac{Y}{m}} \xrightarrow{a}$$

Fig. 6.1 – 44 Acceleration control with angle-of-attack autopilot

But without an acceleration feedback loop, the variation of the aerodynamic parameters in the transfer function $\dfrac{Y(s)}{\alpha(s)}$ will result in an acceleration control error. The pros and cons of this approach in real engineering applications are to be carefully weighed by the designers.

6.1.8 The Acceleration Autopilot Design Method

6.1.8.1 Pole-placement Method

The "state feedback pole assignment method" is to synthesize a state feedback control system for the LTI controlled system with a set of expected poles (eigenvalues) as performance indicators, and to configure the resulting control system eigenvalues to the desired positions on the complex plane. As the author has expounded in another book (reference [1]), for autopilots of different structures, if the influence of actuator dynamics is ignored, when selecting the state variables and output variables to describe the missile body model as shown in Table 6.1-8, the state space model is state-controllable and output-controllable. So the pole-assignment method can be used to simplify the autopilot design. In this section, the idea of pole-assignment and analytical method will be adopted to study the design methods of classic two-loop and three-loop autopilots. For other types of autopilot design methods and derivation of pole-assignment algorithms for specific state feedback and output feedback of autopilots, please refer to reference [1].

Table 6.1-8 Pole assignment of several typical structural autopilots

Autopilot structure	State variables	Output variables	State/output controllability	Observability	Pole-assignment
Classic two-loop autopilot	$[\alpha \ \dot{\vartheta}]^T$	$[\alpha_y \ \dot{\vartheta}]^T$	√	√	√ (2 poles)
Two-loop autopilot with PI compensation	$[\alpha \ \vartheta \ \dot{\vartheta}]^T$	$[\alpha_y \ \int \alpha_y \ \dot{\vartheta}]^T$	√	√	√ (3 poles)
Classic three-loop autopilot	$[\alpha \ \vartheta \ \dot{\vartheta}]^T$	$[\int \alpha_y \ \vartheta \ \dot{\vartheta}]^T$	√	√	√ (3 poles)
Three-loop autopilot with pseudo angle of attack feedback	$[\alpha \ \vartheta \ \dot{\vartheta}]^T$	$[\int \alpha_y \ \alpha \ \dot{\vartheta}]^T$	√	√	√ (3 poles)
Attitude autopilot	$[\alpha \ \vartheta \ \dot{\vartheta}]^T$	$[\vartheta \ \dot{\vartheta}]^T$	√	×	× (3 poles)
Velocity vector autopilot	$[\alpha \ \vartheta \ \dot{\vartheta}]^T$	$[\theta \ \dot{\vartheta}]^T$	√	×	× (3 poles)

1) Design two-loop acceleration autopilot with pole-assignment method

As a preliminary theoretical design, in order to simplify the design process, the hardware dynamics such as accelerometer, angular rate gyro and actuator are not considered temporarily. The accelerometer gain is introduced, and the structure of the two-loop acceleration autopilot is changed into the form of Fig. 6.1-45 based on Fig. 6.1-1. Where k_{ac}, K_g, k_s are accelerometer, angular rate gyro and actuator gains respectively, and K_A, K_g are design parameters—the closed-loop gain adjustment coefficient.

All the feedback in Fig. 6.1-45 is equivalent to the actuator as

$$\delta_e = -k_s k_{ac} K_A (a_y + c\ddot{\vartheta}) - k_s K_g \dot{\vartheta}. \quad (6.1-18)$$

Substitute it into the equation of motion of the missile, and we get

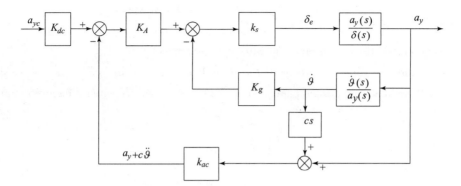

Fig. 6.1-45 Block diagram of two-loop acceleration autopilot

$$\delta_e = -\frac{k_s k_{ac} K_A (Vb_\alpha - ca_\alpha)}{1 + k_s k_{ac} K_A (Vb_\delta - ca_\delta)}\alpha - \frac{k_s K_g - k_s k_{ac} K_A ca_\omega}{1 + k_s k_{ac} K_A (Vb_\delta - ca_\delta)}\dot{\vartheta}. \quad (6.1-19)$$

The gain matrix of state feedback $\boldsymbol{K}$ is

$$\boldsymbol{K} = \begin{bmatrix} k_1 \\ k_2 \end{bmatrix}^T = \begin{bmatrix} \dfrac{k_s k_{ac} K_A (Vb_\alpha - ca_\alpha)}{1 + k_s k_{ac} K_A (Vb_\delta - ca_\delta)} \\ \dfrac{k_s K_g - k_s k_{ac} K_A ca_\omega}{1 + k_s k_{ac} K_A (Vb_\delta - ca_\delta)} \end{bmatrix}^T. \quad (6.1-20)$$

If output feedback is adopted, equation $\boldsymbol{F} = (\boldsymbol{I} - \boldsymbol{KC}^{-1}\boldsymbol{D})^{-1}\boldsymbol{KC}^{-1}$ can be used to get

$$\boldsymbol{F} = \begin{bmatrix} f_1 \\ f_2 \end{bmatrix}^T = \begin{bmatrix} k_s k_{ac} K_A \\ k_s K_g \end{bmatrix}^T = \begin{bmatrix} \dfrac{k_1}{(Vb_\alpha - ca_\alpha) - k_1(Vb_\delta - ca_\delta)} \\ \dfrac{k_1 ca_\omega + k_2(Vb_\alpha - ca_\alpha)}{(Vb_\alpha - ca_\alpha) - k_1(Vb_\delta - ca_\delta)} \end{bmatrix}^T. \quad (6.1-21)$$

Thus, the transformation relationship between design parameters of the two-loop autopilot and state feedback gain and output feedback gain is

$$K_A = \frac{f_1}{k_s k_{ac}} = \frac{k_1/(k_{ac}k_s)}{(Vb_\alpha - ca_\alpha) - k_1(Vb_\delta - ca_\delta)},$$

$$K_g = \frac{f_2}{k_s} = \frac{k_1 ca_\omega/k_s + k_2(Vb_\alpha - ca_\alpha)/k_s}{(Vb_\alpha - ca_\alpha) - k_1(Vb_\delta - ca_\delta)}. \quad (6.1-22)$$

According to the closed-loop transmission function of the two-loop autopilot, the expression of the closed-loop gain of the autopilot K can be obtained as

$$K = \frac{k_s K_A k_{\dot{\vartheta}} V}{1 + k_s k_{\dot{\vartheta}}(K_g + k_{ac}K_A V)} = \frac{-k_s K_A V(a_\delta b_\alpha - a_\alpha b_\delta)}{(a_\alpha + a_\omega b_\alpha) - k_s(K_g + k_{ac}K_A V)(a_\delta b_\alpha - a_\alpha b_\delta)}. \quad (6.1-23)$$

In order to make the steady-state error of the pilot tracking acceleration instruction zero, the system should satisfy the final value theorem

$$\lim_{s \to 0} \frac{a_y}{a_{yc}} = 1. \quad (6.1-24)$$

Therefore, the closed-loop gain adjustment coefficient is

$$K_{dc} = \frac{(a_\alpha + a_\omega b_\alpha) - k_s(K_g + k_{ac}K_AV)(a_\delta b_\alpha - a_\alpha b_\delta)}{-k_sK_AV(a_\delta b_\alpha - a_\alpha b_\delta)}. \qquad (6.1-25)$$

2) Design classic three-loop autopilot with pole-assignment method

Similarly, the structure adjustment of the three-loop autopilot in Section 6.1.4 is shown in Fig. 6.1 – 46. Note that the design parameter variables K_g, ω_i, K_A are different from those in Fig. 6.1 – 27.

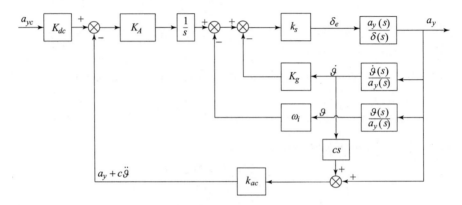

Fig. 6.1 – 46 Block diagram of classic three-loop autopilot

All the feedback in Fig. 6.1 – 45 is equivalent to the actuator as

$$\delta_e = k_s k_{ac} K_A V \alpha - k_s(k_{ac}K_AV + \omega_i)\vartheta - k_s(K_g + ck_{ac}K_A)\dot{\vartheta}. \qquad (6.1-26)$$

The gain matrix of state feedback $\boldsymbol{K}$ is

$$\boldsymbol{K} = \begin{bmatrix} k_1 \\ k_2 \\ k_3 \end{bmatrix}^\mathrm{T} = \begin{bmatrix} -k_s k_{ac} K_A V \\ k_s(\omega_i + k_{ac}K_AV) \\ k_s(K_g + ck_{ac}K_A) \end{bmatrix}^\mathrm{T}. \qquad (6.1-27)$$

The gain matrix of output feedback $\boldsymbol{F}$ is

$$\boldsymbol{F} = \begin{bmatrix} f_1 \\ f_2 \\ f_3 \end{bmatrix}^\mathrm{T} = \begin{bmatrix} k_s k_{ac} K_A \\ k_s \omega_i \\ k_s K_g \end{bmatrix}^\mathrm{T} = \begin{bmatrix} -k_1/V \\ k_1 + k_2 \\ ck_1/V + k_3 \end{bmatrix}^\mathrm{T}. \qquad (6.1-28)$$

Therefore, the design parameters of the three-loop autopilot are

$$\begin{cases} K_A = f_1/(k_s k_{ac}) = -k_1/(k_s k_{ac} V) \\ \omega_i = f_2/k_s = (k_1 + k_2)/k_s \\ K_g = f_3/k_s = (ck_1 + k_3 V)/(k_s V) \end{cases}. \qquad (6.1-29)$$

The closed-loop gain is

$$K = K_A V/(\omega_i + k_{ac}K_AV) \qquad (6.1-30)$$

The above formula shows that the closed-loop gain K is related to the velocity and design parameters. It can be seen that, as described in Section 6.1.5, although the forward channel of the three-loop autopilot contains integral terms, it is not a structure without static error. In order to realize the accurate tracking of acceleration instruction (to ensure zero steady-state error), $K_{dc}K = 1$ is needed to obtain the expression of the pilot's closed-loop gain adjustment coefficient K_{dc}, that is

$$K_{dc} = (\omega_i + k_{ac}K_AV)/(K_AV). \qquad (6.1-31)$$

In the previous autopilot pole-assignment design, the influence of actuator dynamics was ignored, so the corresponding autopilot closed-loop pole assignment can be achieved arbitrarily regardless of state feedback or output feedback. However, an actual actuator cannot be infinitely fast. When the frequency of the actuator is close to the open-loop cut-off frequency of the autopilot, the influence of the actuator on the closed-loop pole of the autopilot will be obvious. Therefore, the dynamics of the actuator cannot be simply ignored. Even if not considered in the design, after the completion of the autopilot design, the actuator dynamics should be brought in for necessary inspections.

As an independent part, the actuator in the autopilot has been "designed." Although the actuator dynamics is introduced in this section, it is only used as the constraint of pole assignment, and it is not redesigned with the missile body for output (state) feedback.

Take the second-order actuator model as an example, as shown in Fig. 6.1 – 47. The transmission function of the actuator is

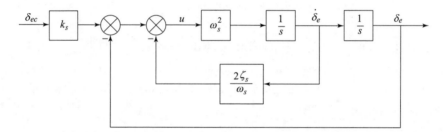

Fig. 6.1 – 47 Block diagram of the second-order actuator model

$$\frac{\delta_e}{\delta_{ec}} = \frac{k_s}{s^2/\omega_s^2 + 2\zeta_s s/\omega_s + 1}. \qquad (6.1-32)$$

In Fig. 6.1 – 46 and Fig. 6.1 – 47, the system state equation considering the steering gear dynamics is constructed, and the feedback is equivalent to u, then the following equation is obtained.

$$u = -k_s[-k_{ac}K_AV\alpha + (\omega_i + k_{ac}K_AV)\vartheta + (K_g + ck_{ac}K_A)\dot{\vartheta}] - \delta_e - (2\zeta_s/\omega_s)\dot{\delta}_e. \qquad (6.1-33)$$

The gain matrix of state feedback K is

$$K = \begin{bmatrix} k_1 \\ k_2 \\ k_3 \\ k_4 \\ k_5 \end{bmatrix}^T = \begin{bmatrix} -k_s k_{ac}K_AV \\ k_s(\omega_i + k_{ac}K_AV) \\ k_s(K_g + ck_{ac}K_A) \\ 1 \\ 2\zeta_s/\omega_s \end{bmatrix}^T. \qquad (6.1-34)$$

The gain matrix of output feedback F is

$$F = \begin{bmatrix} f_1 \\ f_2 \\ f_3 \\ f_4 \\ f_5 \end{bmatrix}^T = \begin{bmatrix} -k_1/V \\ k_1 + k_2 \\ ck_1/V + k_3 \\ k_4 \\ k_5 \end{bmatrix}^T. \tag{6.1-35}$$

The design parameters of the autopilot are

$$\begin{cases} K_A = f_1/(k_s k_{ac}) = -k_1/(k_s k_{ac} V) \\ \omega_i = f_2/k_s = (k_1 + k_2)/k_s \\ K_g = f_3/k_s = (ck_1 + k_3 V)/(k_s V) \end{cases}. \tag{6.1-36}$$

Similarly, in the design process, the constraints $f_4 \approx 1$, $f_5 \approx 2\zeta_s/\omega_s$ need to be satisfied.

6.1.8.2 Analytical Method

This section mainly studies the analytic design method of the classical three-loop autopilot and two-loop autopilot, which is essentially consistent with pole-assignment method. Therefore, autopilots that can be designed by pole-assignment method can also be designed by the analytical method in this section.

1) Design classic three-loop autopilot with analytical method

In order to simplify the design process, dynamics of the actuator is not taken into account temporarily. The open-loop structure diagram of the autopilot which is disconnected at the actuator is shown in Fig. 6.1-48.

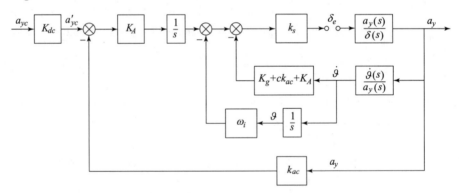

Fig. 6.1-48 Block diagram of the open-loop structure of the classic three-loop autopilot

In Fig. 6.1-48, the open-loop transmission function of the autopilot system is

$$HG(s) = k_s \frac{a_y(s)}{\delta(s)} \left[(K_g + ck_{ac}K_A) \frac{\dot{\vartheta}(s)}{a_y(s)} + \omega_i \frac{\vartheta(s)}{a_y(s)} + \frac{k_{ac}K_A}{s} \right]. \tag{6.1-37}$$

The closed-loop transmission function is

$$\frac{a_y(s)}{a'_{yc}(s)} = \frac{\delta(s)}{a'_{yc}(s)} \cdot \frac{a_y(s)}{\delta(s)} = \frac{k_s K_A/s}{1 + HG(s)} \cdot \frac{a_y(s)}{\delta(s)}. \tag{6.1-38}$$

By introducing intermediate variables M_2, M_1, M_0 and connecting equation (6.1-37) and equation (6.1-38), the open-loop transmission and closed-loop transmission of the autopilot can be expressed as

$$\mathrm{HG}(s) = \frac{M_2 s^2 + M_1 s + M_0}{s(s^2/\omega_m^2 + 2\zeta_m s/\omega_m + 1)}, \quad (6.1-39)$$

$$\frac{a_y(s)}{a'_{yc}(s)} = \frac{k_s k_\vartheta^* K_A V (A_2 s^2 + A_1 s + 1)}{s(s^2/\omega_m^2 + 2\zeta_m s/\omega_m + 1) + M_2 s^2 + M_1 s + M_0}, \quad (6.1-40)$$

$$\begin{cases} M_2 = k_s k_\vartheta^* (K_g T_\alpha + c k_{ac} K_A T_\alpha + k_{ac} K_A V A_2) \\ M_1 = k_s k_\vartheta^* (K_g + c k_{ac} K_A + \omega_i T_\alpha + k_{ac} K_A V A_1) \\ M_0 = k_s k_\vartheta^* (\omega_i + k_{ac} K_A V) \end{cases} \quad (6.1-41)$$

Then introduce intermediate variables $K_{\vartheta A}$, $K_{\vartheta I}$ and $K_{\vartheta G}$, and set

$$\begin{cases} K_{\vartheta A} = k_s k_\vartheta^* K_A \\ K_{\vartheta I} = k_s k_\vartheta^* \omega_i \\ K_{\vartheta G} = k_s k_\vartheta^* K_g \end{cases} \quad (6.1-42)$$

M_2, M_1 and M_0 can be re-expressed as

$$\begin{cases} M_2 = T_\alpha K_{\vartheta G} + (c T_\alpha + V A_2) k_{ac} K_{\vartheta A} \\ M_1 = K_{\vartheta G} + T_\alpha K_{\vartheta I} + (c + V A_1) k_{ac} K_{\vartheta A} \\ M_0 = K_{\vartheta I} + k_{ac} V K_{\vartheta A} \end{cases} \quad (6.1-43)$$

The matrix representation is

$$\begin{bmatrix} M_2 \\ M_1 \\ M_0 \end{bmatrix} = \begin{bmatrix} k_{ac}(c T_\alpha + V A_2) & 0 & T_\alpha \\ k_{ac}(c + V A_1) & T_\alpha & 1 \\ k_{ac} V & 1 & 0 \end{bmatrix} \begin{bmatrix} K_{\vartheta A} \\ K_{\vartheta I} \\ K_{\vartheta G} \end{bmatrix}. \quad (6.1-44)$$

At the same time, the autopilot closed-loop pole parameters τ, ω and ζ are taken as constraint indexes, and the desired characteristic equation of the autopilot is assumed to satisfy

$$(1 + \tau s)\left(\frac{s^2}{\omega^2} + \frac{2\zeta}{\omega} s + 1\right) = \frac{s^3}{M_0 \omega_m^2} + \left(\frac{2\zeta_m}{M_0 \omega_m} + \frac{M_2}{M_0}\right) s^2 + \frac{1 + M_1}{M_0} s + 1, \quad (6.1-45)$$

and we get

$$\begin{bmatrix} M_2 \\ M_1 \\ M_0 \end{bmatrix} = \begin{bmatrix} 1/\tau \omega_m^2 + 2\tau(\zeta \omega - \zeta_m \omega_m)/\tau \omega_m^2 \\ 2\zeta \omega/\tau \omega_m^2 + \tau(\omega^2 - \omega_m^2)/\tau \omega_m^2 \\ \omega^2/\tau \omega_m^2 \end{bmatrix}. \quad (6.1-46)$$

In the above equation, τ, ζ, ω, ω_m and ζ_m are known, so M_2, M_1 and M_0 can also be calculated. In this way, the equation can be expressed as

$$\begin{bmatrix} K_{\vartheta A} \\ K_{\vartheta I} \\ K_{\vartheta G} \end{bmatrix} = \begin{bmatrix} \dfrac{1}{k_{ac} V(A_2 + T_\alpha^2 - A_1 T_\alpha)} & \dfrac{-T_\alpha}{k_{ac} V(A_2 + T_\alpha^2 - A_1 T_\alpha)} & \dfrac{T_\alpha^2}{k_{ac} V(A_2 + T_\alpha^2 - A_1 T_\alpha)} \\ \dfrac{-1}{A_2 + T_\alpha^2 - A_1 T_\alpha} & \dfrac{T_\alpha}{A_2 + T_\alpha^2 - A_1 T_\alpha} & \dfrac{A_2 - A_1 T_\alpha}{A_2 + T_\alpha^2 - A_1 T_\alpha} \\ \dfrac{V T_\alpha - c - V A_1}{V(A_2 + T_\alpha^2 - A_1 T_\alpha)} & \dfrac{c T_\alpha + V A_2}{V(A_2 + T_\alpha^2 - A_1 T_\alpha)} & \dfrac{-(c T_\alpha + V A_2) T_\alpha}{V(A_2 + T_\alpha^2 - A_1 T_\alpha)} \end{bmatrix} \begin{bmatrix} M_2 \\ M_1 \\ M_0 \end{bmatrix}.$$

$$(6.1-47)$$

Combined with equation (6.1-42) and equation (6.1-47), the final equation for calculating autopilot design parameters is obtained as

$$\begin{cases} K_A = \dfrac{M_2 - T_\alpha M_1 + T_\alpha^2 M_0}{k_s k_\vartheta^* k_{ac} V(A_2 + T_\alpha^2 - A_1 T_\alpha)} \\ \omega_i = \dfrac{-M_2 + T_\alpha M_1 + M_0}{k_s k_\vartheta^* (A_2 + T_\alpha^2 - A_1 T_\alpha)} \\ K_g = \dfrac{(VT_\alpha - c - VA_1) M_2 + (cT_\alpha + VA_2) M_1 - (cT_\alpha + VA_2) T_\alpha M_0}{k_s k_\vartheta^* V(A_2 + T_\alpha^2 - A_1 T_\alpha)} \end{cases} \quad (6.1-48)$$

It should be pointed out that although the open-loop crossing frequency index ω_{CR} is not introduced obviously in the design process, it is equivalent to the indirect introduction of ω_{CR} due to the approximate correspondence between ω and ω_{CR} as shown in equation (6.1-49) (see reference [1]).

$$\omega = [\tau(\omega_{CR} + 2\zeta_m \omega_m) - 1]/2\zeta\tau. \quad (6.1-49)$$

At the same time, due to the neglect of actuator dynamics in the design of the autopilot, there is a certain deviation between the open-loop crossover frequency of the autopilot designed according to the above ideas and the real value. In the design with actuator dynamics, if the set open-loop crossover frequency is expected to be consistent with the design value, the set value of ω can be changed, and the design can be repeated until the desired crossover frequency is reached. In this case, the autopilot parameters are the final design result.

2) Design classic two-loop autopilot with analytical method

The above analytical method is used to design the classic two-loop autopilot.

The open loop transmission of the classic two-loop autopilot is

$$HG(s) = \dfrac{M_2' s^2 + M_1' s + M_0'}{s^2/\omega_m^2 + 2\zeta_m s/\omega_m + 1}. \quad (6.1-50)$$

The closed-loop transmission function is

$$\begin{cases} M_2' = k_s k_{ac} k_\vartheta^* K_A (VA_2 + cT_\alpha) \\ M_1' = k_s k_\vartheta^* (k_{ac} K_A VA_1 + ck_{ac} K_A + K_g T_\alpha), \\ M_0' = k_s k_\vartheta^* (K_g + k_{ac} K_A V) \end{cases} \quad (6.1-51)$$

$$\dfrac{a_y(s)}{a_{yc}'(s)} = \dfrac{k_s K_A V k_\vartheta^* (A_2 s^2 + A_1 s + 1)}{(s^2/\omega_m^2 + 2\zeta_m s/\omega_m + 1) + (M_2' s^2 + M_1' s + M_0')}. \quad (6.1-52)$$

Taking ζ, ω as the design indexes of the two-loop autopilot, it is assumed that the desired characteristic equation of the pilot satisfies

$$\left(\dfrac{s^2}{\omega^2} + \dfrac{2\zeta}{\omega} s + 1\right) = \left[\dfrac{1}{\omega_m^2 (1 + M_0')} + \dfrac{M_2'}{1 + M_0'}\right] s^2 + \left[\dfrac{2\zeta_m}{\omega_m (1 + M_0')} + \dfrac{M_1'}{1 + M_0'}\right] s + 1.$$

$$(6.1-53)$$

By the equality of the coefficients of the corresponding terms, we get

$$\begin{aligned} \dfrac{1}{\omega^2} &= \dfrac{1}{\omega_m^2 (1 + M_0')} + \dfrac{M_2'}{1 + M_0'}, \\ \dfrac{2\zeta}{\omega} &= \dfrac{2\zeta_m}{\omega_m (1 + M_0')} + \dfrac{M_1'}{1 + M_0'}. \end{aligned} \quad (6.1-54)$$

Then introduce intermediate variables $K'_{\vartheta A}$ and $K'_{\vartheta G}$, and set

$$\begin{cases} K'_{\vartheta A} = k_s k_{\vartheta}^{\bullet} K_A \\ K'_{\vartheta G} = k_s k_{\vartheta}^{\bullet} K_g \end{cases}. \quad (6.1-55)$$

After further derivation, we get

$$\frac{\omega^2}{\omega_m^2} - 1 = K'_{\vartheta A} k_{ac} (V - VA_2\omega^2 - cT_\alpha\omega^2) + K'_{\vartheta G},$$

$$\frac{2\omega\zeta_m}{\omega_m} - 2\zeta = K'_{\vartheta A} k_{ac} (2\zeta V - VA_1\omega - c\omega) + K'_{\vartheta G}(2\zeta - T_\alpha\omega). \quad (6.1-56)$$

The system in matrix form is

$$\begin{bmatrix} (\omega^2/\omega_m^2 - 1) \\ 2\omega\zeta_m/\omega_m - 2\zeta \end{bmatrix} = \begin{bmatrix} k_{ac}(V - VA_2\omega^2 - cT_\alpha\omega^2) & 1 \\ k_{ac}(2\zeta V - VA_1\omega - c\omega) & 2\zeta - T_\alpha\omega \end{bmatrix} \begin{bmatrix} K'_{\vartheta A} \\ K'_{\vartheta G} \end{bmatrix}. \quad (6.1-57)$$

The design equation of K_A and K_g can be obtained by simultaneous equation (6.1-55) and equation (6.1-57)

$$\begin{bmatrix} K_A \\ K_g \end{bmatrix} = \begin{bmatrix} k_{ac}(V - VA_2\omega^2 - cT_\alpha\omega^2) & 1 \\ k_{ac}(2\zeta V - VA_1\omega - c\omega) & 2\zeta - T_\alpha\omega \end{bmatrix}^{-1} \begin{bmatrix} (\omega^2/\omega_m^2 - 1)/k_s k_{\vartheta}^{\bullet} \\ 2(\omega\zeta_m/\omega_m - \zeta)/k_s k_{\vartheta}^{\bullet} \end{bmatrix}. \quad (6.1-58)$$

6.1.8.3 Autopilot Design Examples

As shown in Table 6.1-9, the dynamic coefficients of a certain air-to-air missile given in reference [12] during mid-air flight are selected as the autopilot-controlled object, where $a_\alpha = -250$ s^{-2} means the missile body is statically unstable; Table 6.1-10 shows a set of on-board hardware parameters and phase lags at different autopilot open-loop crossover frequencies. The data in Table 6.1-9 and Table 6.1-10 are the basis of design and analysis in this section.

Table 6.1-9 Aerodynamic data of the missile

$V/(\text{m} \cdot \text{s}^{-1})$	a_α/s^{-2}	a_δ/s^{-2}	a_ω/s^{-1}	b_α/s^{-1}	b_δ/s^{-1}	c/m
914.4	±250	280	1.5	1.6	0.23	0.68

Table 6.1-10 Hardware parameters and corresponding phase lags

Hardware	Damping ζ	Frequency $\omega/(\text{rad} \cdot \text{s}^{-1})$	Phase lag/(°)			
			35 rad/s	40 rad/s	45 rad/s	50 rad/s
Actuator	0.65	220	12.0	13.7	15.5	17.3
Rate gyro	0.65	300	8.7	10.0	11.3	12.6
Accelerometer	0.65	300	8.7	10.0	11.3	12.6
Filter	0.50	314	6.4	7.4	8.3	9.3
The total phase lag			27.1	31.1	35.1	39.2

It should be noted that since the rate gyro and the accelerometer are in a parallel relationship and their dynamics are completely consistent, only one of the phase lags needs to be calculated when we use table 6.1 – 10 to calculate the total phase lag.

According to the index selection principle in reference [1], the expected performance index of the autopilot is set as follows:

For two-loop acceleration autopilot: $\zeta = 0.9$, $\omega_{CR} = 45$ rad/s;

For three-loop autopilot: $\tau = 0.2$s, $\zeta = 0.9$, $\omega_{CR} = 45$ rad/s.

Table 6.1 – 11 shows the related design results of the two-loop and three-loop acceleration autopilots for statically stable missiles ($a_\alpha = 250$ s^{-2}), statically neutral stable missiles ($a_\alpha = 0$ s^{-2}) and statically unstable missiles ($a_\alpha = -250$ s^{-2}), where: ① means that the autopilot design only considers the actuator dynamics; ② means that the autopilot design considers all the hardware dynamics listed in table 6.1 – 10; $\Delta\phi_K$ represents the phase lead angle that can be provided by the controller transmission function from output a_y to actuator at ω_{CR}; $\Delta\phi_{Total}$ represents the sum of the phase lag angles of the on-board hardware dynamics at ω_{CR} (the last line of table 6.1 – 10); $\Delta\phi_M$ represents the phase lag angle of the missile acceleration transfer function at ω_{CR}.

Table 6.1 – 11 Design results of two kinds of classical structure acceleration autopilots

(a) Classic two-loop acceleration autopilots										
Items		$K_A/$ (rad·ms^{-2})	K_g /s	K_{dc}	$\omega_{CR}/$ (rad·s^{-1})	ΔL /dB	$\Delta\phi/$ (°)	$\Delta\phi_K/$ (°)	$\Delta\phi_{Total\,M}/$ (°)	$\Delta\phi_{Total\,M}/$ (°)
Statically stable missiles	①	0.000,67	0.14	2.38	45	16.3	68.6	80.6	-15.5	-176.5
	②	0.000,63	0.14	2.35	45	7.8	49.3	80.9	-35.1	-176.5
Statically neutral stable missiles	①	0.001,4	0.15	1.12	45	14.5	58.1	70.6	-15.5	-177.0
	②	0.001,4	0.15	1.12	45	6.6	38.7	70.8	-35.1	-177.0
Statically unstable missiles	①	0.001,9	0.16	0.81	45	13.3	50.2	63.1	-15.5	-177.4
	②	0.001,9	0.16	0.81	45	5.6	30.8	63.3	-35.1	-177.4

(b) Classic three-loop acceleration autopilots											
Items		$K_A/$ (rad·ms^{-2})	ω_i /s	K_g /s	K_{dc}	$\omega_{CR}/$ (rad·s^{-1})	ΔL /dB	$\Delta\phi/$ (°)	$\Delta\phi_K/$ (°)	$\Delta\phi_{Total}/$ (°)	$\Delta\phi_M/$ (°)
Statically stable missiles	①	0.004,9	1.13	0.14	1.25	45	16.8	66.8	78.8	-15.5	-176.5
	②	0.004,8	1.08	0.14	1.25	45	7.8	47.6	79.2	-35.1	-176.5
Statically neutral stable missiles	①	0.003,8	2.35	0.15	1.67	45	15.7	57.3	69.8	-15.5	-177.0
	②	0.003,7	2.30	0.15	1.68	45	6.8	37.9	70.0	-35.1	-177.0
Statically unstable missiles	①	0.002,6	3.51	0.16	2.46	45	14.9	50.0	62.9	-15.5	-177.4
	②	0.002,5	3.44	0.16	2.49	45	6.0	30.5	63.0	-35.1	-177.4

Since the phase lag of the missile $a_y(s)/\delta(s)$ at the open-loop crossover frequency of the autopilot is close to $-180°$, in order to achieve a phase margin of more than $30°$, the feedback correction loop of the autopilot needs to provide an advanced phase of more than $75°$ for statically stable missiles, and an advanced phase of more than 60° for statically unstable missiles. At the same time, the introduction of angular rate gyro, accelerometer and structural filter dynamics has a great impact on the magnitude and phase margin of the autopilot, which cannot be simply ignored in the design.

By observing the magnitude and phase margin in Table 6.1 – 11, it can be seen that the magnitude margin of the classical three-loop autopilot is better than that of the classical two-loop autopilot for statically unstable missiles.

Fig. 6.1 –49 shows the unit step response curves of the two-loop and three-loop acceleration autopilots when a_α is 250 s^{-2}, 0 s^{-2}, and -250 s^{-2}, respectively. The results show that, as described in section 6.1.5, although the control ability of the typical damping loop to the statically unstable missile is very weak, the control ability of the autopilot to the statically unstable missile can be greatly enhanced by the introduction of the attitude angle feedback or the acceleration feedback (that is, the angular velocity feedback). For three-loop acceleration autopilots, the introduction of the angular velocity feedback reduces the autopilot's ability to control the statically unstable missile slightly, but the reduction is extremely limited, so it can be considered that the introduction of the angular velocity feedback maintains the autopilot's ability to control the statically unstable missile.

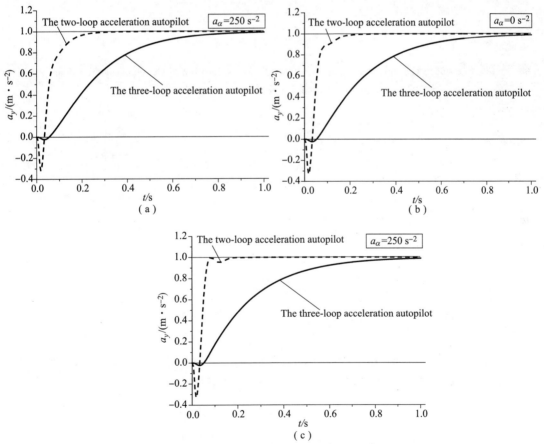

Fig. 6.1 –49 Unit step response of a classical structure acceleration autopilot
(a) The statically stable missile; (b) The statically neutral stable missile; (c) The statically unstable missile

Fig. 6.1 – 50, Fig. 6.1 – 51 and Fig. 6.1 – 52 show the comparison curves of unit step responses of the classical three-loop autopilot with different stable missiles for acceleration, rudder angle and angle of attack respectively. Fig. 6.1 – 52 shows that the steady-state rudder angle required by a statically unstable missile is slightly less than a statically stable missile. A statically stable missile needs a large rudder angle of negative value at a steady state. In the initial process, the statically unstable missile and the static neutral stable missile should hit the negative rudder angle first. In the steady state, the statically unstable missile is stable at the positive rudder angle, and the rudder angle required by the static neutral stable missile is approximately zero.

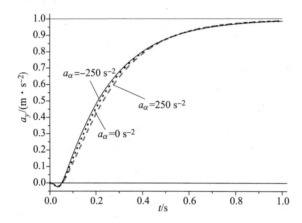

Fig. 6.1 – 50 The unit step response curve of the acceleration of the classical three-loop autopilot (missiles with different stabilities)

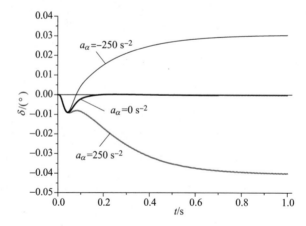

Fig. 6.1 – 51 The unit step response curve of the rudder angle of the classical three-loop autopilot (missiles with different stabilities)

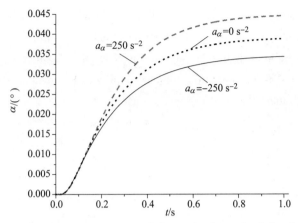

Fig. 6.1 – 52 The unit step response curve of the angle of attack of the classical three-loop autopilot (missiles with different stabilities)

§ 6.2 Pitch/Yaw Attitude Autopilot

The pitch/yaw attitude autopilot design has two possible structures. Structure I takes the output of an attitude gyro as the main feedback, and uses a lead compensator to generate the damping signal required for the damping loop (Fig. 6.2 – 1).

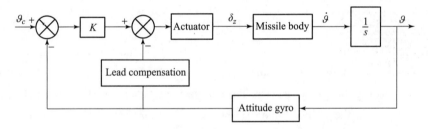

Fig. 6.2 – 1 Block diagram of an early attitude autopilot

This is the common solution for early attitude autopilots.

In Structure II, the attitude ϑ and angular velocity $\dot{\vartheta}$ signals are respectively used to form the attitude feedback and angular velocity feedback. As most missiles currently have a strapdown inertial navigation system onboard, this is the current scheme adopted by most missiles (Fig. 6.2 – 2).

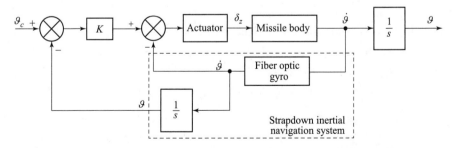

Fig. 6.2 – 2 Block diagram of the attitude autopilot with the fiber optic gyro

The attitude autopilot has been used in the boosting phase of the early ballistic missile programs for trajectory control. With the help of missile angle of attack and normal acceleration generated by the attitude control, an indirect flight path angle θ control can be obtained (Fig. 6.2-3). Since no flight path angle feedback is used in this approach, all the parameter variations in the transfer function from ϑ to θ will affect the trajectory control accuracy.

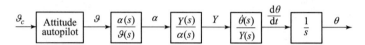

Fig. 6.2-3 θ **indirect control with the help of an attitude autopilot**

At present, as the flight path angle θ can be given by the strapdown integrated navigation system on a missile, the control of the θ angle can be accurately completed by the flight path angle autopilot without the help of attitude autopilots. Therefore, the attitude autopilot is used less often at present.

Currently several possible applications of attitude autopilots are:

(1) In the air-to-air missile application, when the missile is launched from the aircraft, the attitude of the missile is often stabilized by means of an attitude autopilot to prevent the missile attitude change. Since a variation of the missile body attitude will lead to a missile thrust direction change, its corresponding missile trajectory variations could have a direct effect on the aircraft safety.

(2) In order to ensure the penetration effect, an air-to-ground penetration missile requires that the missile inertial angle of attack be as small as possible when it hits the target. Note that the missile inertial angle of attack is the angle between the missile inertial velocity direction and missile x-axis which can be given by the onboard inertial navigation system. Therefore, it is required that an attitude autopilot be used to ensure that in the final moment of flight, the missile attitude tracks the missile inertial velocity direction. This will eliminate the inertial angle of attack of the missile before penetration. Here it is necessary to note that for simplification some penetration missiles achieve the penetration effect with the help of an acceleration autopilot. As the missile normal acceleration is zero, a zero missile wind-angle of attack (the angle between the missile wind-affected velocity and missile body axis) can be achieved. When wind exists, this control approach will not be perfect.

(3) In order to ensure the maximum electromagnetic damage effect when the electromagnetic missile approaches the target, it is required to stabilize the missile direction to the electromagnetic warhead maximum damage effect direction.

Here an example is given to illustrate the design features of this type of autopilot.

Table 6.2-1 shows the parameters of the missile body and the actuator parameters of the attitude autopilot required to be designed. The design block diagram is shown in Fig. 6.2-4.

6 Missile Autopilot Design

Table 6.2-1 Parameters of the example missile and actuator

Velocity	Parameters of missile dynamics					Actuator dynamics	
$V/(\mathrm{m \cdot s^{-1}})$	a_α/s^{-2}	a_δ/s^{-2}	a_ω/s^{-1}	b_α/s^{-1}	b_δ/s^{-1}	T_S	k_S
306	62.3	71.6	1.17	1.02	0.17	0.025	-0.070,3

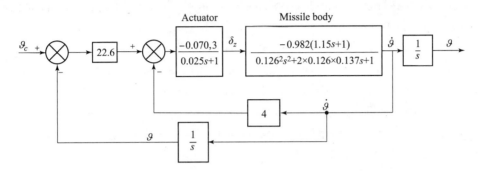

Fig. 6.2-4 Block diagram of the attitude autopilot

Fig. 6.2-5 shows the missile attitude ϑ response and flight path angle θ response for this attitude autopilot when a unit attitude angle command ϑ_c is given. When θ changes, an angle of attack α will be generated, and the related lift will change the missile flight path angle θ. According to the above sequence, θ will always be lagging behind ϑ response.

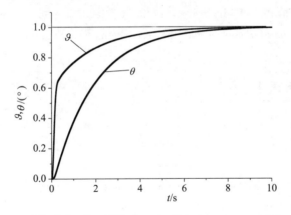

Fig. 6.2-5 Attitude autopilot step response

It should be noted that the angular velocity loop of an attitude autopilot is to improve the damping of the attitude loop, so its phase angle compensation frequency area is designed close to the attitude loop crossover frequency. Because the bandwidth difference between the attitude loop and the damping loop is not very large, the two loops are not suitable for independent design. The current applications of effective nonlinear programming optimization design methods have made it possible to design these two loops simultaneously (see the section on roll attitude autopilot design).

It should be noted that the relative relationship between the angular velocity loop and the angle loop frequency bandwidth could have three structures as follows:

(1) When the angular velocity loop is used to improve the damping of the angle loop (such as

the attitude autopilot in this section), the frequency bandwidths of the two are not much different and it is not suitable for independent design.

(2) The angle loops are of a type II structure (e. g., for ground guidance radars, see Chapter 4). Because a type II system has two integrators and $-180°$ phase lag, in order to ensure that the angular velocity loop does not have a large phase shift at the crossover frequency of the angle loop, generally the angular velocity loop frequency bandwidth is at least set to be about five times that of the angle loop. In this case, the two loops can be designed independently.

(3) When the angular velocity loop is designed to resist the influence of the disturbance moment generated by the angular motion of the missile on the seeker angle tracking loop (see Chapter 5), to have strong suppression on this type of disturbance, the seeker's angular velocity loop bandwidth is generally designed to be ten to fifteen times the frequency bandwidth of the seeker's angular tracking loop. Therefore, the two loops can also be designed independently for the missile seeker.

§ 6.3 Flight Path Angle Autopilot

At present, flight path angle autopilots are often used to control the initial programmed turn of the ballistic type missiles or rockets or used for velocity pursuit guidance. There are two structures that can be adopted for flight path angle autopilots (Fig. 6.3 - 1 and Fig. 6.3 - 2). The structure of Fig. 6.3 - 2 is superior to that of Fig. 6.3 - 1 as the acceleration autopilot inner loop of Fig. 6.3 - 2 has a better normal acceleration control than the aerodynamic acceleration control loop of Fig. 6.3 - 1.

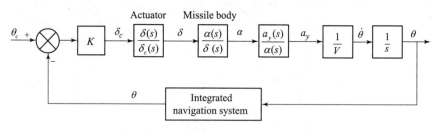

Fig. 6.3 - 1 Flight path angle autopilot, structure I

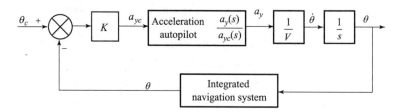

Fig. 6.3 - 2 Flight path angle autopilot, structure II

Fig. 6.3 - 3 shows the flight path angle transient difference, when both attitude autopilots and flight path angle autopilots are used to generate a flight path angle unit step response. With the attitude autopilot, its attitude overshoot suppression function has greatly limited the angle of attack value which is needed for flight path angle change. In the flight path angle autopilot case, a large attitude overshoot is allowed. Its related large angle of attack could lead to a much faster flight path angle response and its flight path angle feedback also makes it more accurate in flight path angle control.

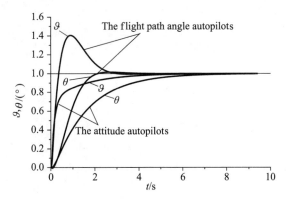

Fig. 6.3 – 3   Step responses of the velocity vector autopilot and the attitude autopilot

§ 6.4   Roll Attitude Autopilot

The difference between the roll attitude autopilot and the pitch/yaw attitude autopilot is that the larger disturbance moment possibly existing in the roll channel must be taken into account in its design. The reason for this is that when the missile total angle of attack α_T plane (the missile maneuvering plane) is not the symmetrical plane of the missile body, the missile body aerodynamic asymmetry relative to the maneuvering plane will generate a roll disturbance moment as shown in Fig. 6.4 – 1.

The design of the autopilot should ensure that the steady state roll angle error of the roll autopilot does not exceed its permissible value when a maximum roll disturbance moment is applied.

The aerodynamic block diagram of the roll missile body is shown in Fig. 6.4 – 2. Here, $M_x^{\delta_x}$ is the actuator roll moment derivative, $M_x^{\omega_x}$ is the roll damping moment derivative ($M_x^{\omega_x} < 0$), and J_x is the moment of inertia of the missile around x-axis.

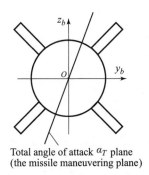

Fig. 6.4 – 1   Sketch of the maneuvering surface of an axial symmetrical missile

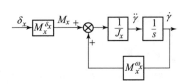

Fig. 6.4 – 2   Aerodynamic block diagram of the missile body

Therefore, the roll aerodynamic transfer function of the missile body will be

$$\frac{\dot{\gamma}(s)}{M_x(s)} = \frac{1}{M_x^{\omega_x}\left(\dfrac{1}{-M_x^{\omega_x}/J_x}s + 1\right)} = \frac{1}{M_x^{\omega_x}(T_r s + 1)}, \quad (6.4-1)$$

where $T_r = \dfrac{J_x}{-M_x^{\omega_x}}$.

The transfer function from the actuator angle δ_x to roll angular velocity $\dot{\gamma}$ is

$$\frac{\dot{\gamma}(s)}{\delta_x(s)} = \frac{k_r}{T_r s + 1}, \quad (6.4-2)$$

where $k_r = \dfrac{-M_x^{\delta_x}}{M_x^{\omega_x}}$.

Fig. 6.4-3 shows the typical structure of a roll attitude autopilot. Here, K_A and $K_{\omega x}$ are the autopilot design parameters, ω_s and μ_s are the actuator natural frequency and damping. The value of K_A and $K_{\omega x}$ can be determined by selecting an appropriate objective function and design constraints and using an available nonlinear programming optimization software to search for their optimum solutions.

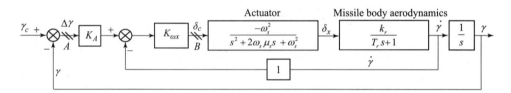

Fig. 6.4-3 Typical structure of the roll attitude autopilot

An example of the above design process will now be explained in detail.

Suppose that for the example missile $J_x = 0.96 \text{kg} \cdot \text{m}^2$, $M_x^{\delta_x} = -13,500$ ((N·m)/rad), $M_x^{\omega_x} = -37.3$ ((N·m)/(rad·s^{-1})), $T_r = 0.025,7\text{s}$. Therefore

$$\frac{\dot{\gamma}(s)}{M_x(s)} = \frac{1}{37.3(0.025,7s + 1)}. \quad (6.4-3)$$

If the actuator second order model has $\omega_s = 180\text{rad/s}$, $\mu_s = 0.7$, Fig. 6.4-4 shows the block diagram of the roll channel autopilot to be designed. The value of $K_A, K_{\omega x}$ of the roll autopilot can then be determined by the following optimization algorithm.

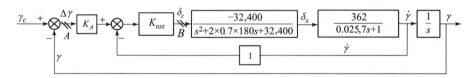

Fig. 6.4-4 Block diagram of the example roll autopilot

Assume a phase margin of more than 55° constraint and a gain margin of more than 8 dB constraint are given for this design. And an optimum transient to a step command γ_c will be obtained by minimizing the following objective function. Here T is the autopilot transient settling time. The mathematical description of this optimization problem is

$$\begin{cases} \text{Constraint}: \Delta\phi(K_A, K_{\omega x}) > 55° \text{ and } \Delta L(K_A, K_{\omega x}) > 8 \text{ dB}, \\ \min J(K_A, K_{\omega x}) = \min \int_0^T |\gamma(t) - \gamma_{\text{steady state}}| dt (\text{take } \gamma_c = 1, \gamma_{\text{steady state}} = 1, T = 0.4 \text{ s}). \end{cases}$$

Fig. 6.4-5 shows the constraint boundaries of gain margin ΔL and phase margin $\Delta\phi$, and the contour line of the objective function for design parameters K_A and $K_{\omega x}$. It can be seen that the minimum value of the objective function is 0.034, and the corresponding optimal design results are $K_A = 50.0$, $K_{\omega x} = 0.003,99$ with the satisfaction of the above-mentioned gain and phase margin constraints. It is seen that for this design the system response speed is restricted by its phase margin constraint.

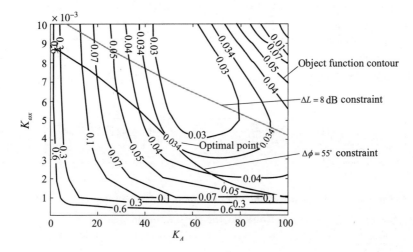

Fig. 6.4-5 Contour line of the time domain objective function and frequency domain constraints

Fig. 6.4-6 shows the time domain step response of the roll autopilot.

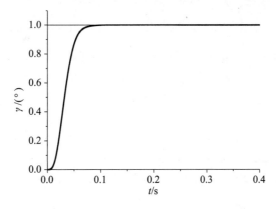

Fig. 6.4-6 Roll autopilot step response

A roll autopilot is usually affected by a disturbance moment L as the missile is maneuvering in the missile asymmetric plane and is required to have sufficient anti-disturbance ability. Given that the maximum allowable roll angle error under the action of the disturbance moment L is $\Delta\gamma$, then the autopilot gain from the allowable steady state error to the counter-balance control moment must be no less than the following value K. As the autopilot gain value is constrainted by stability considerations, a lag compensator in the form $\beta\left(\dfrac{T_b s + 1}{\beta T_b s + 1}\right)$ could be added to increase the autopilot low frequency gain by a factor of β. Here the value of β should be chosen to satisfy the constraint K and the value of T_b should be taken to have the roll error reduced to the allowed value at the end of the autopilot transient. The Bode diagrams of the roll autopilot with a lag compensator are shown in Fig. 6.4 −7 and Fig. 6.4 −8.

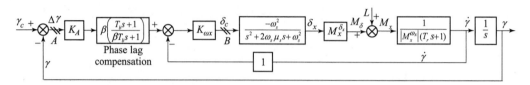

Fig. 6.4 −7 Structure of the roll attitude autopilot with phase lag compensation

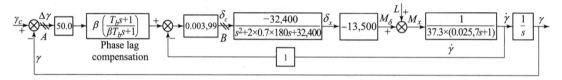

Fig. 6.4 −8 Structure of the roll attitude autopilot with lag compensation
network for the given example

Assuming that the maximum disturbance moment acting on the missile body is $L = 1,000$ N · m, and the maximum allowable roll angle error is $\Delta\gamma = 0.05$ rad $= 2.9°$, then the transfer ratio of the allowed roll angle error to the autopilot countering balancing control moment should be $K = \dfrac{1,000}{0.05} = 20,000$. However, the gain of the roll angle to the control moment of Fig. 6.4 −4 is only 2,692.4. In order to make the previous system have a sufficient anti-disturbance gain in the low frequency range, the value of β for the lag compensator should be no less than the following value

$$\beta = \dfrac{20,000}{2,692.4} = 7.43. \qquad (6.4-4)$$

In order to have the stability margin of the transient overshoot and the roll error elimination speed of the designed autopilot to meet their design requirements, the time constant of the phase lag network is selected as $T_1 = 0.12$ s. Therefore, the lag network parameter could be taken as

$$\beta\left(\dfrac{T_b s + 1}{\beta T_b s + 1}\right) = 7.43 \times \left(\dfrac{0.12s + 1}{0.892s + 1}\right). \qquad (6.4-5)$$

The open-loop Bode diagrams of the inner and outer loops of the roll autopilot with and without the lag compensator are given in Fig. 6.4 −9 and Fig. 6.4 −10.

Table 6.4 −1 shows the cross-over frequencies of the inner and outer loops with and without the lag compensation, and their related phase and gain margins.

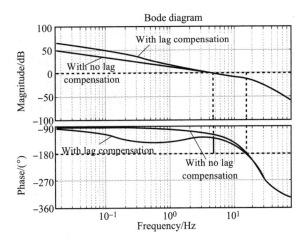

Fig. 6.4 – 9   Open loop Bode diagram of the outer loop

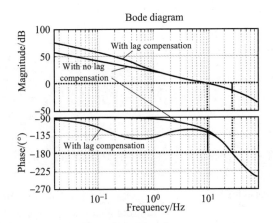

Fig. 6.4 – 10   Open loop Bode diagram of the inner loop

Table 6.4 – 1   Characteristics of system magnitude frequency

Items	System	ΔL /dB	Magnitude cross-over frequency/Hz	$\Delta \varphi$ /(°)	Phase cross-over frequency /Hz
Outer loop	With no lag compensation	11.3	16.4	66.6	4.71
Outer loop	With lag compensation	10.8	15.7	52.6	4.87
Inner loop	With no lag compensation	12.2	27.5	55	9.7
Inner loop	With lag compensation	12.2	27.4	53.1	9.2

Fig. 6.4 – 11 shows the closed loop Bode diagram of the system after lag compensation.

Fig. 6.4 – 12 and Fig. 6.4 – 13 respectively show the step response of the autopilot with and without lag compensation, and the transition process of the autopilot with disturbance. It can be seen from these figures that with the help of the lag compensation the roll error introduced by the disturbance moment L can still be reduced to the allowable value 2.9° at the end of the autopilot transient.

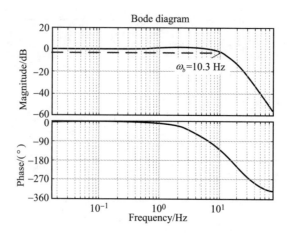

Fig. 6.4 – 11 Closed loop Bode diagram with lag compensation

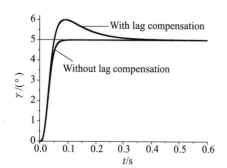

Fig. 6.4 – 12 Step response for a command $\gamma_c = 5°$

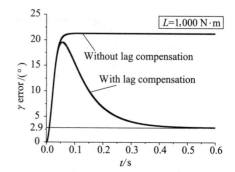

Fig. 6.4 – 13 System response under the action of the disturbance moment L with command $\gamma_c = 0°$

§ 6.5 BTT Missile Autopilot

The aerodynamic layout characteristic of a BTT missile is surface symmetrical. Generally, its lift surface must be large in one direction and small in the other direction. The maneuvering of a missile is obtained by the control of the missile roll and pitch simultaneously to have the angle of attack produce lift force in the expected maneuvering plane. The possible aerodynamic layouts of a BTT controlled missile are given in Fig. 6.5 – 1.

Fig. 6.5 – 1 (d) is the layout for air-to-air missile "meteor," which has a ramjet rocket to keep a high missile speed even at the end of the engagement.

However, there are also many tactical missiles that perform their roll control by the rear control surface but not the aileron.

Usually for axis-symmetric missile the aerodynamic couplings of the yaw, roll and pitch channels are not large and can generally be ignored in the autopilot design phase. Their effects are evaluated only in the final six-degree-of-freedom simulations. However, for BTT missiles due to their special non-axisymmetric aerodynamic configuration, the aerodynamic coupling between the yaw and

6 Missile Autopilot Design

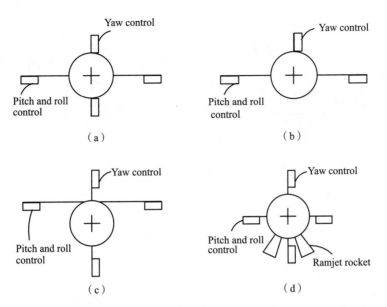

Fig. 6.5-1 Possible schemes of the BTT missile aerodynamic layout

roll is large and cannot be ignored. Table 6.5-1, Table 6.5-2 and Table 6.5-3 show the definitions of all the dynamic coefficients of BTT missiles with yaw and roll coupling effect included. Here they are defined respectively as: δ_e (the elevator angle), δ_r (the rudder angle) and δ_a (the aileron angle).

Table 6.5-1 Dynamic coefficients of the non-coupling model

Items	Symbol	Expression	Significance
Pitch	$a_\alpha = -M_z^\alpha/J_z$	$-m_z^\alpha qSL/J_z$	The effect of pitch moment caused by the angle of attack m_z^α, for static stable missile $m_z^\alpha < 0$
	$a_{\omega_z} = -M_z^{\omega_z}/J_z$	$-m_z^{\omega_z}qSL^2/(J_zV)$	The effect of pitch damping moment caused by pitch angular velocity $m_z^{\omega_z} < 0$
	$a_{\delta_e} = -M_z^{\delta_e}/J_z$	$-m_z^{\delta_e}qSL/J_z$	The effect of pitch moment caused by the elevator $m_z^{\delta_e} < 0$
	$b_\alpha = (P+Y^\alpha)/(mV)$	$(P+c_y^\alpha qS)/(mV)$	The effect of pitch force caused by the angle of attack $c_y^\alpha > 0$
	$b_{\delta_e} = Y^{\delta_e}/(mV)$	$c_y^{\delta_e}qS/(mV)$	The effect of pitch force caused by the elevator $c_y^{\delta_e} > 0$
Yaw	$a_\beta = -M_y^\beta/J_y$	$-m_y^\beta qSl/J_y$	The effect of yaw moment caused by the sideslip angle $m_y^\beta < 0$
	$a_{\omega_y} = -M_y^{\omega_y}/J_y$	$-m_y^{\omega_y}qSl^2/(J_yV)$	The effect of yaw damping moment caused by the yaw anglar velocity $m_y^{\omega_y} < 0$
	$a_{\delta_r} = -M_y^{\delta_r}/J_y$	$-m_y^{\delta_r}qSl/J_y$	The effect of yaw moment caused by the rudder $m_y^{\delta_r} < 0$
	$b_\beta = -Z^\beta/(mV)$	$-c_z^\beta qS/(mV)$	The effect of yaw force caused by the sideslip angle $c_z^\beta < 0$
	$b_{\delta_r} = -Z^{\delta_r}/(mV)$	$-c_z^{\delta_r}qS/(mV)$	The effect of yaw force caused by the rudder $c_z^{\delta_r} < 0$

Continued

Items	Symbol	Expression	Significance
Roll	$c_{\omega_x} = -M_x^{\omega_x}/J_x$	$-m_x^{\omega_x}qSl^2/(2J_xV)$	The effect of roll moment caused by the roll angular velocity $m_x^{\omega_x} < 0$
	$c_{\delta_a} = -M_x^{\delta_a}/J_x$	$-m_x^{\delta_a}qSl/J_x$	The effect of roll moment caused by the aileron $m_x^{\delta_a} < 0$

Table 6.5-2 Dynamic coefficients of the yaw coupling to roll

Symbol	Expression	Significance
$c_\beta = -M_x^\beta/J_x$	$-m_x^\beta qSl/J_x$	The effect of roll moment caused by the yaw sideslip angle $m_x^\beta < 0$
$c_{\omega_y} = -M_x^{\omega_y}/J_x$	$-m_x^{\omega_y}qSl^2/(J_xV)$	The effect of roll moment caused by the yaw angular velocity $m_x^{\omega_y} < 0$
$c_{\delta_r} = -M_x^{\delta_r}/J_x$	$-m_x^{\delta_r}qSl/J_x$	The effect of roll moment caused by the yaw rudder $m_x^{\delta_r} < 0$

Table 6.5-3 Dynamic coefficients of the roll coupling to yaw

Symbol	Expression	Significance
$a_{\omega_x} = M_y^{\omega_x}/J_y$	$m_y^{\omega_x}qSL^2/(2J_yV)$	The effect of yaw moment caused by the roll angular velocity $m_y^{\omega_x} > 0$
$a_{\delta_a} = -M_y^{\delta_a}/J_y$	$-m_y^{\delta_a}qSl/J_y$	The effect of yaw moment caused by the roll aileron $m_y^{\delta_a}$

The following gives the physical mechanism illustrations for several important yaw and roll coupling moments:

1) The roll moment M_x^β produced by the coupling of the sideslip angle

The aerodynamic coefficient c_β related to the roll moment generated by the coupling of the sideslip angle is defined as

$$c_\beta = -m_x^\beta qSl/J_x = -M_x^\beta/J_x. \qquad (6.5-1)$$

As shown in Fig. 6.5-2, when the missile has a sideslip angle and an angle of attack α for a swept wing configuration missile, the velocity components V_R, V_L perpendicular to the leading edges of the two wings are different $(V_R > V_L)$, and the lift of the right wing will be greater than that of the left wing, thus creating a coupling roll moment.

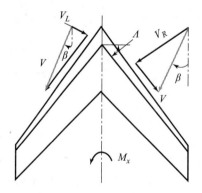

Fig. 6.5-2 Effect of wing sweep angle on the roll/yaw coupling

2) The rolling moment $M_x^{\omega_y}$ caused by the yaw angular velocity

The aerodynamic coefficient c_{ω_y} associated with the coupling rolling moment due to yaw angular velocity is defined as

$$c_{\omega_y} = - m_x^{\omega_y} q S l^2 / (J_x V). \qquad (6.5-2)$$

As shown in Fig. 6.5 – 3, in the case of angle of attack $\alpha > 0$, a yaw angular velocity exists. Here the right wing will turn forward, which will make its relative airspeed and lift force increase. At the same time, the left wing turns backward, its relative airspeed decreases. So its airspeed and lift force decrease, resulting in a negative roll moment.

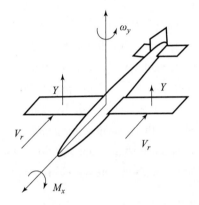

Fig. 6.5 – 3 Sketch of the roll moment produced by yaw angular velocity

3) The yaw moment $M_y^{\omega_x}$ produced by the roll angular velocity

As shown in Fig. 6.5 – 4, if a roll angular velocity exists and $\alpha = 0°$, the roll angular velocity will make the right wing have a positive angle of attack and left wing a negative angle of attack. The difference of the two lift horizontal projection forces will introduce a coupling yaw moment.

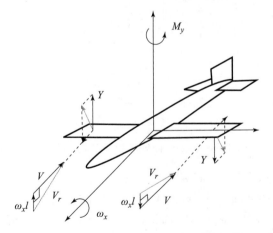

Fig. 6.5 – 4 Sketch of the yaw moment generated by roll angular velocity

The linear time-invariant mathematical model for a BTT missile with coupling included is shown in the following equations.

Pitch:
$$\dot{\alpha} = -b_\alpha \alpha + \omega_z - b_{\delta_e}\delta_e - \omega_x \beta$$
$$\dot{\omega}_z = -a_\alpha \alpha - a_{\omega_z}\omega_z - a_{\delta_e}\delta_e \quad ; \quad (6.5-3)$$

Yaw:
$$\dot{\beta} = -b_\beta \beta + \omega_y - b_{\delta_\gamma}\delta_\gamma + \omega_x \alpha - b_{\delta_a}\delta_a$$
$$\dot{\omega}_y = -a_\beta \beta - a_{\omega_y}\omega_y - a_{\delta_\gamma}\delta_\gamma + a_{\omega_x}\omega_x - a_{\delta_a}\delta_a \quad ; \quad (6.5-4)$$

Roll:
$$\dot{\gamma} = \omega_x$$
$$\dot{\omega}_x = -c_{\omega_x}\omega_x - c_{\delta_a}\delta_a - c_\beta \beta - c_{\omega_y}\omega_y - c_{\delta_\gamma}\delta_\gamma \quad (6.5-5)$$

In the above equations, the $\omega_x \beta$ term in pitch and the $\omega_x \alpha$ term in yaw are kinematic coupling related terms. Since in BTT control, angular velocity ω_x is often large, it is not possible to consider these two items as second order small quantity and to ignore them. However, if they are taken into consideration in the autopilot design, the model will become a nonlinear model and the linear control theories will not be applied. For this reason, the general design method is still to omit the kinematic coupling in the initial autopilot design, and the linear theories are adopted. Finally, the non-linear effects of these kinematic couplings are carefully evaluated in a six-degree-of-freedom simulation.

The following is a brief description of the kinematic coupling mechanism. Assume that the missile has a sudden roll $\Delta \gamma$ and a constant angle of attack α_0, and its polar graph is given in Fig. 6.5-5.

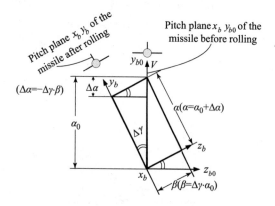

Fig. 6.5-5 Polar graph of the BTT missile kinematic coupling

The point x_b on the graph is the missile-axis direction, the plane $x_b y_{b0}$ is the pitch symmetrical plane of the missile before rolling, the point V is the direction of the velocity axis, and the angle between x_b axis and V axis is the angle of attack α_0. Since the inertia of the missile velocity vector is much larger than that of the missile attitude, the velocity axis can be supposed to be in its original direction when the missile is rolling by an angle $\Delta \gamma$. It can be seen that a sideslip angle $\beta = \Delta \gamma \cdot \alpha_0$ and an angle of attack increment $\Delta \alpha = -\Delta \gamma \cdot \beta$ are formed relative to the pitch symmetrical plane of the missile $x_b y_b$ after a roll angle $\Delta \gamma$. As $\omega_x = \dfrac{\Delta \gamma}{\Delta t}$, the two terms related to the $\dot{\alpha}$ and $\dot{\beta}$ should be $\omega_x \beta$ and $\omega_x \alpha$ respectively in the system equations.

In the following, a model of a plane-symmetric air-to-air missile with a ramjet rocket will be used to analyze the characteristics of the BTT-controlled missile coupling model. The dynamic equations of this missile are as follows.

Pitch channel:
$$\begin{aligned} \dot{\alpha} &= \omega_z - 3.028,6\alpha - 0.778,5\delta_e - \omega_x \beta \\ \dot{\omega}_z &= -0.028\omega_z + 149\alpha - 701\delta_e \end{aligned} ; \qquad (6.5-6)$$

Yaw channel:
$$\begin{aligned} \dot{\beta} &= \omega_y - 0.853\beta - 0.8\delta_r + \alpha \omega_x \\ \dot{\omega}_y &= 0.002,8\omega_x - 0.028\omega_y - 295\beta - 85\delta_a - 695\delta_r \end{aligned} ; \qquad (6.5-7)$$

Roll channel:
$$\begin{aligned} \dot{\gamma} &= \omega_x \\ \dot{\omega}_x &= -2.61\omega_x - 0.014\omega_y - 6,438\beta - 11,113\delta_a \end{aligned} . \qquad (6.5-8)$$

First, the effect of the roll coupling on the yaw is analyzed below.

In the yaw angular acceleration equation, the dynamic coefficient of the roll aileron coupling to yaw a_{δ_a} is 85, but the yaw rudder dynamic coefficient a_{δ_y} is 695. Therefore, the roll aileron coupling effect is very small compared to the rudder. In addition, the dynamic coefficient of the roll angular velocity coupling a_{ω_x} to yaw is 0.002,8, while the dynamic coefficient of the yaw angular velocity effect a_{ω_y} is 0.028. Therefore, compared with the yaw angular velocity effect, the roll angular velocity coupling is also weak. A conclusion can be made that the roll coupling to the yaw is a weak coupling.

Next, the coupling effect of the yaw on the roll is discussed below. The coupling dynamic coefficient of the yaw angular velocity to the roll c_{ω_y} is 0.014, but the roll angular velocity coefficient c_{ω_x} is 2.61, so the influence of the yaw angular velocity on the roll is also very weak. Then let us analyze the influence of the sideslip angle β on the roll. The dynamic coefficient of the sideslip angle to the roll coupling c_β is 6,438, and the dynamic coefficient of the roll aileron c_{δ_a} is 11,113, which indicates that to balance the disturbance roll moment generated by a $1°$ sideslip angle, a roll aileron deflection angle $0.6°$ should be applied. Obviously, the roll coupling moment caused by the sideslip angle is a strong coupling term in the roll equation. We can conclude that the coupling of the yaw sideslip to the roll is a strong coupling.

The response of the sideslip angle β to a step yaw disturbance moment is given in Fig. 6.5-6.

This yaw sideslip response will be coupled to the roll channel owing to the coupling moment m_x^β effect and its related coupling transfer function is given as follows.

$$\frac{\gamma(s)}{\beta(s)} = \frac{-2,466.8(0.001,2s+1)}{s(0.383s+2.61)(0.001,2s+1)}. \qquad (6.5-9)$$

At the yaw oscillation frequency 2.73 Hz (also the coupling roll oscillation frequency), the

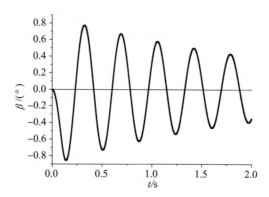

Fig. 6.5-6　The β response to a step yaw disturbance moment

transfer function $\dfrac{\gamma(s)}{\beta(s)}$ has a gain of 21.6. It means that a very strong coupling effect exists from yaw to roll. This is because the missile roll moment of inertia is small. Fig. 6.5-7 shows the roll response to the coupling moment m_x^β.

Normally, the yaw and roll coupling motion mode is called the Dutch roll mode.

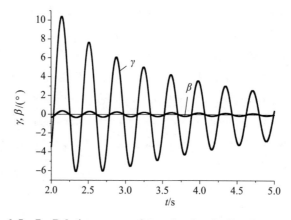

Fig. 6.5-7　Relation curves of β and γ for the Dutch roll mode

From the above analysis, we can see that for a BTT missile, due to its yaw and roll constituting a strong coupling system, it is known from the control system theory that for the design of a strong-coupling system, the multi-input multi-output (MIMO) control theory should be used. But BTT missile control has its own particularity. Because for a BTT missile to maneuver in a certain direction only roll and pitch channel controls are needed, and the yaw channel is not required. The yaw sideslip coupling roll moment is only an interference for the roll channel control. Therefore, we can independently design a yaw autopilot to suppress the sideslip angle β to nearly zero during the roll and pitch channel controls and to reduce interference from the yaw to roll to weak coupling. With this approach, the BTT missile roll and pitch control can also be designed separately.

The above-mentioned aircraft BTT maneuver with a sideslip-suppressing yaw autopilot is generally known as BTT coordinated turn.

The structure and parameters of the roll autopilot designed for the above BTT missile example are shown in Fig. 6.5 – 8. Here, the actuator has a second order model with an undamped natural frequency $\omega_n = 150$ rad/s $= 24$ Hz and damping coefficient $\xi = 0.7$.

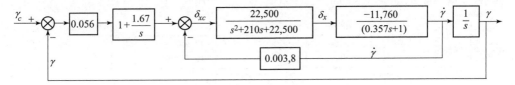

Fig. 6.5 – 8 The roll autopilot block diagram of the BTT missile

Fig. 6.5 – 9 shows the missile roll autopilot response to a 20° step roll command for an initial angle of attack $\alpha_0 = 3°$ with and without the yaw coupling effect. In Fig. 6.5 – 10 yaw sideslip angle β response introduced by the roll control is given.

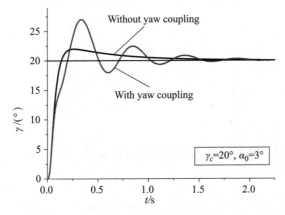

Fig. 6.5 – 9 Comparison of the roll step responses with and without yaw coupling

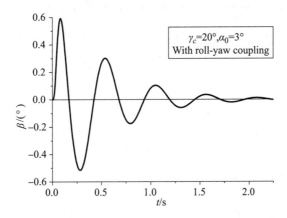

Fig. 6.5 – 10 Roll control introduced to sideslip angle response

The structure and parameters of the yaw autopilot designed to achieve a coordinated turn are shown in Fig. 6.5 – 11. The main feedback of the autopilot was taken as the sideslip angle β. In order to improve stability, a lead compensation network was added to the main loop of the autopilot. More specifically, the angular velocity ω_y damping loop also has a high-pass filter $\dfrac{\tau s}{\tau s + 1}$ ($\tau = 0.15$ s) added. Its role is to let the yaw oscillation high frequency signal pass through but the coordinated turn low frequency signal is blocked.

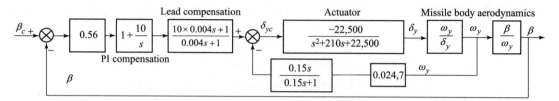

Fig. 6.5 – 11 Block diagram of the yaw autopilot with the sideslip angle feedback

Fig. 6.5 – 12 gives the transient of the yaw autopilot to an initial 1° sideslip angle β disturbance. It shows the effectiveness of the yaw autopilot for depressing the yaw sideslip angle disturbance.

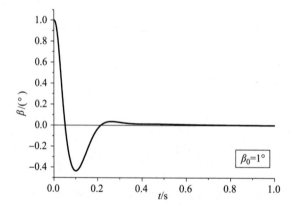

Fig. 6.5 – 12 Suppression curve of initial perturbation by the yaw autopilot

Fig. 6.5 – 13 shows the roll angle response for the roll autopilot (Fig. 6.5 – 8) under the same condition as shown in Fig. 6.5 – 9 with and without yaw autopilot coupling suppression.

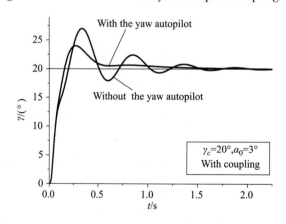

Fig. 6.5 – 13 Time domain transient process of the roll autopilot with and without a yaw autopilot

Fig. 6.5 – 14 shows the suppression effect of the above yaw autopilot on the sideslip angle.

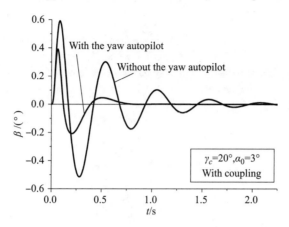

Fig. 6.5 – 14 Time domain transient process of the sideslip angle with and without the yaw autopilot

Another possible yaw autopilot structure for achieving a coordinated turn takes the yaw acceleration as its feedback. Since the missile lateral acceleration is mainly generated by the yaw sideslip angle, clearly this structure can also be used to approximately complete the task of sideslip angle suppression.

In theory, to completely suppress the existence of the sideslip angle in BTT control, it is required that the yaw autopilot should be faster than the roll autopilot. However, this is impossible as the missile yaw moment of inertia is much greater than the roll moment of inertia. For this reason, BTT control is primarily used for missile midcourse guidance control, where low maneuverability can be acceptable. While for terminal guidance that requires missile rapid response, STT control is usually used.

§ 6.6 Thrust Vector Control and Thruster Control

To obtain a fast change of flight path direction (high value of $\dot{\theta}$) at low missile speeds and high attitudes (low air densities), thrust vector control and thruster control are often used for tactical missiles. The classification of thrust vector control and thruster control for tactical missiles is given in Table 6.6 – 1.

Table 6.6 – 1 The classification of thrust vector control and thruster control

Items	Type	Application
Endoatmosphere application	Jet vane (thruster vector control)	To rapidly change missile flight direction to incoming target for surface-to-air missile at low initial missile speeds. Example: Patriot surface-to-air missile; S-300. To rapidly change air-to-air dogfight missile flight direction to incoming enemy target. The required missile $\dot{\theta}$ value could be as high as 100°/s. Example: AIM – 9X; Archer AA – 11; Mica; IRIS – T

Continued

Items	Type	Application
Endoatmosphere application	Pulse thruster (thruster control)	From Chapter 8, it is known that fast missile autopilot response will lead to smaller interception miss distance. For this reason, some modern surface-to-air missiles at high attitude end game phase use pulse thrusters to increase missile autopilot response speed. Example: PAC-3 surface-to-air missile
	Trajectory control thruster (thruster control)	In this case, the control thrusters are arranged near the missile center of gravity to directly produce the trajectory control force, which greatly increases the lateral control force and speeds up the autopilot speed. Example: Aster surface-to-air missile
Exoatmosphere application	Attitude thruster control plus trajectory thruster control (thruster control)	Exoatmosphere ground-to-air missile application. Example: SM-3 anti-ballistic missile

1) Jet vane

Jet vanes are allocated at the missile rocket engine exhaust (Fig. 6.6-1). Its deflection δ_V will change the rocket thrust direction leading to the generation of lateral control force F and control moment M. Their mathematical models are often simply given as $F = k\delta_V$ and $M = kL_V\delta_V$.

Here L_V is the distance between the jet vane and the missile center of gravity and the parameter k in the control force expression can be measured by rocket engine firing test on the 6-degree-of-freedom test bench.

The dynamic coefficients related to missile rotational control a_δ and flight path control b_δ for aerodynamic control and jet vane control are given in Table 6.6-2.

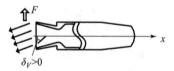

Fig. 6.6-1 Jet vane

Table 6.6-2 The dynamic coefficients of aerodynamic control and jet vane control

Items	Missile rotational control efficiency a_δ	Flight path control efficiency b_δ
Aerodynamic control	$a_{\delta_{ae}} = \dfrac{-m_z^{\delta_{ae}}\rho V^2 \cdot SL}{2J_z}$	$b_{\delta_{ae}} = \dfrac{c_y^{\delta_{ae}}\rho VS}{2m}$
Jet vane control	$a_{\delta_V} = \dfrac{-kL_V}{J_z}$	$b_{\delta_V} = \dfrac{k}{mV}$

It can be seen from Table 6.6−2 that for aerodynamic control $a_{\delta_{ae}}$ is proportional to ρV^2 and $b_{\delta_{ae}}$ is proportional to ρV, so it is clear that the value of its rotational control efficiency $a_{\delta_{ae}}$ and flight path control efficiency $b_{\delta_{ae}}$ are low at low missile speeds and high attitudes (low air densities).

But for jet vane control its a_{δ_V} is not related to V and ρ, and b_{δ_V} is proportional to $1/V$. For this reason, jet vane rotational control efficiency is not effected by low missile speeds and high attitudes, and its flight path control efficiency is even higher at low missile speeds. That is why jet vane control is so popular in modern surface-to-air and air-to-air missile control.

The block diagram for a typical aerodynamic and jet vane hybrid control autopilot is shown in Fig. 6.6−2.

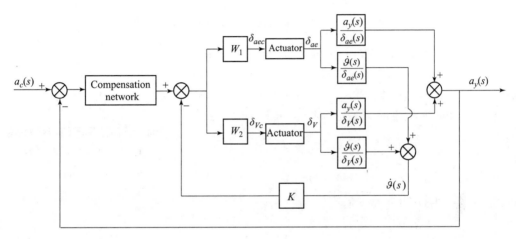

Fig. 6.6−2 The aerodynamic and jet vane hybrid control acceleration autopilot block diagram

For aerodynamic control, the transfer function $a_y(s)/\delta_{ae}(s)$ from the aerodynamic actuator deflection angle δ_{ae} to the missile lateral acceleration a_y is

$$\frac{a_y(s)}{\delta_{ae}(s)} = -V \cdot \frac{-b_{\delta_{ae}}s^2 - a_\omega b_{\delta_{ae}}s + (a_{\delta_{ae}}b_\alpha - a_\alpha b_{\delta_{ae}})}{s^2 + (a_\omega + b_\alpha)s + (a_\alpha + a_\omega b_\alpha)}. \tag{6.6-1}$$

And the transfer function $\dot{\vartheta}(s)/\delta_{ae}(s)$ from the aerodynamic actuator deflection angle δ_{ae} to the missile angular velocity $\dot{\vartheta}$ is

$$\frac{\dot{\vartheta}(s)}{\delta_{ae}(s)} = -\frac{a_{\delta_{ae}}s + (a_{\delta_{ae}}b_\alpha - a_\alpha b_{\delta_{ae}})}{s^2 + (a_\omega + b_\alpha)s + (a_\alpha + a_\omega b_\alpha)}. \tag{6.6-2}$$

For jet vane control, the transfer function $a_y(s)/\delta_V(s)$ from the jet vane deflection angle δ_V to the missile lateral acceleration a_y is

$$\frac{a_y(s)}{\delta_V(s)} = -V \cdot \frac{-b_{\delta_V}s^2 - a_\omega b_{\delta_V}s + (a_{\delta_V}b_\alpha - a_\alpha b_{\delta_V})}{s^2 + (a_\omega + b_\alpha)s + (a_\alpha + a_\omega b_\alpha)}. \tag{6.6-3}$$

And the transfer function $\dot{\vartheta}(s)/\delta_V(s)$ from jet vane deflection angle δ_V to missile angular velocity $\dot{\vartheta}$ is

$$\frac{\dot{\vartheta}(s)}{\delta_V(s)} = -\frac{a_{\delta_V}s + (a_{\delta_V}b_\alpha - a_\alpha b_{\delta_V})}{s^2 + (a_\omega + b_\alpha)s + (a_\alpha + a_\omega b_\alpha)}. \tag{6.6-4}$$

2) Pulse thruster control

To speed up autopilot response at high attitudes, some modern surface-to-air missiles adopt a pulse thruster control scheme. The typical missile used in this scheme is the American PAC – 3 surface-to-air missile. It contains 180 solid propellant pulse thrusters allocated perpendicular to the centerline of the missile to provide pitch and yaw control. These thrusters are spaced evenly around the centerline of the missile in rings containing 18 motors. There are 10 rings of the thrusters in the longitudinal direction (Fig. 6.6 – 3).

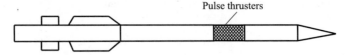

Fig. 6. 6 – 3 American PAC – 3 surface-to-air missile.

In this case, the control variable n is the number of pulse thrusters fired in the sampling interval, its average force generated in time interval Δt is nf (Fig. 6. 6 – 4).

Here f is the control force generated by one pulse thruster firing.

Fig. 6. 6 – 4 The control force generated by pulse thrusters

If the distance between the pulse thruster and the missile center of gravity is L_P, the pulse thruster control related dynamic coefficients will be given as

$$a_n = \frac{-fL_P}{J_z}, \qquad (6.6-5)$$

$$b_n = \frac{f}{mV}. \qquad (6.6-6)$$

The block diagram for a typical aerodynamic and pulse thruster hybrid control autopilot is shown in Fig. 6. 6 – 5.

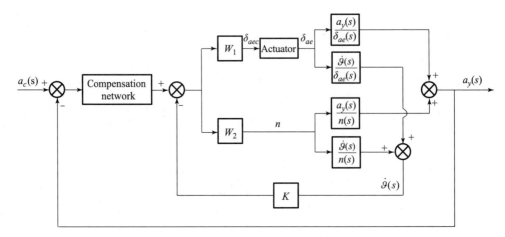

Fig. 6. 6 – 5 The aerodynamic and pulse thruster hybrid control acceleration autopilot block diagram

For pulse thruster control, the transfer function $a_y(s)/n(s)$ from the fired pulse thruster number n to the missile lateral acceleration a_y is

$$\frac{a_y(s)}{n(s)} = -V \cdot \frac{-b_n s^2 - a_\omega b_n s + (a_n b_\alpha - a_\alpha b_n)}{s^2 + (a_\omega + b_\alpha)s + (a_\alpha + a_\omega b_\alpha)}. \tag{6.6-7}$$

And the transfer function $\dot{\vartheta}(s)/n(s)$ from the fired pulse thruster number n to missile angular velocity $\dot{\vartheta}$ is

$$\frac{\dot{\vartheta}(s)}{n(s)} = -\frac{a_n s + (a_n b_\alpha - a_\alpha b_n)}{s^2 + (a_\omega + b_\alpha)s + (a_\alpha + a_\omega b_\alpha)}. \tag{6.6-8}$$

The Russia reference [4] [ПРОЕКТИРОВАНИЕ ЗЕНИТНЫХ УПРАВЛЯЕМЫХ РАКЕТ] has studied the American PAC-3 pulse thruster controlled autopilot performance by using the following parameters.

Pulse thruster operation time $\Delta t = 16\text{ms}$, pulse force $f = 2,500\text{ N}$, missile roll angle velocity $\dot{\gamma} = 3\text{Hz} = 1,080°/\text{s}$, total pulse thruster number 180, the distance between the pulse thruster and missile center of gravity $L_P = 1\text{m}$. The pulse force direction change in 16ms is $\Delta r = 1,080 \times 0.016 = 17.3°$, which is acceptable. The result of analysis showed that at high attitudes the acceleration autopilot can have a response time $t_{63} = 0.05\text{ s}$.

3) Trajectory control thruster

In the pulse thruster control scheme, the distance between the pulse thruster and the missile center of gravity is relatively large. Its main control effect is the generation of a larger control moment to speed up autopilot response. If the adopted control thruster has more power and the distance between the control thruster and the missile center of gravity is small, the goal of this design will be to generate a large guidance required trajectory control force and put the control moment benefit second.

Given in Fig. 6.6-6 is the France Aster surface-to-air missile trajectory control thruster arrangement on the missile.

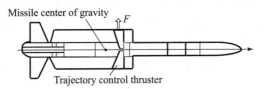

Fig. 6.6-6 France Aster surface-to-air missile

Suppose the trajectory control thruster control variable is taken as u, then the control force will be $F = k_1 u$, and the related dynamic coefficients will be

$$a_u = \frac{-k_1 L_u}{J_z}, \tag{6.6-9}$$

$$b_u = \frac{k_1}{mV}. \tag{6.6-10}$$

where L_u is the distance between the trajectory thruster and the missile center of gravity.

This design gives much enhanced missile lateral acceleration capability and faster autopilot response but the price to pay is the increased missile weight, dimension and costs.

The block diagram for a typical aerodynamic and trajectory thruster hybrid control autopilot is shown in Fig. 6.6 – 7.

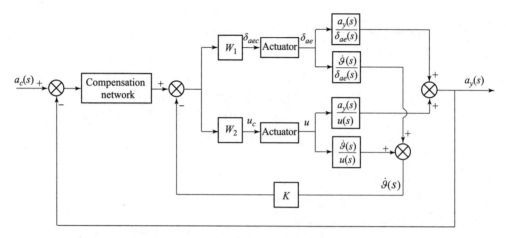

Fig. 6.6 – 7 The aerodynamic and trajectory hybrid control thruster acceleration autopilot block diagram

For trajectory thruster control, the transfer function $a_y(s)/u(s)$ from u to the missile lateral acceleration a_y is

$$\frac{a_y(s)}{u(s)} = -V \cdot \frac{-b_u s^2 - a_\omega b_u s + (a_u b_\alpha - a_\alpha b_u)}{s^2 + (a_\omega + b_\alpha)s + (a_\alpha + a_\omega b_\alpha)}. \tag{6.6-11}$$

And the transfer function $\dot{\vartheta}(s)/u(s)$ from u to missile angular velocity $\dot{\vartheta}$ is

$$\frac{\dot{\vartheta}(s)}{u(s)} = -\frac{a_u s + (a_u b_\alpha - a_\alpha b_u)}{s^2 + (a_\omega + b_\alpha)s + (a_\alpha + a_\omega b_\alpha)}. \tag{6.6-12}$$

4) Attitude thruster control plus trajectory thruster control

In exoatmospheric flight, there is no aerodynamic force that could be used for missile control. For this reason, in exoatmospheric flight both missile attitude control and trajectory control have to use thruster control. Fig. 6.6 – 8 gives an example for this type of missile in its end phase configuration and Fig. 6.6 – 9 shows the end phase engagement trajectories.

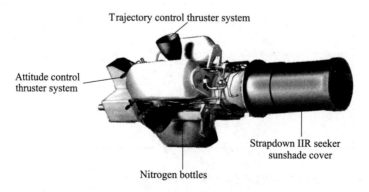

Fig. 6.6 – 8 An example for the attitude control thruster plus trajectory control thruster

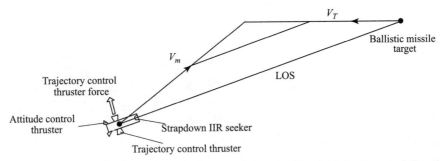

Fig. 6.6 – 9 The example for the attitude control thruster plus trajectory control thruster

In this application, the attitude control thruster could keep the missile centerline always pointing to the target (by way of the strapdown seeker angular error ε feedback). The guidance required LOS rate $\dot{q}$ signal could be obtained by the following relation $\dot{q} = \dot{\vartheta} + \dot{\varepsilon}$. Since the guidance acceleration generated by the trajectory control thruster is perpendicular to the missile centerline, it must also be perpendicular to the missile-target line-of-sight (LOS). This direction is just the control direction desired for proportional navigation guidance law.

§ 6.7 Spinning Missiles Control

Due to the significant advantages of excellent robustness to some disturbances, such as thrust asymmetries, fin misalignments, etc. , missiles of spinning airframe require less on the manufacturing technology than nonspinning ones, so that they are widely applied in engineering, such as some low-cost gun-launched terminal guided projectiles or mortars. However, unlike the nonspinning missiles, cross-coupling effects would be induced by spinning airframe between the pitch and yaw, which may lead to an undesirable coning motion. In this subsection, we establish the model of spinning missiles, and describe the unique phenomenon of the coning motion.

For convenience to separate from continuous spinning, we perform the derivation in a nonspinning frame, which is defined as follows:

In the nonspinning body coordinate system $Ox_{nb}y_{nb}z_{nb}$, Ox_{nb} is coincident with the velocity vector, Oy_{nb} is perpendicular to Ox_{nb} in the vertical plane, and Oz_{nb} is determined by the right-hand rule.

Note that the new frame can be obtained by rotating the body-fixed coordinate system $Ox_b y_b z_b$. The time-varing spinning angle γ about the Ox_b axis is shown in Fig. 6.7 – 1.

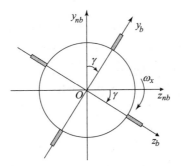

Fig. 6.7 – 1 The relationship between nonspinning frame and spinning frame

There exist relations between two frames as

$$\sigma_b = L(\gamma)\sigma_{nb}, \sigma_{nb} = L^{-1}(\gamma)\sigma_b, \quad (6.7-1)$$

where the coordinate transformation matrix $L(\gamma)$ is

$$L(\gamma) = \begin{bmatrix} \cos\gamma & \sin\gamma \\ -\sin\gamma & \cos\gamma \end{bmatrix}. \quad (6.7-2)$$

The governing equations of dynamics should be obtained according to the theorem of angular momentum. However, the moments and forces acting on spinning missles are slightly different to nonspinning ones as to couplings. The model of spinning missiles must take the extra specific moments and force originated from the coupling effect into account.

The cross coupling effects induced by aerodynamics consist of gyroscopic effect and Magnus effect, control coupling effect. Thereafter, we introduce them in details.

6.7.1 Aerodynamic Coupling

The Magnus effect, arising from asymmetric vortices shed, is proposed by Magnus early in 1852, which is always compared to Achilles Heel of the vehicle with regard to cross coupling. It has been proved in applications that owing to the Magnus effect, the spinning missile would deviate from its firing direction, even evolve to be unstable. The extra terms for the dynamical equations are Magnus moments and forces, and gyroscopic moments.

The Magnus moments in nonspinning frame can be expressed as

$$\begin{aligned} M_{y\_Mag} &= a_M \alpha \\ M_{z\_Mag} &= -a_M \beta \end{aligned} \quad (6.7-3)$$

Nevertheless, as the spinning rate ω_x is high, the terms $J_x \omega_x \omega_z$ and $J_x \omega_x \omega_y$ are significant enough to be ignored

$$\begin{aligned} M_{y\_Gyro} &= -a_G \dot{\vartheta} \\ M_{z\_Gyro} &= a_G \dot{\psi} \end{aligned}, \quad (6.7-4)$$

where

$$a_G = \frac{J_x}{J_y}\omega_x. \quad (6.7-5)$$

Fig. 6.7-2 is the aerodynamics of the spinning missile.

6.7.2 Control Coupling

For spinning missiles, dynamic of actuator in nonspinning coordinate would result in control coupling effect. In order to achieve high control precision, it is necessary to analyze the control coupling induced by the dynamic of actuator.

Original dynamic of actuator $G_b(s)$ in the body-fixed frame can be expressed as a typical second-order element

$$G_b(s) = \frac{\delta_{yb}}{\delta_{ybc}} = \frac{\delta_{zb}}{\delta_{zbc}} = \frac{1}{T_a^2 s^2 + 2T_a \mu_a s + 1}, \quad (6.7-6)$$

where T_a, μ_a stand for the time constant and damping ratio of the actuator, respectively.

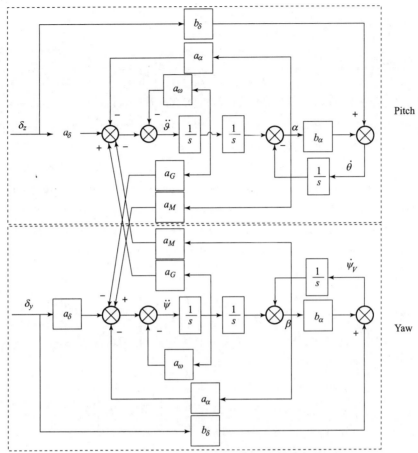

Fig. 6.7−2 The aerodynamics for the spinning missile

The relation between the command vector and responses vector of control surface deflections in the body-fixed frame can be rewritten as

$$\begin{bmatrix} \delta_{zb} \\ \delta_{yb} \end{bmatrix} = \begin{bmatrix} G_b(s) & 0 \\ 0 & G_b(s) \end{bmatrix} \begin{bmatrix} \delta_{zbc} \\ \delta_{ybc} \end{bmatrix}. \tag{6.7-7}$$

Likewise,

$$\begin{bmatrix} \dot{\delta}_{zb}(t) \\ \dot{\delta}_{yb}(t) \end{bmatrix} = \begin{bmatrix} \cos\gamma & -\sin\gamma \\ \sin\gamma & \cos\gamma \end{bmatrix} \begin{bmatrix} \dot{\delta}_z(t) \\ \dot{\delta}_y(t) \end{bmatrix} + \omega_x \begin{bmatrix} -\sin\gamma & -\cos\gamma \\ \cos\gamma & -\sin\gamma \end{bmatrix} \begin{bmatrix} \delta_z(t) \\ \delta_y(t) \end{bmatrix}, \tag{6.7-8}$$

$$\begin{bmatrix} \ddot{\delta}_{zb}(t) \\ \ddot{\delta}_{yb}(t) \end{bmatrix} = \begin{bmatrix} \cos\gamma & -\sin\gamma \\ \sin\gamma & \cos\gamma \end{bmatrix} \begin{bmatrix} \ddot{\delta}_z(t) \\ \ddot{\delta}_y(t) \end{bmatrix} + 2\omega_x \begin{bmatrix} -\sin\gamma & -\cos\gamma \\ \cos\gamma & -\sin\gamma \end{bmatrix} \begin{bmatrix} \dot{\delta}_z(t) \\ \dot{\delta}_y(t) \end{bmatrix} + \omega_x^2 \begin{bmatrix} -\cos\gamma & \sin\gamma \\ -\sin\gamma & -\cos\gamma \end{bmatrix} \begin{bmatrix} \delta_z(t) \\ \delta_y(t) \end{bmatrix}. \tag{6.7-9}$$

The transfer function matrix of equivalent dynamic of equation (6.7−6) is

$$\boldsymbol{G}_a(s) \triangleq \begin{bmatrix} G_f(s) & G_{co}(s) \\ -G_{co}(s) & G_f(s) \end{bmatrix}, \tag{6.7-10}$$

where

$$G_f(s) = \frac{T_a^2 s^2 + 2T_a\mu_a s + 1 - T_a^2\omega_x^2}{(T_a^2 s^2 + 2T_a\mu_a s + 1 - T_a^2\omega_x^2)^2 + 4\omega_x^2(T_a^2 s + T_a\mu_a)^2}$$

$$G_{co}(s) = \frac{2\omega_x(T_a^2 s + T_a\mu_a)}{(T_a^2 s^2 + 2T_a\mu_a s + 1 - T_a^2\omega_x^2)^2 + 4\omega_x^2(T_a^2 s + T_a\mu_a)^2} \quad (6.7-11)$$

$G_f(s)$ and $G_{co}(s)$ are transfer functions of forward channel and coupling channel, respectively. When ω_x equals zeros, $G_{co}(s) = 0$, the transfer function matrix is diagonal with no cross coupling, which can refer to the non-spinning case.

For the spinning missiles, the characteristics of diagonal element are different from the original dynamics, meanwhile, the non-diagonal elements symbolizing cross coupling won't equal zero.

The poles of the equivalent dynamics for first-order elements will be doubled, which are, according to theorem 1, all lying on the left half-plane. However, the location of the extra poles would vary with gains of the autopilot, thus, they may affect the stability of the system, which will be studied below.

$$\begin{bmatrix} \ddot{\vartheta}(t) \\ \ddot{\varphi}(t) \\ \dot{\alpha}(t) \\ \dot{\beta}(t) \end{bmatrix} = \begin{bmatrix} a_\omega & a_G & a_\alpha & -a_M \\ -a_G & a_\omega & a_M & a_\alpha \\ 1 & 0 & -b_\alpha & 0 \\ 0 & 1 & 0 & -b_\alpha \end{bmatrix} \begin{bmatrix} \dot{\vartheta}(t) \\ \dot{\varphi}(t) \\ \alpha(t) \\ \beta(t) \end{bmatrix} + \begin{bmatrix} a_\delta & 0 \\ 0 & a_\delta \\ -b_\delta & 0 \\ 0 & -b_\delta \end{bmatrix} \begin{bmatrix} \delta_z(t) \\ \delta_y(t) \end{bmatrix} \quad (6.7-12)$$

$$\begin{bmatrix} \dot{\vartheta}_m \\ \dot{\varphi}_m \\ a_{ym} \\ a_{zm} \end{bmatrix} = \begin{bmatrix} 1 & 0 & 0 & 0 \\ 0 & 1 & 0 & 0 \\ 0 & 0 & Vb_\alpha & 0 \\ 0 & 0 & 0 & Vb_\alpha \end{bmatrix} \begin{bmatrix} \dot{\vartheta} \\ \dot{\varphi} \\ \alpha \\ \beta \end{bmatrix} + \begin{bmatrix} 0 & 0 \\ 0 & 0 \\ Vb_\delta & 0 \\ 0 & Vb_\delta \end{bmatrix} \begin{bmatrix} \delta_z \\ \delta_y \end{bmatrix} \quad (6.7-13)$$

where, the subscript m denotes the measurements of the missile obtained by IMU. Applying Laplace transformation on equation $(6.7-12)$, the transformation function matrix from commands in the nonspinning frame to the states is obtained as

$$G_\delta^x(s) = \frac{\begin{bmatrix} N_{11}(s) & N_{12}(s) \\ N_{21}(s) & N_{22}(s) \\ N_{31}(s) & N_{32}(s) \\ N_{41}(s) & N_{42}(s) \end{bmatrix}}{D(s)} \quad (6.7-14)$$

where

$$D(s) = a_\alpha^2 + a_M^2 + 2[a_\alpha(s + a_\omega) - a_M a_G](s + b_\alpha) + [a_G^2 + (s + a_\omega)^2](s + b_\alpha)^2 \quad (6.7-15)$$

$$N_{11}(s) = a_\delta(s + b_\alpha)[a_\alpha + (s + a_\omega)(s + b_\alpha)] - b_\delta[(a_\alpha s + a_\alpha a_\omega - a_M a_G)(s + b_\alpha) + a_\alpha^2 + a_M^2] \quad (6.7-16)$$

$$N_{21}(s) = (s + b_\alpha)[a_\delta(a_G s + a_G b_\alpha - a_M) - b_\delta(a_M s + a_G a_\alpha + a_M a_\omega)] \quad (6.7-17)$$

$$N_{31}(s) = -a_\delta[(s + a_\omega)(s + b_\alpha) + a_\alpha] - b_\delta[(s + b_\alpha)((s + a_\omega)^2 + a_G^2) + a_\alpha(s + a_\omega) - a_M a_G] \quad (6.7-18)$$

$$N_{41}(s) = a_\delta[a_G(s + b_\alpha) - a_M] - b_\delta[a_M(s + a_\omega) + a_G a_\alpha] \quad (6.7-19)$$

It is worth noticing that the state matrix with respect to equation (6.7 – 12) is a skew – symmetric matrix, thus, the rest of the transformation function matrix elements exist:

$$N_{22}(s) = N_{11}(s), N_{12}(s) = -N_{21}(s)$$
$$N_{42}(s) = N_{31}(s), N_{32}(s) = -N_{41}(s)$$
(6.7 – 20)

According to the equation (6.7 – 13), the transformation function matrix from commands in the nonspinning frame to outputs can be obtained as

$$\boldsymbol{G}_{\delta}^{a_y,s}(s) = \frac{\begin{bmatrix} N_{y11}(s) & N_{y12}(s) \\ N_{y21}(s) & N_{y22}(s) \end{bmatrix}}{D(s)}$$
(6.7 – 21)

where

$$N_{y11}(s) = N_{y22}(s) = V(b_\alpha N_{31} + b_\delta D)$$
$$N_{y12}(s) = -N_{y21}(s) = Vb_\alpha N_{32}$$
(6.7 – 22)

The dynamics of the pitch and yaw should be associated with each other on account of the couplings, consequently, the model of spinning missiles comes out to be a Multiple-Input-Multiple-Output system, and the system cannot be separated for design of the autopilot. Meanwhile, the cross-coupling effects of rolling missiles result in a coning motion and degrade the system stability.

7

Line of Sight Guidance Methods

§ 7.1 LOS Guidance System

There are many types of line-of-sight guidance systems, but the operational principle of all the systems is the same to guide the missile to flight along the guidance radar to the target line of sight (LOS). Therefore, such a guidance method is often referred to as the three-point guidance. Since the deviations of the missile line from the line of sight are measured by the guidance radar and uploaded to the missile in the form of commands, and corrected by the missile control system, this type of guidance is also referred to as the command guidance.

Line-of-sight guidance is often divided into semi-automatic command guidance and automatic command guidance. The difference is that the tracking of the target in the semi-automatic command guidance is accomplished by the operator, while the tracking of the target in the automatic command guidance is automatically completed by a guidance radar or an optoelectronic tracking device.

Most of the current anti-tank guided missiles that use command guidance adopt the semi-automatic command guidance. In this system, the operator tracks the target through an optical or inferred sight, and the deviation of the missile infrared beacon from the goniometer axis is measured by the goniometer. The related guidance command will be uploaded to the missile by a wire or radio link to complete the guidance loop. In this scheme, the misalignment $\Delta\theta$ of the optic axis of the operator sight and the goniometer axis will influence the performance of the guidance (Fig. 7.1 – 1).

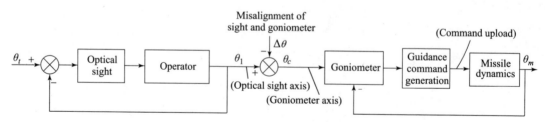

Fig. 7.1 – 1 Block diagram Ⅰ of the semi-automatic anti-tank missile guidance with the help of the operator sight and goniometer loop

When the operator sight adopts an infrared imaging system, the infrared imaging sight can simultaneously observe the target and measure the error angle of the infrared beacon of the missile. At this time, since the sight and the goniometer are the same device, the misalignment error does not exist (Fig. 7.1 – 2).

7 Line of Sight Guidance Methods

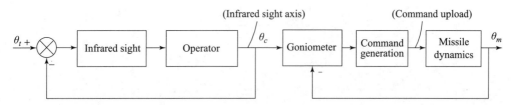

Fig. 7.1-2 Block diagram Ⅱ of the infrared semi-automatic anti-tank missile guidance loop

For an anti-tank missile with laser beam riding system, the deviation of the missile from the line of sight is automatically measured by sensing the coded laser signal field which is generated by the missile operator controlled laser guidance device. The guidance command is generated on the missile and there is no need with commands uploading. At this time, since the sight and the laser guidance device are not the same device, the misalignment of the two devices will still affect the guidance accuracy. The guidance block diagram of this type of system is shown in Fig. 7.1-3.

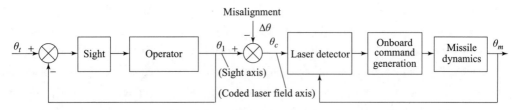

Fig. 7.1-3 Block diagram of the anti-tank missile with the laser beam riding system

Current ground-to-air missiles with LOS (line of sight) guidance usually adopt the automatic command guidance scheme, in which the automatic tracking of the target is accomplished by the tracking radar or the optoelectronic tracking device. Since the radar can achieve the target tracking and the angle measurement simultaneously through the target echo and the missile radio beacon downloaded signal, this type of system no longer has the misalignment problem (Fig. 7.1-4). A detailed description of the loop analysis and design of the guidance radar can be found in Chapter 4.

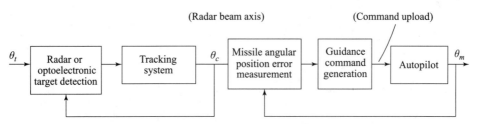

Fig. 7.1-4 Guidance block diagram of the automatic command guidance scheme

§ 7.2 Required Acceleration for the Missile with LOS Guidance

Fig. 7.2-1 shows the trajectory diagrams of the target and the missile in a line-of-sight guidance.

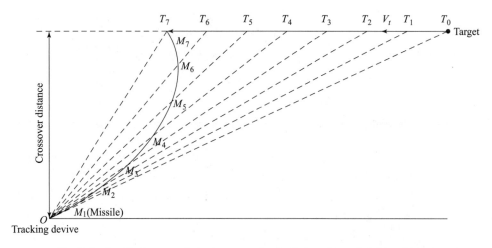

Fig. 7.2-1 Trajectory of the target and missile in a line of sight guidance

In Fig. 7.2-1, suppose that the target T makes a constant velocity straight line flight, and the missile is launched at a speed higher than the target at the initial time T_0. The plane containing the missile speed vector and the target speed vector is called the missile flight plane. The lines of sights at the moments 1, 2 and 3 after launch are respectively given as OT_1, OT_2, OT_3 and so on. Since the three-point guidance law always tries to force the missile to stay on the line of sight, the actual trajectory path must be a curved one, and the curvature of the trajectory will increase sharply in the terminal phase of the guidance. In addition, since the direction of the missile velocity at any point of the flight path will be consistent with the direction of the tangent of the instantaneous trajectory, the velocity direction will not be the same as the direction of the line of sight, and in the terminal phase, the two will differ by a larger angle.

Since the line of sight usually refers to the beam direction of the tracking system, the angle between the missile velocity direction and the line of sight is often called the "trajectory-beam angle," and the angle between the missile body x-axis and the beam direction, usually more concerned in the guidance system design, is referred to as the "missile body-beam angle," which differs from the "trajectory-beam angle" by the angle of attack (Fig. 7.2-2). It is known from the figure that the lateral acceleration of the missile must be to the left of the line of sight in the

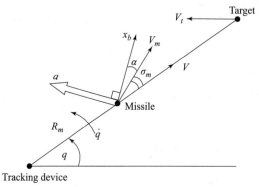

Fig. 7.2-2 Definitions of the symbols related to the LOS guidance

engagement scenario shown, that is, the velocity direction will be on the right side of the missile, and head of the missile must be on the left side of the velocity.

Generally, the LOS guidance missile of a ground-to-air system is equipped with a radio beacon, which can transmit the deviation of the missile and track radar axis. However, the beacon has a limited beam width. Once the "missile body-beam angle" is greater than this width, the tracking device will lose the return signal of the missile beacon. In general, the angle of the "missile body-beam angle" is limited to about 40°. In addition, during the curved flight a necessary lateral acceleration which is perpendicular to the LOS must be acted upon the missile to ensure the missile to fly against from leaving the LOS. Of course, the larger the "missile body-beam angle" is, the less effective lateral acceleration is generated, or the larger normal acceleration command is required to accomplish the LOS guidance flight. From this, we can see that the actual value of the "missile-beam angle" in the flight is a significant parameter that needs our close attention.

To simplify the analysis, suppose that the missile's angle of attack during the guidance is not significant and may be omitted and the missile has a lateral acceleration a perpendicular to the missile's velocity V_m direction (Fig. 7.2 – 3). Taking the trajectory-beam angle as σ_m, the LOS rotation angular velocity and acceleration as $\dot{q}$ and $\ddot{q}$, the velocity of the missile along the line of sight should be $V = V_m \cos\sigma_m$. In addition, due to the presence of the thrust and aerodynamic drag, the missile will have an acceleration component $\dot{V}_m$ in the direction of the velocity vector.

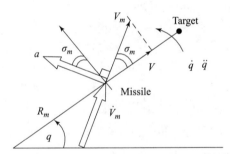

Fig. 7.2 – 3 Definitions of related parameters on the flight plane

In order to realize the flight of the missile along the rotating line of sight, the acceleration projection perpendicular to the LOS direction should be

$$a\cos\sigma_m + \dot{V}_m\sin\sigma_m = 2V_m\cos\sigma_m \dot{q} + R_m\ddot{q}. \quad (7.2-1)$$

In equation (7.2 – 1), the right hand side of the equation is the acceleration of the missile perpendicular to the line of sight required to fly along the rotating line of sight, and the left hand side is the source of this acceleration (the combined action of the missile's normal acceleration a and the missile axis acceleration $\dot{V}_m$).

That is, to achieve the motion of the missile along the rotation LOS, the required missile normal acceleration will be

$$a = 2V_m\dot{q} + \frac{R_m}{\cos\sigma_m}\ddot{q} - \dot{V}_m\tan\sigma_m. \quad (7.2-2)$$

The values of $\dot{q}$ and $\ddot{q}$ depend on the velocity and position of the target. According to the geometric relationship shown in Fig. 7.2 −4,

$$\dot{q} = V_m \frac{\sin\sigma_m}{R_m} = V_t \frac{\sin\sigma_t}{R_t}. \qquad (7.2-3)$$

That is

$$\sin\sigma_m = \sin\sigma_t \frac{R_m}{R_t} \cdot \frac{V_t}{V_m}. \qquad (7.2-4)$$

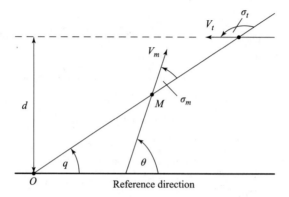

Fig. 7.2 −4 **Geometric relationship of the flight**

If the target flies parallel to the reference direction, its cross distance d will be a constant. Therefore, $\dot{q}$ can also be expressed as

$$\dot{q} = \frac{V_t \sin^2\sigma_t}{d}. \qquad (7.2-5)$$

Note that $\ddot{q}$ can be obtained from the equations $\dot{q} = -\dot{\sigma}_t$ and $dq = -d\sigma_t$

$$\ddot{q} = \frac{d^2 q}{dt^2} = \left[\frac{d}{dq}\left(\frac{dq}{dt}\right)\right] \cdot \left(\frac{dq}{dt}\right) = -\frac{d\dot{q}}{d\sigma_t} \cdot \dot{q} = -\frac{2V_t^2}{d^2}\sin^3\sigma_t\cos\sigma_t. \qquad (7.2-6)$$

Inserting the expressions of $\dot{q}$ and $\ddot{q}$ into the expression of the normal acceleration a, and assuming that the missile is flying at a constant speed ($\dot{V}_m = 0$), we have

$$a = \frac{2V_m V_t}{d}\left(\sin^2\sigma_t - \frac{\sin\sigma_t \cos\sigma_t \sin\sigma_m}{\cos\sigma_m}\right)$$

$$= \frac{2V_m V_t}{d}(\beta_1 + \beta_2). \qquad (7.2-7)$$

In the above equation, $\beta_1 = \sin^2\sigma_t$, $\beta_2 = -\frac{\sin\sigma_t \cos\sigma_t \sin\sigma_m}{\cos\sigma_m}$, and β_1 is the Coriolis acceleration factor corresponding to the line-of-sight rotation angular velocity $\dot{q}$. β_2 is the translational acceleration factor corresponding to the line-of-sight rotation angular acceleration $\ddot{q}$.

It is seen from the expression of the normal acceleration a that its value is proportional to $V_m V_t$ and inversely proportional to d. That is, the larger the target velocity V_t and the missile velocity V_m are, the greater the normal acceleration is required. When the target is flying over the tracking device, the smaller the cross distance d is, the greater the missile's required normal acceleration is. Fig. 7.2 −5 shows the relative magnitudes of Coriolis acceleration factor β_1 caused by $\dot{q}$ and the translational acceleration factor β_2 caused by $\ddot{q}$ at different target positions σ_t when $V_m/V_t = 2.5$ and at the hit target time (as $R_m/R_t = 1$). It is clear that the Coriolis acceleration caused by $\dot{q}$ is the primary factor of the normal acceleration required when hitting the target in the line-of-sight guidance.

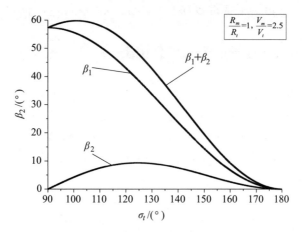

Fig. 7.2 – 5 Variation of the two normal acceleration components with the change of σ_t when $R_m/R_t = 1$, and $V_m/V_t = 2.5$

From the expression $\beta_1 = \sin^2\sigma_t$, it is known that when $\dot{q}$ does not change, Coriolis acceleration has nothing to do with the relative position R_m/R_t of the missile on the line of sight, but when the aircraft is an incoming target, the line-of-sight angle q will gradually increase from zero to 90°. Since $\dot{q} = \dfrac{V_t \sin^2 q}{d}$, $\dot{q}$ increases as the target approaches. When the missile hits the target, $\dot{q}$ will be the largest in comparison with its other positions on the LOS, that is, the lateral acceleration of the missile required in the line-of-sight guidance is the maximum as the missile meets the target, which is the biggest drawback of this guidance law.

Fig. 7.2 – 6 shows the variation of the trajectory-beam angle σ_m of the missile as the missile at different LOS positions R_m/R_t and the target coming at different σ_t positions for $V_m/V_t = 2.5$. It is

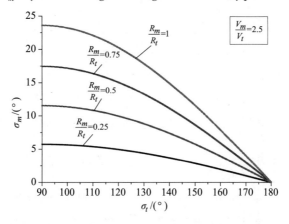

Fig. 7.2 – 6 Variation of the trajectory-beam angle σ_m with the change of σ_t and R_m/R_t

known that as the target approaches, the value of σ_m is the greatest when the missile encounters the target ($R_m/R_t = 1$) (if the angle of attack is included, the actual missile body-beam angle will be greater than the trajectory-beam angle σ_m). In other words, the missile normal acceleration efficiency is the lowest at the end of the attack, and the direction of the signal emitted by the missile beacon has the largest deviation from the LOS. These are the disadvantages of line-of-sight guidance.

§ 7.3 Analysis of the LOS Guidance Loop

Fig. 7.3 – 1 shows the block diagram of the line-of-sight guidance loop, in which $\Delta\theta$ is the angular deviation of the missile from the line-of-sight, h_c is the linear deviation of the missile and R is the distance between the target tracking device and the missile, Δy_m is the missile linear displacement response and θ_m is the angular response of the missile.

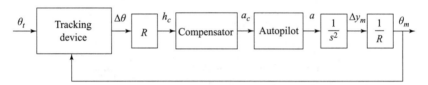

Fig. 7.3 – 1   Block diagram of the line-of-sight guidance loop

The loop of Fig. 7.3 – 1 is an angular tracking loop, but R and $1/R$ in the loop can cancel each other. Therefore, the effect of R can be neglected in the control system loop design and only the control loop of the linear deviation h needs to be considered. That is, Fig. 7.3 – 1 can be simplified as Fig. 7.3 – 2.

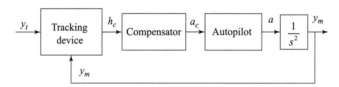

Fig. 7.3 – 2   Control loop of the line-of-sight guidance with only linear deviation considered

Since there are two integrators in this loop, which brings a $-180°$ phase shift, the system is unstable without a compensation network after the autopilot lag is added. After the lead compensation design is introduced, the phase margin of the system should be the compensation lead phase $\Delta\phi_{compensator}$ minus the phase shift of the autopilot at the system crossover frequency.

The compensation network could be a single lead compensation $\left(\dfrac{\alpha Ts + 1}{Ts + 1}\right)$ or a cascaded lead compensation $\left(\dfrac{\alpha^* Ts + 1}{Ts + 1}\right)\left(\dfrac{\alpha^* Ts + 1}{Ts + 1}\right)$. When the lead compensator is providing a positive angle compensation, its gain will increase at higher frequencies, which will cause the crossover frequency to shift to a high frequency. This is detrimental to the system stability and noise depression. Therefore, to maintain the same gain increase effect, the single lead compensator and cascade lead compensator parameters α and α^* should have the following relation

$$\alpha^* = \sqrt{\alpha}. \tag{7.3-1}$$

Fig. 7.3-3 shows the difference between the gain and the phase characteristics of the transfer function of the single lead compensator and the cascaded lead compensator when $\alpha = 10$ and $\alpha^* = \sqrt{10}$. Note that the maximum phase compensation of the cascaded lead compensator is 62.6°, which is higher than the maximum phase compensation of the single lead compensator 55°, but its compensation bandwidth is slightly narrower. However, they have the same gain value 10 dB at the maximum phase compensation and 20 dB at the high-frequency band.

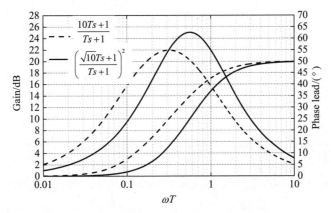

Fig. 7.3-3 **Gain and phase characteristics of the two lead compensator transfer functions**

The following gives an example of the design of a line-of-sight guidance control loop. The missile's acceleration autopilot transfer function selected in this system is of the second order, in which $\omega_m = 12$ rad/s = 1.91 Hz and damping coefficient $\mu = 0.6$ (it should be noted that the transient speed of the line-of-sight guidance loop is almost completely determined by the acceleration autopilot bandwidth). If a cascaded lead compensation network is adopted in this design with $\alpha^* = \sqrt{10}$, it is required to select the proper system open loop gain K and the lead compensation network time constant T, so that the loop gain margin is greater than 6 dB and the phase margin is greater than 45°. Fig. 7.3-4 shows the block diagram of this loop.

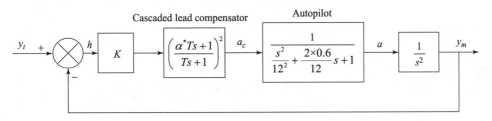

Fig. 7.3-4 **Block diagram of the line-of-sight guidance control loop**

Taking $K = 3$ and $T = 0.15$, the open loop Bode diagrams of the system with the compensation network and without a compensation network are shown in Fig. 7.3-5. The design results are a phase margin of 45° and a gain margin of 9.68 dB.

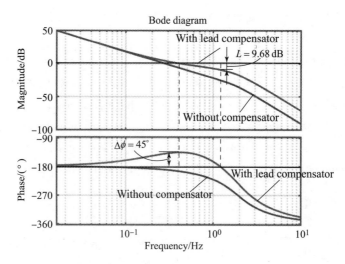

Fig. 7.3-5 Open loop Bode diagram of the line-of-sight guidance loop above

The crossover frequency in the design is 0.41 Hz. At the crossover frequency, the autopilot lag is $-14°$, the double-integrator's lag is $-180°$, and the lead compensation phase lead is $59°$. So, the final design has a phase margin of $\Delta\phi = 59° - 14° = 45°$.

Fig. 7.3-6 shows the Bode diagram of the selected cascaded lead compensator. It is shown that this compensation network can provide a phase lead at the system crossover frequency and it is quite close to its maximum possible value.

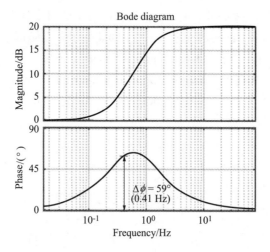

Fig. 7.3-6 Bode diagram of the lead compensator $\left(\dfrac{\sqrt{10}\times 0.15s+1}{0.15s+1}\right)^2$

Fig. 7.3-7 shows the time response of the system with a unit step command h_c input. Since the line-of-sight guidance control loop is a type II system, it is difficult for its response to be fast, and its overshoot of the transition process is often large.

The line-of-sight control loop of the above design is used to intercept a flyover aircraft target. Take the target speed as $V_t = 250$ m/s and the flyover distance $d = 4,000$ m.

7 Line of Sight Guidance Methods

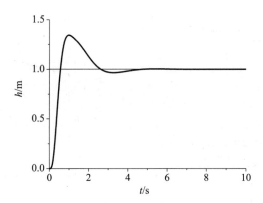

Fig. 7.3-7 Time response of the example line-of-sight guidance loop

Suppose that the system accomplishes the accurate target tracking and launches the missile in the direction of the target when $q = 45°$, and the missile velocity is $V_m = 600$ m/s. When the tracking error of the tracking radar is neglected, the flight trajectories of the target and the missile for this given scenario are shown in Fig. 7.3-8.

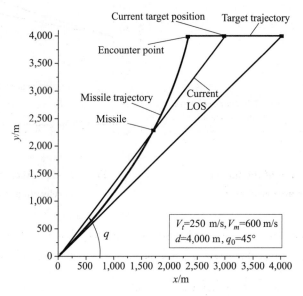

Fig. 7.3-8 Flight paths of the missile trajectory and the target

$q(t), \dot{q}(t)$ and $\ddot{q}(t)$ for this example are respectively shown in Fig. 7.3-9, Fig. 7.3-10 and Fig. 7.3-11. It can be seen from the $\dot{q}(t)$ plot that the line-of-sight angular velocity is increasing when the missile is coming toward to the guidance station, and its value reaches the maximum at the end of the missile-target engagement.

Fig. 7.3-12 shows the corresponding lateral Coriolis acceleration, translational acceleration and total lateral acceleration when the missile flies along the line-of-sight. It is indicated that the required lateral acceleration of the missile is mainly to meet the demand of the Coriolis acceleration corresponding to the line-of-sight angular velocity $\dot{q}$.

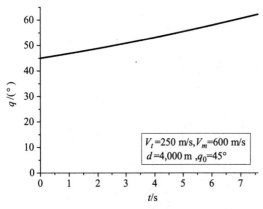

Fig. 7.3-9 Curve of the line-of-sight angle q

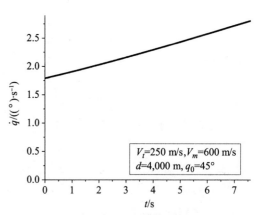

Fig. 7.3-10 Curve of the line-of-sight angular velocity $\dot{q}$

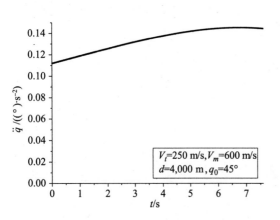

Fig. 7.3-11 Curve of the line-of-sight angular acceleration $\ddot{q}$

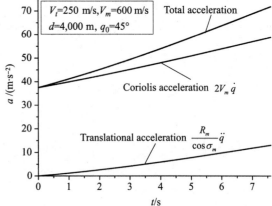

Fig. 7.3-12 Acceleration curve and acceleration components

Fig. 7.3-13 shows the variation of the missile trajectory-beam angle σ_m during this guidance process. It can be observed that this angle reaches its maximum at the end of missile-target engagement.

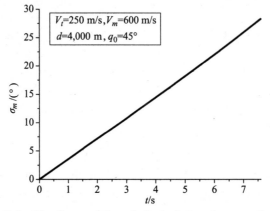

Fig. 7.3-13 Curve of the missile trajectory-beam angle σ_m

In the following, the line-of-sight motion $q(t)$ corresponding to the above scenario is used as the command to the designed missile guidance loop (Fig. 7.3 – 14). Fig. 7.3 – 15 shows the definition of the position error h and the lateral acceleration a. Here, the positive h represents the missile lagging behind the line-of-sight, and the positive a corresponds to positive guidance error h. Fig. 7.3 – 16 shows the $h(t)$ curve of the guidance process and Fig. 7.3 – 17 shows the normal acceleration command $a_c(t)$ and the autopilot acceleration output $a(t)$.

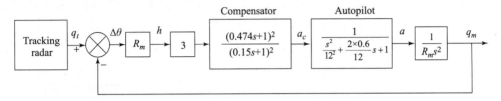

Fig. 7.3 – 14 Block diagram of the line-of-sight guidance system

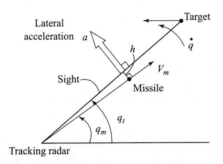

Fig. 7.3 – 15 Definitions of relative variables of the line-of-sight guidance

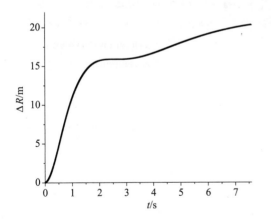

Fig. 7.3 – 16 Curve of the guidance error $h(t)$

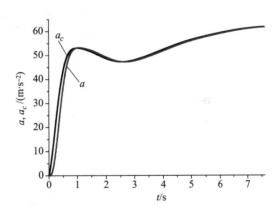

Fig. 7.3 – 17 Curves of the guidance command $a_c(t)$ and the autopilot acceleration output $a(t)$

It is seen from the figures above that the required command $a_c(t)$ is increasing with the increase of the line-of-sight angular velocity $\dot q$ during the guidance, and it reaches the maximum at the missile-target encounter with a value of 6.2 g. Due to the stability constraints of the type II system, the open loop gain value of the line-of-sight guidance loop is low (in this case $K = 3$). For this reason the required guidance acceleration command can only be generated by a large guidance error h. In this case, the miss distance at the encounter can reach 20.4 m.

Since the tracking radar can provide the line-of-sight angular velocity $\dot{q}$ output when tracking the target, the line-of-sight guidance system can adopt a feedforward strategy to improve its guidance accuracy. That is, the tracking radar can use its measured line-of-sight angular velocity $\dot{q}$ and the missile velocity V_m to generate the Coriolis acceleration feedforward command, and superimpose this feedforward command on the error command (Fig. 7.3 – 18). In Fig. 7.3 – 18, the time-varying block in the feedforward channel is used to smooth the initial sudden change in the feedforward signal. In this way, since most of the required acceleration commands for the missile are provided by the feedforward signal, it greatly reduces the contribution of the tracking error Δq in the command a_c, that is to say, the guidance accuracy can be greatly improved. Fig. 7.3 – 19 shows the guidance error curves before and after the feedforward is adopted, and Fig. 7.3 – 20 shows the related acceleration command curve. As seen from these figures, the terminal miss distance with feedforward can be reduced from the original 20.4 m to 0.68 m (Fig. 7.3 – 19), but the demand for terminal acceleration has not been changed (Fig. 7.3 – 20).

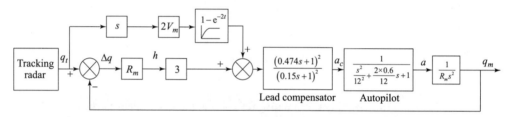

Fig. 7.3 – 18　Line-of-sight guidance ($\dot{q}$ related Coriolis acceleration feedforward compensator is introduced)

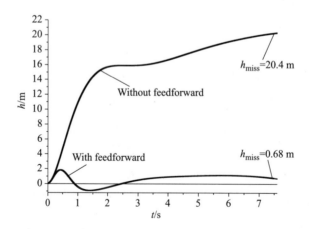

Fig. 7.3 – 19　Position error curves before and after the feedforward is introduced

7 Line of Sight Guidance Methods

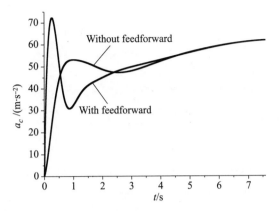

Fig. 7.3 − 20 Curves of the acceleration command before and after the feedforward is introduced

§ 7.4 Lead Angle Guidance Method

In the line of sight guidance, the missile trajectory becomes more and more curved with the increase of time, and its lateral acceleration reaches the maximum at the end of the missile-target engagement. Obviously, if the missile is allowed to fly at a more straight trajectory ahead of the line-of-sight, its required terminal acceleration will inevitably be reduced. Moreover, since the guidance error is proportional to the acceleration command, the miss distance will be reduced simultaneously with the decrease of the required terminal acceleration. With this approach, the missile will no longer fly along the actual LOS, but track a virtual LOS ahead of the real one. Let us denote the virtual LOS as q^* (Fig. 7.4 − 1):

$$q^* = q + \Delta q. \tag{7.4-1}$$

The expression of the lead angle Δq is taken as

$$\Delta q = c(R_t - R_m) = c \cdot \Delta R. \tag{7.4-2}$$

Clearly, $R_t - R_m = \Delta R = 0$ when the missile encounters the target. Therefore, the virtual LOS will coincide with the actual LOS automatically at the end of the missile-target engagement ($\Delta q = 0$), that is, this approach also ensures that the missile hits the target at the end of the engagement.

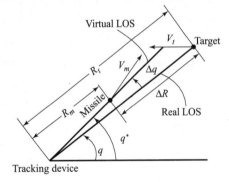

Fig. 7.4 − 1 Lead angle method

In addition, since the missile's required acceleration is mainly related to the LOS angular velocity, a constraint $\dot{q}^* = 0$ as $\Delta R = 0$ can be taken to derive the coefficient expression. In this way, the terminal acceleration required will certainly be reduced.

Since

$$\dot{q}^* = \dot{q} + \dot{c} \cdot \Delta R + c \cdot \Delta \dot{R}, \tag{7.4-3}$$

take $\dot{q}^* = 0$ as $\Delta R = 0$, therefore

$$\dot{q}^* = \dot{q} + c \cdot \Delta \dot{R} = 0. \tag{7.4-4}$$

As c is taken as $c = -\dfrac{\dot{q}}{\Delta \dot{R}} = \dfrac{\dot{q}}{|\Delta \dot{R}|}$, this will make $\dot{q}^* = 0$ when $\Delta R = 0$. The final lead angle expression will be

$$\Delta q = \dfrac{\dot{q}}{|\Delta \dot{R}|} \Delta R. \qquad (7.4-5)$$

Since both $\dot{q}$ and ΔR can be obtained from the ground tracking device, this scheme is not very difficult to realize. However, due to the constraint on the limited beam angle width of the tracking device, a half lead angle approach is more widely used. That is

$$q^* = q + \dfrac{1}{2}\left(\dfrac{\dot{q}}{|\Delta \dot{R}|}\right)\Delta R. \qquad (7.4-6)$$

Differentiating the above expression with respect to t with $\Delta R = 0$ gives the following relation

$$\dot{q}^* = \dfrac{1}{2}\dot{q}. \qquad (7.4-7)$$

In other words, the half lead angle approach can help reduce the required acceleration command by about a half when hitting the target.

Fig. 7.4 – 2 shows the variation of q and q^* with the change of ΔR in the three-point method, the half lead angle method and the lead angle method for the previous example. It can be seen that when $\Delta R = 0$, that is when the missile encounters the target, the value of q is the same for the three methods and all the missiles will hit the target. However, the lead angle in the half lead angle method is small, which reduces the risk of the missile deviating from the beam. Fig. 7.4 – 3 shows the variation of $\dot{q}$ and $\dot{q}^*$ for the three methods. It can be seen that the LOS angular velocity $\dot{q}^*$ is zero when $\Delta R = 0$ in the lead angle method, and the line-of-sight angular velocity of the half lead angle method is half that of the three-point method.

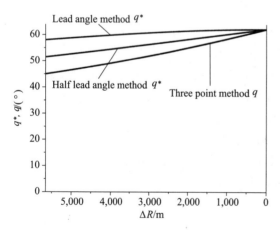

Fig. 7.4 – 2 Variation of q and q^* for the three-point method, half lead angle method and lead angle method

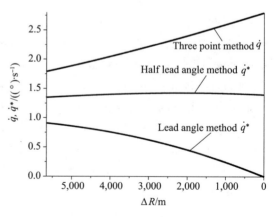

Fig. 7.4 – 3 Variation of $\dot{q}$ and $\dot{q}^*$ for the three-point method, half lead angle method and lead angle method

As mentioned before, the feedforward compensation cannot change the required terminal acceleration value, but it can greatly improve the guidance accuracy. The lead angle method can reduce the terminal acceleration requirement and properly improve the guidance accuracy. If the two methods are used together, we can have the guidance accuracy improved and the terminal acceleration reduced at the same time.

8

Proportional Navigation and Extended Proportional Navigation Guidance Laws

§ 8.1 Proportional Navigation Guidance Law

8.1.1 Proportional Navigation Guidance Law (PN)

From maritime experience throughout history, it is known that if the ship navigator's observation line to another ship is found to be non-rotating, the two ships while continuing to sail will collide eventually, as shown in Fig. 8.1 − 1. In order to avoid such accidents, the navigator will have to change course and keep the observation line rotating while continuing to move forward.

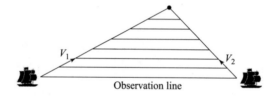

Fig. 8.1 − 1 Sketch of a sea accident

Missile guidance is the opposite problem of the above navigational experience. The guidance law designer only needs to keep the missile-target line (or line of sight, LOS) non-rotating in the inertial space when the missile is flying forward. In this way, the missile will be sure to hit the target at the end of the flight. To implement this approach, people skillfully designed the seeker, which can measure the angular velocity $\dot{q}$ of the missile-target line of sight in the inertial space. When the measured $\dot{q}$ is not zero, the missile changes its flight path direction through the proportional navigation guidance law $\dot{\theta} = N\dot{q}$ to control $\dot{q}$ towards zero so as to hit the target. This is the basic idea used in the proportional navigation guidance.

Fig. 8.1 − 2 shows the schematic of the target and missile trajectories as the proportional navigation guidance law is implemented, in which the proportional guidance constant is taken as $N = 4$ and the ratio of the missile velocity and the target velocity is 2 ($V_m/V_t = 2$).

Many years later, scholars have discovered that proportional navigation law is actually an optimal guidance law that can be rigorously proved. Below we will give the mathematical model of this optimal guidance problem.

8 Proportional Navigation and Extended Proportional Navigation Guidance Laws

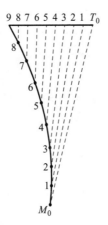

Fig. 8.1-2　Proportional guidance target and missile trajectories ($V_m/V_t = 2$, proportional guidance constant $N = 4$)

Suppose that the missile and the target have already been in the nominal engagement trajectories (Fig. 8.1-3) and V_m and V_t are the missile and the target velocities. At the time t moment, the missile will be at point A and the target will be at point B. We take the origin of the coordinate system to be at point B, and the direction x is in the nominal direction of the missile-target line of sight. Since the xBy coordinate system is moving in the inertial space with a constant velocity, it can be considered as an inertial system, where Newton's Law applies.

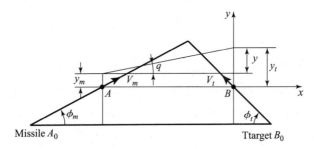

Fig. 8.1-3　Relative relationship between the missile and the target

Suppose that a disturbance deviation occurs in the missile and target flight in relation to the nominal trajectory coordinate system xBy and the disturbances are y_t and y_m, in which y_t is the position disturbance of the target perpendicular to the LOS and y_m is the position disturbance of the missile perpendicular to the LOS. Therefore, the relative position error of the two is $y = y_t - y_m$, the relative velocity error is $V = \dot{y}$, and the relative acceleration error $\ddot{y}$ is $a_t - a_m$.

It should be noted that y and V are respectively the relative position and velocity of the missile and target, while a_t, a_m are the absolute accelerations of the target and the missile perpendicular to the LOS with respect to inertial space, and only their difference is the relative acceleration. Therefore, the dynamic model of this problem can be established as

$$\begin{cases} \dot{y} = V \\ \dot{V} = a_t - a_m \end{cases} \quad (8.1-1)$$

In studying the basic model, a simple situation like the following is often assumed: the target does not maneuver (i.e., $a_t = 0$), and the required missile maneuver acceleration a_m in the inertial space is taken as the guidance command a_c ($a_m = a_c$). Then the following basic dynamic model can be obtained

$$\begin{cases} \dot{y} = V \\ \dot{V} = -a_c \end{cases} \tag{8.1-2}$$

Set the objective function for deriving the optimal guidance law as $J = \left[S \dfrac{y^2(T)}{2} + \dfrac{1}{2} \int_0^T a_c^2(t) \, dt \right]$, then the optimization problem will be given as

$$\min J = \min \left[S \frac{y^2(T)}{2} + \frac{1}{2} \int_0^T a_c^2(t) \, dt \right], \tag{8.1-3}$$

where T is the engagement time, and $\phi(T) = S \dfrac{y^2(T)}{2}$ is the penalty function of the miss distance at the moment of T. When the value of the coefficient S is set as ∞, the miss distance will be forced to become zero at the end of the engagement. The integral term of the objective function represents the integrated square of the missile acceleration over the missile flight time interval of T, which means that the design tries to hit the target with minimum control costs.

After the above optimization problem is solved, the optimal control solution will give the following guidance law

$$a_c = \frac{3}{(T-t)^2} y(t) + \frac{3}{(T-t)} V(t). \tag{8.1-4}$$

Please refer to the end of the section for the detailed mathematical derivation process.

It should be pointed out that the above dynamic equation (8.1-2) is a linear time-invariant model. It is known from the basic control theory that, when the upper limit of the integral of the objective function is taken as ∞, the optimal control solution obtained is the state feedback of the normal optimal control. It is a linear time-invariant control law. However, when the integral time T of the objective function is finite, such an optimal control solution will become a linear time-varying control law, because the guidance law is related to time $(T-t)$.

The above-mentioned closed-loop guidance law expressed in terms of the state variables $y(t)$ and $V(t)$ requires the missile to provide the relative position and velocity of the current missile and target perpendicular to the LOS. The guidance law coefficient is related to time. Such a guidance law is not suitable for engineering application. But it has been found that when the seeker output $\dot{q}$ is used for guidance feedback, the guidance law can be greatly simplified.

In the case of small disturbances, the LOS angle of the missile to target can be expressed as $q = \dfrac{y(t)}{V_r \cdot (T-t)}$, in which V_r is the relative velocity of the missile and target along the LOS, that is $V_r = V_m \cos\phi_m + V_t \cos\phi_t$ (Fig. 8.1-3).

Taking the derivative of q with respect to time, we have

8 Proportional Navigation and Extended Proportional Navigation Guidance Laws

$$\dot{q} = \frac{y(t)}{V_r \cdot (T-t)^2} + \frac{V(t)}{V_r \cdot (T-t)}. \tag{8.1-5}$$

Another form of the proportional navigation law can be acquired by substituting the above expression of $\dot{q}$ into the optimal guidance law, that is

$$a_c = 3V_r \dot{q}. \tag{8.1-6}$$

So, the general form of proportional navigation law that is more commonly used in engineering implementation is

$$a_c = NV_r \dot{q}. \tag{8.1-7}$$

Here N is known as the proportional navigation constant.

The benefits of converting the guidance law feedback variable from the missile-target relative position y and the relative velocity V to the seeker output $t_{go} = T - t$ are:

(a) The number of the feedback variables is changed from y and V, which are hard to measure, to measurable seeker output $t_{go} = T - t$.

(b) The feedback guidance law is changed from a time-varying guidance law to a time-invariant guidance law, that is to say, to implement the proportional navigation guidance law, the missile does not need to know $t_{go} = (T-t)$. This greatly simplifies the engineering application of the proportional navigation guidance law.

Since the above acceleration command a_c of the proportional navigation guidance law refers to the acceleration command perpendicular to the LOS, when there is an angle of ϕ_m between the missile velocity direction and the LOS (Fig. 8.1-4), the acceleration autopilot command should be modified as the following

$$a_c = \frac{NV_c}{\cos\phi_m} \dot{q}. \tag{8.1-8}$$

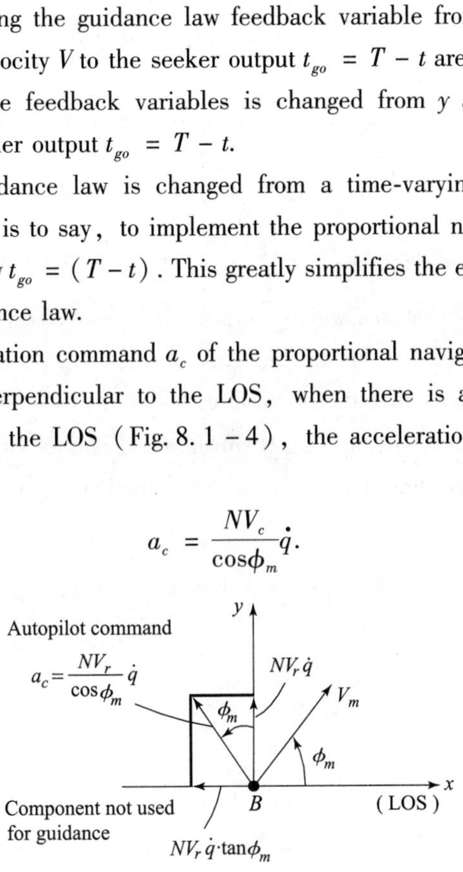

Fig. 8.1-4 Acceleration command modification when a seeker gimbal angle ϕ_m exists

When the guidance law is implemented, the angle ϕ_m can be approximated by the seeker gimbal angle.

Proof of the Proportional Navigation Law with the Help of Optimal Control Theory

Under the assumption of small disturbances, the simplified model of the missile attacking a constant velocity flight target can be represented by Fig. 8.1-5. Its corresponding state equation is

$$\begin{cases} \dot{y} = V \\ \dot{V} = -a_c \end{cases} \tag{8.1-9}$$

In the above equation, y is the relative position of the missile-target perpendicular to the LOS (line

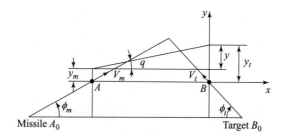

Fig. 8.1-5 Simplified model of the missile attacking a target

of sight), V is the relative velocity of the missile-target perpendicular to the LOS and a_c is the absolute acceleration command of the missile-target perpendicular to the LOS. We are given that $a_t = 0$, $a_m = a_c$.

Taking the objective function for deriving the optimal guidance law as

$$J = S \frac{y(T)^2}{2} + \frac{1}{2} \int_0^T a_c^2(t) \, dt, \qquad (8.1-10)$$

where T is the engagement time and $\phi(T) = S \frac{y^2(T)}{2}$ is the penalty function of the miss distance at the time moment T. The miss distance will be zero when $S \to \infty$. The integral term of the objective function requires the minimization of the integral of the missile acceleration square over the engagement time interval. Obviously, the purpose of this optimal model is to seek an optimal control law $a_c(t)$ so that the missile can hit the target with minimal control costs.

According to the optimal control theory, the Hamiltonian function of this problem should be given first, that is

$$H = \frac{1}{2} a_c^2 + \lambda_y V + \lambda_V(-a_c), \qquad (8.1-11)$$

and its adjoint equations

$$\begin{aligned} \dot{\lambda}_y &= -\frac{\partial H}{\partial y} = 0, \\ \dot{\lambda}_V &= -\frac{\partial H}{\partial V} = -\lambda_y. \end{aligned} \qquad (8.1-12)$$

The optimal control a_c can be given by the stationarity of this problem

$$\frac{\partial H}{\partial a_c} = a_c - \lambda_V = 0, \qquad (8.1-13)$$

$$a_c = \lambda_V. \qquad (8.1-14)$$

It is known that the boundary value of the state equation is given as the initial value $y(0)$, $V(0)$ of the state. The boundary value of the adjoint equation is given by the final values $\lambda_y(T)$ and $\lambda_V(T)$ of λ_y and λ_V, and it can be derived from the penalty function $\phi(T)$.

$$\begin{aligned} \lambda_y(T) &= \frac{\partial \phi(T)}{\partial y(T)} = S \cdot y(T), \\ \lambda_V(T) &= \frac{\partial \phi(T)}{\partial V(T)} \equiv 0. \end{aligned} \qquad (8.1-15)$$

The differential equations of the boundary value problem can be acquired as follows by substituting the optimal control expression $a_c = \lambda_V$ into the state and adjoint equations.

$$\begin{cases} \dot{y} = V \\ \dot{V} = -\lambda_V \\ \dot{\lambda}_y = 0 \\ \dot{\lambda}_V = -\lambda_y \end{cases} \quad (8.1-16)$$

The boundary conditions for solving this problem are:
The state initial values

$$y(0), V(0),$$

and the adjoint final values

$$\lambda_y(T) = S \cdot y(T),$$
$$\lambda_V(T) = 0. \quad (8.1-17)$$

Generally, there are no analytical solutions for dual boundary value differential equations, but there are analytical solutions expressed with $y(0)$, $V(0)$, $\lambda_y(T)$ and t for this problem, and the open loop optimal control law for this problem can be obtained by substituting the expression of λ_V into a_c, that is

$$a_c(t) = (T-t)\frac{S}{1+\frac{S}{3}T^3}[y(0) + T \cdot V(0)]. \quad (8.1-18)$$

Suppose $S \to \infty$ and the open loop guidance law with a zero miss distance is

$$a_c(t) = \frac{3(T-t)}{T^3}[y(0) + T \cdot V(0)]. \quad (8.1-19)$$

Consider the current time t as the initial time, then the engagement time T becomes $(T-t)$, and the initial state value will be $y(t)$ and $V(t)$. The time variable after the initial time t can be expressed with τ (Fig. 8.1-6).

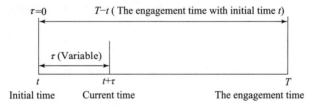

Fig. 8.1-6 Time variable transformation diagram

Then the open loop guidance law with the initial time (the time t) and the time variable τ is

$$a_c(t+\tau) = \frac{3(T-t-\tau)}{(T-t)^3}[y(t) + (T-t) \cdot V(t)]. \quad (8.1-20)$$

Let $\tau=0$ for each moment t, the desired proportional navigation law is obtained as follows

$$a_c(t) = \frac{3}{(T-t)^2}y(t) + \frac{3}{(T-t)}V(t), \quad (8.1-21)$$

where the missile-target LOS angle q is small, that is

$$q \approx \tan q = \frac{y(t)}{(T-t) \cdot V_r}. \tag{8.1-22}$$

Take q by the derivative of time,

$$\dot{q} = \frac{y(t)}{(T-t)^2 \cdot V_r} + \frac{V(t)}{(T-t) \cdot V_r}. \tag{8.1-23}$$

The proportional navigation guidance law with feedback $\dot{q}$ can be obtained as follows

$$a_c(t) = 3V_r \cdot \dot{q}. \tag{8.1-24}$$

8.1.2 PN Analysis without Guidance System Lag

1) Effect of the initial heading error on the proportional navigation guidance

The optimal guidance law derived from the above model can be used to hit the target with minimal control cost under any initial state $(y(0), V(0))$ disturbance.

When there is an initial missile heading error ε, the relative velocity $V(0)$ disturbance (Fig. 8.1-7) that is perpendicular to the LOS will be

$$V(0) = V_m \varepsilon \cos \phi_m. \tag{8.1-25}$$

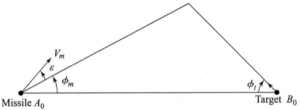

Fig. 8.1-7 Proportional guidance model under the $V(0)$ disturbance

Under the disturbance $V(0)$, the block diagram of the proportional navigation guidance closed-loop control is shown in Fig. 8.1-8. Here again, it is emphasized that y_m, y_t are the absolute displacements of the missile and the target perpendicular to the LOS, y is the relative displacement between the missile and the target in the same direction, and V_m is the flight velocity of the missile.

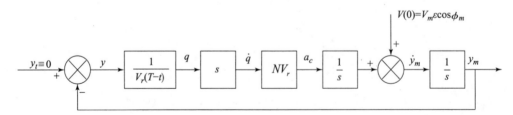

Fig. 8.1-8 Block diagram of the proportional navigation guidance under the $V(0)$ disturbance

Fig. 8.1-8 can be simplified to an equivalent block diagram as shown in Fig. 8.1-9. The transfer function from $V_m \varepsilon \cos \phi_m$ to y_m is

$$\frac{y_m}{V_m \varepsilon \cos \phi_m} = \frac{\frac{1}{s}}{1 + \left(\frac{1}{s}\right)\left(\frac{N}{T-t}\right)} = \frac{1}{s + \frac{N}{T-t}}, \tag{8.1-26}$$

8 Proportional Navigation and Extended Proportional Navigation Guidance Laws

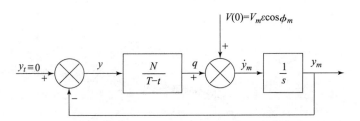

Fig. 8.1-9 A simplified block diagram of the proportional navigation guidance under the $V(0)$ disturbance

or it can be more precisely expressed as a linear time-varying differential equation

$$\dot{y}_m + \frac{N}{T-t} y_m = V_m \varepsilon \cos\phi_m. \quad (8.1-27)$$

The solution of the equation is

$$y_m = V_m \varepsilon \cos\phi_m \frac{T\left(1-\frac{t}{T}\right)}{N-1}\left[1-\left(1-\frac{t}{T}\right)^{N-1}\right], \quad (8.1-28)$$

and the non-dimensional missile displacement will be

$$\frac{y_m}{V_m \varepsilon T \cos\phi_m} = \frac{1-\frac{t}{T}}{N-1}\left[1-\left(1-\frac{t}{T}\right)^{N-1}\right]. \quad (8.1-29)$$

Obviously, for a simplified lag-free proportional navigation guidance model, there is $y = 0$ when $t = T$, i.e., the missile can certainly hit the target.

The dimensionless model of $\dot{y}_m$ is as follows by taking the derivative of y_m with respect to time t

$$\frac{\dot{y}_m}{V_m \varepsilon \cos\phi_m} = -\frac{1}{N-1}\left[1 - N\left(1-\frac{t}{T}\right)^{N-1}\right]. \quad (8.1-30)$$

Then taking the above equation by the derivative of time t, we have

$$-\frac{\ddot{y}_m T}{V_m \varepsilon \cos\phi_m} = N\left(1-\frac{t}{T}\right)^{N-2}. \quad (8.1-31)$$

Since $\ddot{y}_m = a_c \cos\phi_m$, the missile acceleration can be non-dimensionalized as

$$-\frac{a_c T}{V_m \varepsilon} = N\left(1-\frac{t}{T}\right)^{N-2}. \quad (8.1-32)$$

Obviously, for the given N and t/T, there are $y \propto V_m \varepsilon T$, $\dot{y}_m \propto V_m \varepsilon$ and $a_c \propto \frac{V_m \varepsilon}{T}$.

Fig. 8.1-10, Fig. 8.1-11 and Fig. 8.1-12 show the non-dimensional relative position, velocity and absolute missile acceleration time curves for proportional navigation guidance under the $V(0)$ disturbance with the different proportional navigation constant N.

It is indicated from the missile acceleration curve in Fig. 8.1-12 that the required missile acceleration diverges and cannot be used when $N < 2$; it is a constant when $N = 2$; the missile acceleration curve is a straight line when $N = 3$. As $N > 3$, the initial required missile acceleration will increase with an increase of N, but the required missile acceleration will decrease later. In order to ensure that the missile acceleration at the terminal phase of guidance is not saturated, the

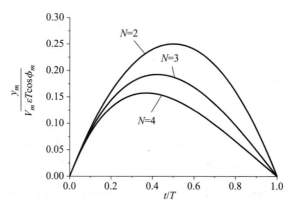

Fig. 8.1−10 Non-dimensional displacement of the proportional navigation guidance system when $N = 2, 3, 4$

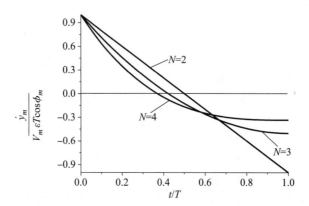

Fig. 8.1−11 Non-dimensional velocity of the proportional navigation guidance system when $N = 2, 3, 4$

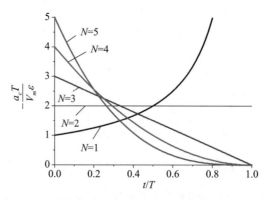

Fig. 8.1−12 Non-dimensional missile acceleration of the proportional navigation guidance system when $N = 1, 2, 3, 4, 5$

required acceleration at this phase should be as small as possible. In addition, it is known from the following analysis that when the target is maneuvering, the proportional navigation constant N is required to be not less than 3, so in practical proportional navigation guidance implementation, N is often taken as 4. Therefore in real applications, even if the variation of N could reach $\pm 25\%$, the N value used could still be in the range 3 to 5, which will not greatly affect the guidance performance.

From the displacement curve of Fig. 8.1 – 10, the trajectory is a circular arc when $N = 2$. Since the required acceleration is large at the early stage of guidance with the increase of N and then it decreases afterwards, the trajectory turns to be closer to a straight line before interception of the target.

2) Effects of constant target maneuver on proportional navigation guidance performance

The performance of the proportional guidance control when the target has a constant maneuver acceleration a_t perpendicular to the LOS will be discussed in the following. Since the above studied model of the proportional navigation law is supposed to be with no target maneuver, it is optimal only when the disturbance is the system initial state $y(0)$ or $V(0)$. So, it should be pointed out that the proportional navigation guidance is not an optimal guidance law for attacking maneuvering targets. However, since it is a closed-loop guidance law, it can still be used to attack maneuvering targets, but we must study its effectiveness for this application with rigorous theoretical analysis.

When $y(0)$ and $V(0)$ are zero and the target has an absolute maneuver acceleration a_t perpendicular to the LOS, the block diagram of the proportional navigation guidance control is as shown in Fig. 8.1 – 13.

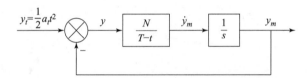

Fig. 8.1 – 13 Block diagram of the proportional navigation guidance under the disturbance of constant target maneuver $y_t = \frac{1}{2}a_t t^2$

Its equivalent control block diagram is as shown in Fig. 8.1 – 14.

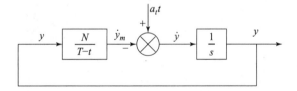

Fig. 8.1 – 14 Simplified block diagram of the proportional navigation guidance under the disturbance of the constant target maneuver $y_t = \frac{1}{2}a_t t^2$

Then, the corresponding linear time-varying differential equation will be

$$\dot{y} + \left(\frac{N}{T-t}\right) y = a_t t. \qquad (8.1-33)$$

With

$$y_m = \frac{1}{2}a_t t^2 - y, \qquad (8.1-34)$$

by solving the above differentiation equation and substituting y with the expression of y_m, we arrive at

$$\frac{y_m}{a_t T^2} = \frac{1}{2}\left(\frac{t}{T}\right)^2 - \frac{1 - \frac{t}{T}}{(N-1)(N-2)}\left[(N-1)\frac{t}{T} - 1 + \left(1 - \frac{t}{T}\right)^{N-1}\right]. \quad (8.1-35)$$

It is clear that as $t = T$, $y_m = \frac{1}{2}a_t T^2$ and $y = 0$.

The solution of its non-dimensional acceleration command will be

$$\frac{a_c}{a_t} = \frac{N}{N-2}\left[1 - \left(1 - \frac{t}{T}\right)^{N-2}\right]. \quad (8.1-36)$$

Fig. 8.1 – 15 and Fig. 8.1 – 16 are the non-dimensional trajectory curves and acceleration curves when the missile attacks a constant acceleration a_t maneuvering target.

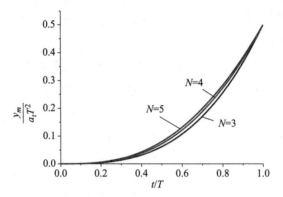

Fig. 8.1 – 15 Non-dimensional displacement curves of the missile when a constantly maneuvering target is attacked with $N = 3, 4, 5$

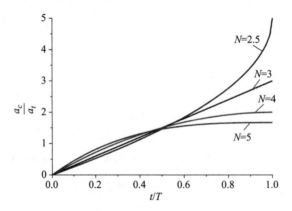

Fig. 8.1 – 16 Non-dimensional acceleration of the missile when a constantly maneuvering target is attacked with $N = 2.5, 3, 4, 5$

As shown in Fig. 8.1 – 16, it is known from the non-dimensional acceleration curve of the missile when a constantly maneuvering target is attacked that the required acceleration will diverge when $N < 3$. Therefore, when a maneuvering target is attacked, the proportional navigation constant N should not be less than 3. The missile's initial acceleration will increase with the increase of N when $N > 3$, but the terminal acceleration will decrease, e. g., $\dfrac{a_c}{a_t} = 3$ at the end of guidance when $N = 3$; $\dfrac{a_c}{a_t} = 2$ at the end of guidance when $N = 4$.

In order to ensure that the missile has sufficient acceleration to intercept a maneuvering target, a US missile designer suggested that the missile available acceleration be designed as 5 times the target maneuver acceleration, that is $a_c = 5a_t$. While Russian references suggested it to be $a_c = 3a_t + 10g$.

It is known from the equation (8.1 – 19) that $y_m = \dfrac{1}{2}aT^2$ when $t = T$. That is to say, although it is not an optimal guidance law for the proportional navigation guidance to be used for attacking a maneuvering target, as long as the missile has enough acceleration capability, the task of hitting a maneuvering target can still be accomplished.

8.1.3 PN Characteristics Including the Missile Guidance Dynamics

In the previous section, we have assumed that the entire signal flow from the measurement of the LOS angular velocity $\dot{q}$ to the generation of the missile acceleration a_m is instantaneous without any dynamics involved. However, in real engineering implementation, the guidance loop involves many hardware elements (e.g., the seeker, the guidance filter and the autopilot) and they all have dynamics. To simplify the system analysis, it is often supposed that the seeker and the guidance filter both have a first order dynamics, the autopilot has a second order dynamics and the four first order dynamic components all have the same time constant $T_g/4$ (Fig. 8.1 – 17). So the whole guidance dynamics will have the form of $\left(\dfrac{T_g}{4}s + 1\right)^{-4}$.

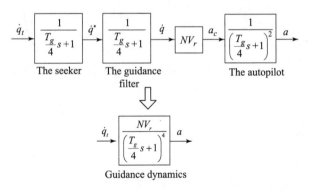

Fig. 8.1 – 17 Simplified model of the guidance dynamics

It can be seen from the Bode diagram (Fig. 8.1 – 18, here taking $T_g = 0.5\text{s}$) of the guidance dynamics transfer function $\dfrac{1}{\left(\dfrac{T_g}{4}s + 1\right)^4} = \dfrac{1}{\dfrac{T_g^4}{256}s^4 + \dfrac{T_g^3}{16}s^3 + \dfrac{3T_g^2}{8}s^2 + T_g s + 1}$ that its low frequency dynamics characteristics are quite similar to a first order system $\dfrac{1}{T_g s + 1}$, in which T_g is often known as the guidance time constant. However, it should be noted that its phase shift and gain roll-off rate at the higher frequency are totally different from a lower order system and their effects on the guidance loop stability, and guidance miss distance must be carefully considered.

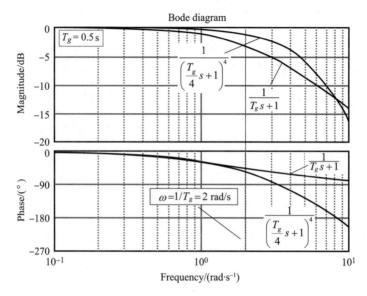

Fig. 8.1–18 Bode diagram of the guidance dynamics $\left(\dfrac{T_g}{4}s+1\right)^{-4}$ and first order dynamics $(T_g s+1)^{-1}$

The block diagram of the proportional navigation guidance loop with guidance dynamics considered is shown in Fig. 8.1–19.

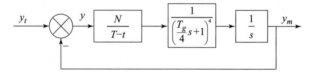

Fig. 8.1–19 Proportional guidance loop with guidance dynamics

The above block diagram can be made non-dimensional with respect to the guidance time constant T_g. This will greatly simplify its analysis. For this we take $\bar{t} = \dfrac{t}{T_g}$ and the corresponding non-dimensional frequency $\bar{\omega} = T_g \omega$ and $\bar{s} = T_g s$. Substituting the above non-dimensional time domain and frequency domain expressions into Fig. 8.1–19, the proportional navigation guidance loop can be simplified as Fig. 8.1–20 and Fig. 8.1–21, in which $\bar{t}_{go} = \dfrac{T-t}{T_g} = \dfrac{t_{go}}{T_g}$.

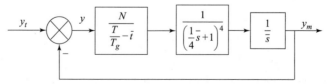

Fig. 8.1–20 Non-dimensional proportional navigation guidance model with guidance dynamics (1)

8 Proportional Navigation and Extended Proportional Navigation Guidance Laws

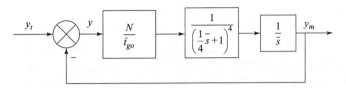

Fig. 8.1 – 21 Non-dimensional proportional navigation guidance model with guidance dynamics (2)

It can be seen from Fig. 8.1 – 21 that with the missile approach to the target, the $\bar{t}_{go}$ will decrease and the guidance loop gain $K = \dfrac{N}{\bar{t}_{go}}$ could finally go to ∞.

Fig. 8.1 – 22 shows the changes of the guidance loop gain $K = \dfrac{N}{\bar{t}_{go}}$ with the decrease of $\bar{t}_{go}$ when $N = 3, 4, 5$.

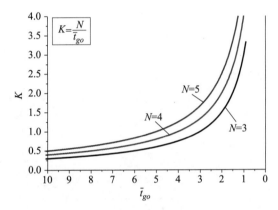

Fig. 8.1 – 22 Variation of the guidance open loop gain K with the decrease of $\bar{t}_{go}$

Take $N = 4$ and draw the open loop Bode diagram (Fig. 8.1 – 23) of the proportional navigation guidance with different $\bar{t}_{go}$. It can be seen that the gain margin and phase margin of the proportional navigation guidance are decreasing when the missile approaches the target ($\bar{t}_{go}$ decreases). Fig. 8.1 – 24 shows the variation of the gain margin and phase margin with the decrease of $a_m = a_c$. From Fig. 8.1 – 23 and Fig. 8.1 – 24, it can be seen that as $N = 4$ and $\bar{t}_{go} = \dfrac{t_{go}}{T_g} = 1.76$, the phase shift of the loop has reached $-180°$, and the actual control direction has been reversed in comparison with the required feedback control direction. This means that the guidance loop tends to be unstable after $\bar{t}_{go} < 1.76$.

It should be pointed out that since the proportional guidance is a linear time-varying system, the loop stability analysis described above with a linear time-invariant model is not completely correct in theory, but qualitatively, the conclusion that a problem of stability occurs in the guidance loop at the later stage of proportional navigation guidance is still correct.

The effects of different disturbance sources on the performance of proportional navigation guidance will be studied in the following.

1) Analysis of the effect of the initial heading error ε on the proportional navigation guidance

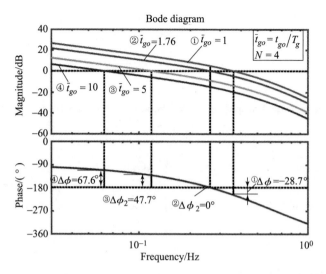

Fig. 8.1-23 Open loop diagram of the guidance system with different $\bar{t}_{go}$ parameters

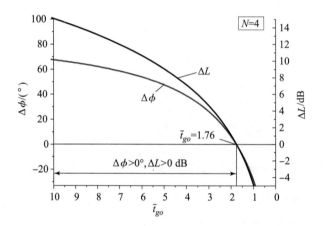

Fig. 8.1-24 Gain margin and phase margin of the guidance system with the variation of $\bar{t}_{go}$

Fig. 8.1-25 shows the block diagram of the non-dimensional proportional navigation guidance control with guidance dynamics under the initial heading error disturbance ε.

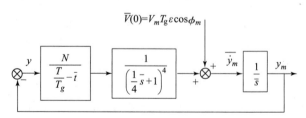

Fig. 8.1-25 Block diagram of the non-dimensional proportional navigation guidance control with fourth order guidance dynamics under the initial heading error disturbance ε

The variation curve (Fig. 8.1-26) of the non-dimensional miss distance $\dfrac{y_{miss}}{V_m T_g \varepsilon \cos\phi_m}$ with the non-dimensional engagement time T/T_g under the initial heading error disturbance ε can be obtained by using the adjoint method (Section 8.1.4). Note that y_{miss} is defined as the guidance terminal miss distance.

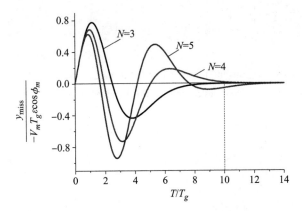

Fig. 8.1－26 Miss distance curves for the guidance system with guidance dynamics considered for different engagement time T/T_g and N

It can be observed from Fig. 8.1－26 that in the presence of the fourth order guidance dynamics, the duration of the engagement time has a great influence on the miss distance. To ensure the convergence of the miss distance, the engagement time should not be less than 10 times the guidance time constant T_g.

Next, the proportional navigation guidance acceleration under the initial heading error disturbance ε will be discussed. Fig. 8.1－27 shows the block diagram for this investigation. Here, the missile acceleration is taken as the output.

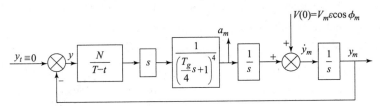

Fig. 8.1－27 Block diagram of the guidance system with fourth order lag dynamics

When studying the missile acceleration, it is more reasonable to take the engagement time T to non-dimensionalize the loop. For this reason, take $\bar{\bar{t}} = \dfrac{t}{T}, \bar{\bar{\omega}} = T\omega, \bar{\bar{s}} = Ts$, that is, $t = T\bar{\bar{t}}$ and $s = \dfrac{1}{T}\bar{\bar{s}}$. The block diagram after non-dimensionalization is shown in Fig. 8.1－28.

At this time, the expression of the non-dimensional acceleration is

$$\bar{\bar{a}}_m = \dfrac{T}{V_m \varepsilon \cos\phi_m} a_m. \tag{8.1-37}$$

Fig. 8.1－29 shows the variation curves of the non-dimensional acceleration with the non-dimensional time t/T when different engagement time T/T_g is taken as the parameter and $N = 4$.

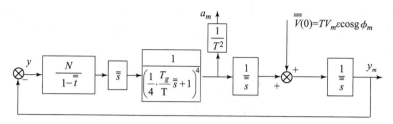

Fig. 8.1 – 28 Block diagram of the fourth order dynamics non-dimensional proportional navigation guidance under the initial heading error disturbance ε

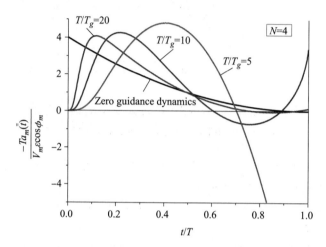

Fig. 8.1 – 29 Variation curves of the non-dimensional acceleration with the non-dimensional time for different engagement time T/T_g when $N = 4$

From Fig. 8.1 – 29 we know, the acceleration curve is closer to the zero guidance dynamics system as the engagement time T/T_g is longer. When the engagement time is $T < 10T_g$, the terminal guidance acceleration will be much larger than the zero guidance dynamics system, which will certainly lead to a larger miss distance.

2) The effect of the target maneuver a_t on the proportional guidance performance

The miss distance caused by the target maneuver a_t will be analyzed in the following. Still taking the guidance time constant T_g to make the non-dimensionalization of the time t, $\bar{t} = \dfrac{t}{T_g}$ ($\bar{\omega} = T_g \omega$, $\bar{s} = T_g s$), and the resulting non-dimensional block diagram is shown in Fig. 8.1 – 30.

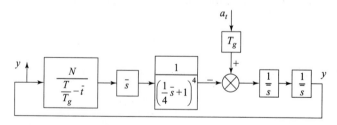

Fig. 8.1 – 30 Guidance dynamics non-dimensional system under the action of the target maneuver

The non-dimensional miss distance is defined as

$$\bar{y}_{miss} = \frac{y_{miss}}{a_t T_g^2}. \qquad (8.1-38)$$

Fig. 8.1-31 shows the variation curves of the non-dimensional miss distance over the non-dimensional engagement time $\bar{T} = T/T_g$. It can be seen from the figure that when a maneuvering target is involved, the proportional navigation guidance constant N should not be lower than 3 to ensure the convergence of miss distance. At the same time, the missile's engagement time T should, again, not be less than 10 times the missile's guidance time constant T_g.

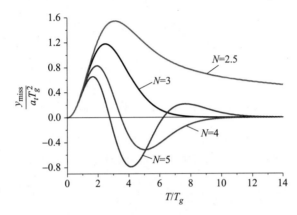

Fig. 8.1-31 Miss distance curves corresponding to different N with the guidance dynamics taken into account (under the action of the target maneuver)

Similarly, in the analysis of the missile acceleration, we take the missile's engagement time T to make the non-dimensionalization of t. Take $\bar{\bar{t}} = \frac{t}{T}, \bar{\bar{\omega}} = T\omega, \bar{\bar{s}} = Ts$, and the guidance block diagram after the non-dimensionalization is shown in Fig. 8.1-32.

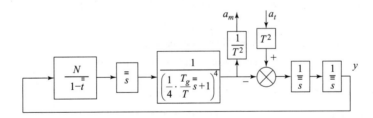

Fig. 8.1-32 Guidance block diagram after the non-dimensionalization with respect to the engagement time T

From the above figure, the expression of the non-dimensional acceleration can be given as follows

$$\bar{\bar{a}}_m = \frac{a_m}{a_t}. \qquad (8.1-39)$$

Fig. 8.1-33 shows the missile acceleration over time t/T with different engagement time T/T_g when $N = 4$ where one can observe that the missile acceleration gradually deviates from the ideal proportional navigation guidance law when the engagement time T is less than 10 times the guidance time constant T_g, which also leads to a larger miss distance. Therefore, when attacking a maneuvering target in the presence of guidance dynamics, the engagement time T should not be less than 10 times the guidance time constant T_g in order to obtain an acceptably small miss distance.

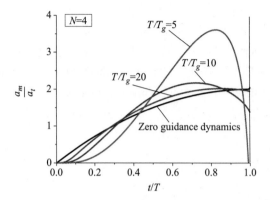

Fig. 8.1-33 Non-dimensional acceleration curves caused by the target maneuver

Fig. 8.1-34 shows the variation of the acceleration corresponding to the different proportional navigation constant N over the non-dimensional time t/T when $T = 10T_g$.

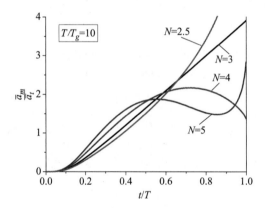

Fig. 8.1-34 Variation of the acceleration corresponding to the different proportional guidance coefficient N under the action of the target maneuver

3) Analysis of the influence of the radar seeker receiver thermal noise

Since the thermal noise of the seeker receiver has a wide bandwidth, it can be regarded as a white noise input in comparison with the guidance loop bandwidth. Suppose that the power spectrum density of the white noise signal u_{RN} is $S(\omega) = \phi_{RN}$ (rad^2/Hz) $= \phi_{RN}$ (rad^{2}s) when the missile-target distance is R_0. Since the echo signal of the target increases with the decrease of the missile-target distance, the seeker has been designed with an automatic gain adjustment function to maintain a constant signal level for the subsequent circuit when the missile-target distance decreases and the echo signal increases. For this, the automatic gain adjustment coefficient of the seeker is R^2/R_0^2, in which R is the current missile-target distance. Therefore, the actual effective power spectrum of the thermal noise u_{RN} is $S(\omega, R) = \dfrac{R^2}{R_0^2}\phi_{RN} = \left(\dfrac{(T-t)V_r}{R_0}\right)^2 \phi_{RN}$ (rad^{2}s). Accordingly, the guidance loop diagram under the influence of thermal noise is as shown in Fig. 8. 1 – 35.

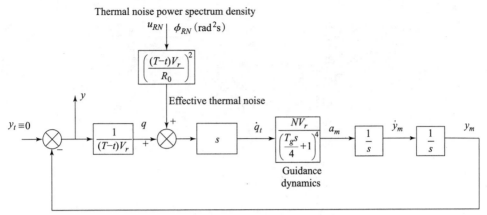

Fig. 8. 1 – 35 Block diagram of the guidance system under the influence of the thermal noise (including the fourth order guidance dynamics)

Take the above block diagram to make the non-dimensionalization with respect to the guidance time constant T_g. The result is shown in Fig. 8. 1 – 36, in which $\bar{u}_{RN}$ is the non-dimensional thermal noise signal, and the non-dimensional power spectrum of its corresponding disturbance white noise power spectrum is $\bar{\phi}_{RN}$ $\left(\bar{\phi}_{RN} = \dfrac{\phi_{RN}}{T_g}\right)$, and its unit is rad^2.

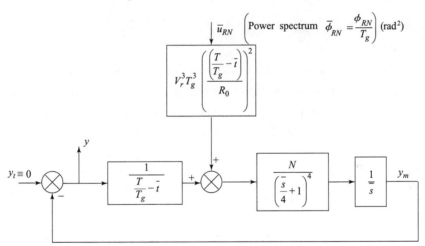

Fig. 8. 1 – 36 Non-dimensional block diagram of the guidance system under the influence of the thermal noise

Next, the adjoint function method can be used to solve this problem (see Section 8. 1. 4 of this book for the specific solution procedure). The expression of the non-dimensional miss distance standard deviation $\bar{\sigma}_{\text{miss}(RN)}$ caused by the thermal noise is as follows

$$\bar{\sigma}_{\text{miss}(RN)} = \dfrac{\sigma_{\text{miss}(RN)} R_0^2}{V_r^3 T_g^3 \sqrt{\phi_{RN}}}, \qquad (8.1-40)$$

where $\sigma_{\text{miss}(RN)}$ is the miss distance standard deviation caused by the thermal noise, and its unit is in meters. The non-dimensional miss distance of the system under the influence of the thermal noise is shown in Fig. 8.1 –37.

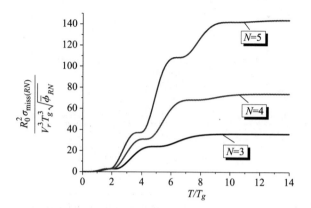

Fig. 8.1 –37 Non-dimensional miss distance of the system under the influence of the thermal noise

It can be seen that the miss distance caused by the thermal noise does not converge to "0" no matter how long the engagement time T is. However, in the case where the usual engagement time $T/T_g > 10$ is adopted, the miss distance caused by the thermal noise is basically stable and flat. However, when the proportional guidance coefficient N increases, the bandwidth of the guidance loop will also increase, which will certainly increase the miss distance response to the high-frequency thermal noise input.

4) The influence of target glint

As mentioned in Chapter 4, the target glint noise is a low frequency disturbance noise. Generally, it can be simulated with a white noise with a power spectrum density ϕ_{GL} (m²s) passing a first order low pass filter with the time constant T_{gl}. It should be noted that the output colored noise signal is the target position glint, which can be converted into the target angle glint and added to the guidance system block diagram (Fig. 8.1 –38).

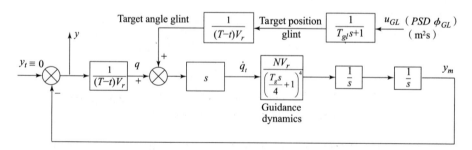

Fig. 8.1 –38 Block diagram of the guidance system under the influence of the target glint

Taking non-dimensionalization with respect to the guidance time constant T_g the system can be simplified as Fig. 8.1 – 39, in which $\bar{u}_{GL}$ is the non-dimensional white noise signal, and its corresponding non-dimensional power spectrum is $\bar{\phi}_{GL}$ $\left(\bar{\phi}_{GL} = \dfrac{\varphi_{GL}}{T_g}\right)$, with a unit m².

8 Proportional Navigation and Extended Proportional Navigation Guidance Laws

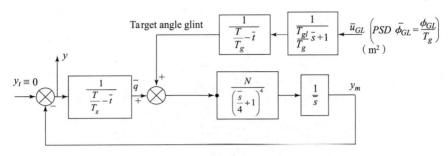

Fig. 8.1-39 Block diagram of the non-dimensional guidance system under the influence of the target glint

Using the adjoint method (see Section 8.1.4 for details), the expression of the non-dimensional miss distance standard deviation caused by the target glint can be obtained as

$$\bar{\sigma}_{\text{miss}(GL)} = \frac{\sigma_{\text{miss}(GL)}}{\sqrt{\bar{\phi}_{GL}}}. \tag{8.1-41}$$

According to the theory of random process, when the white noise with the power spectrum ϕ_{GL} (m²s) passes through a first order low pass filter with a time constant T_{gl} (Fig. 8.1-40), the output colored noise variance will be

$$\sigma_{GL}^2 = \frac{\phi_{GL}}{2T_{gl}}(\text{m}^2). \tag{8.1-42}$$

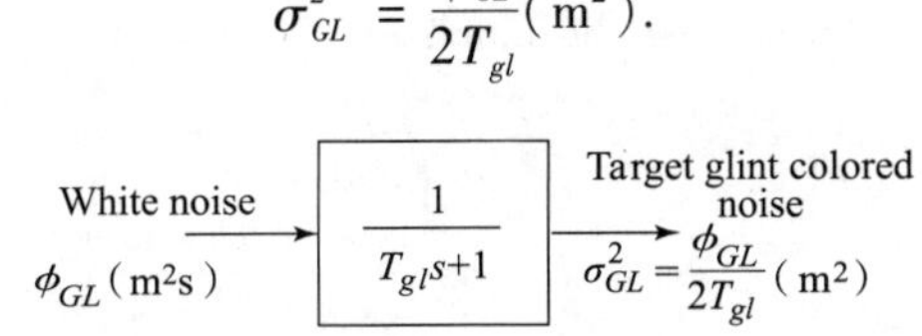

Fig. 8.1-40 Model of the colored noise

Therefore, the relationship between the power spectrum of the target glint power spectrum density and its standard deviation is

$$\sqrt{\bar{\phi}_{GL}} = \sqrt{\frac{\phi_{GL}}{T_g}} = \sqrt{\frac{2T_{gl}}{T_g}}\sigma_{GL}. \tag{8.1-43}$$

Taking the above equation into the expression $\bar{\sigma}_{\text{miss}(GL)}$ obtained via the adjoint method, we obtain

$$\bar{\sigma}_{\text{miss}(GL)} = \frac{\sigma_{\text{miss}(GL)}}{\sqrt{2\frac{T_{gl}}{T_g}}\sigma_{GL}}. \tag{8.1-44}$$

Thus, the ratio $\sigma_{\text{miss}(GL)}/\sigma_{GL}$ of the root-mean-squares (RMS) of the miss distance to the glint noise is

$$\frac{\sigma_{\text{miss}(GL)}}{\sigma_{GL}} = \sqrt{2\frac{T_{gl}}{T_g}}\bar{\sigma}_{\text{miss}(GL)}. \tag{8.1-45}$$

Fig. 8.1-41 shows the variation of the ratio $\sigma_{\text{miss}(GL)}/\sigma_{GL}$ of the miss distance to the glint noise over the non-dimensional engagement time. It can be seen from the figure that in the case of the common engagement time $T/T_g > 10$, the target glint miss distance has converged to a constant error.

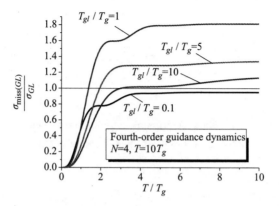

Fig. 8.1-41 Variation of the ratio $\sigma_{miss(GL)}/\sigma_{GL}$ of the miss distance to the glint noise over the non-dimensional engagement time

Fig. 8.1-42 shows the effect of T_{gl}/T_g on $\sigma_{miss(GL)}/\sigma_{GL}$. From this figure, it can be known that when the glint frequency is lower in comparison with the guidance dynamics frequency (i.e., T_{gl}/T_g is high), the missile can completely track the glint, and thus $\sigma_{miss(GL)}/\sigma_{GL} \approx 1$. When the glint frequency is high, the guidance loop fails to respond to the target glint (T_{gl}/T_g is very small), and the value of $\sigma_{miss(GL)}/\sigma_{GL}$ can be much less than 1.

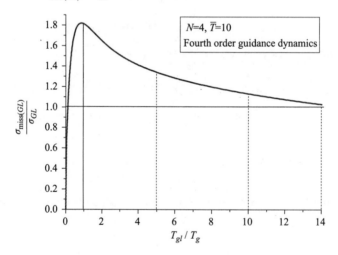

Fig. 8.1-42 Variation of the ratio $\dfrac{\sigma_{miss(GL)}}{\sigma_{GL}}$ of the miss distance with respect to T_{gl}/T_g when $\overline{T} = 10$

It also should be noted that when the glint frequency is close to the guidance dynamics frequency ($T_{gl}/T_g \approx 1$), a resonance will occur, so that the value of $\sigma_{miss(GL)}/\sigma_{GL}$ could be greater than 1, that is to say, the miss distance variance could be greater than the target glint variance in certain cases.

8.1.4 Adjoint Method

The adjoint method is an effective tool for analyzing the linear time-varying system final state $y(T)$ (here T is the final control time) under deterministic disturbance or random disturbance.

The linear time-varying system loops in which the adjoint method can be used should have the following characteristics: its time-varying blocks should only contain $(T-t)$ form element, that is t_{go}, where T is the total control time.

The transformation of the above linear time-varying system loop to the adjoint system should follow the steps below:

(1) Replace the original time variable t with the new time variable $\tau = T - t$. So that all time-varying structure $(T-t)$ can be converted to τ and the new system no longer contains T.

(2) Invert the signal flow of the original loop. Transform the branch nodes of the original system to summing junctions and the original summing junctions into branch nodes, as shown in table 8.1-1.

Table 8.1-1 Interchange the branch nodes and summing junctions for the adjoint method

Item	Original system	Adjoint system
Summing junctions	$G_2(s)$, $G_1(s)$	$G_2(s)$, $G_1(s)$
Branch nodes	$G_2(s)$, $G_1(s)$	$G_2(s)$, $G_1(s)$

(3) Transform the original system.

(a) For deterministic step disturbance input.

Transform the original system output to the adjoint system impulse input and transform the original step input as the impulse response output of the adjoint system. With deterministic step disturbance input, the linear time-varying system final state $y(T)$ is the integration of the impulse response output of the adjoint system (Fig. 8.1-43).

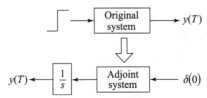

Fig. 8.1-43 The transformation of the original system to the adjoint system with deterministic step disturbance input

(b) For white noise input with power spectrum of ϕ.

It should be noticed that the unit of power spectrum density ϕ for angle noise is rad^2/Hz, and the unit of power spectrum density ϕ for position noise is m^2/Hz.

Transform the original system output to the impulse input of the adjoint system. Treat the original system input as the impulse response of the adjoint system. To obtain the linear time-varying system final state variance $E[y^2(T)]$, the adjoint system impulse response should be squared, integrated and multiplied by the power spectrum density ϕ (Fig. 8.1-44).

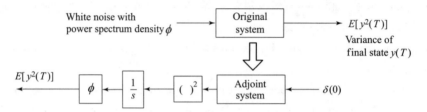

Fig. 8.1-44 The transformation of the original system to the adjoint system under white noise input

Notice that the adjoint system time variable is taken as τ. when $\tau = T$, the adjoint system output $y(\tau)$ will be $y(T)$ or $E[y^2(T)]$. The benefit of the adjoint method is that the original system final state $y(T)$ or its variance $E[y^2(T)]$ for different disturbance and different control time T can be obtained by only one simulation.

In the following, two examples are given to show the applicability of the adjoint method.

1) Example for deterministic disturbance

The original proportional navigation guidance loop under deterministic disturbance of target maneuver with acceleration a_T (m/s^2) is given in Fig. 8.1-45, where a fourth order guidance dynamic $\left(\frac{T_g}{4}s + 1\right)^{-4}$ is considered.

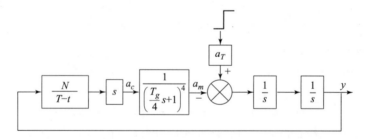

Fig. 8.1-45 Original block diagram with target maneuver a_T

The adjoint block diagram of the proportional guidance system under target maneuvering acceleration a_T disturbance is shown in Fig. 8.1-46.

Suppose the target maneuvering acceleration a_T is $5g$ ($50\ m/s^2$) and the system guidance time constant T_g is 0.5 s. The miss distance y_{miss} obtained by the adjoint method is given in Fig. 8.1-47 for different proportional constant N and control time T.

The adjoint method analysis result shows that to cope with target maneuver, the proportional navigation constant N should be greater than 3 and the non-dimensional guidance time T/T_g more than 10.

8 Proportional Navigation and Extended Proportional Navigation Guidance Laws

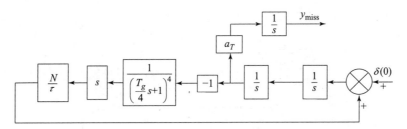

Fig. 8.1 – 46 Adjoint block diagram of the corresponding system

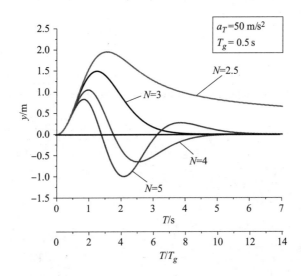

Fig. 8.1 – 47 Miss distance curves corresponding to different N for the proportional guidance system (under the target maneuver disturbance)

2) Example for random disturbance

Due to the seeker automatic gain control, the thermal noise generated effective angular variation will be proportional to the square of missile-target distance R^2 and its effective angular disturbance power spectrum density will be $\dfrac{R^2}{R_0^2}\phi_{RN} = \left(\dfrac{(T-t)V_r}{R_0}\right)^2 \phi_{RN}$ (rad^2/Hz), where ϕ_{RN} is the thermal noise power spectrum density at distance R_0 and V_r is the relative missile-target velocity.

The original proportional navigation guidance loop under random disturbance of seeker receiver thermal noise is given in Fig. 8.1 – 48, where a fourth order guidance dynamic $\left(\dfrac{T_g}{4}s+1\right)^{-4}$ is considered.

The adjoint block diagram of the proportional guidance system under thermal noise disturbance is shown in Fig. 8.1 – 49.

Suppose the initial power spectrum density of the thermal noise signal at R_0 is $\phi_{RN} = 1.20 \times 10^{-9}\,\text{rad}^2/\text{Hz}$, the relative velocity of the missile and target is $V_r = 800$ m/s and the initial missile-target distance is $R_0 = 10$ km. The miss distance $\sigma_{\text{miss}(RN)}$ generated by the adjoint method is given in Fig. 8.1 – 50 for different proportional constant N and control time T.

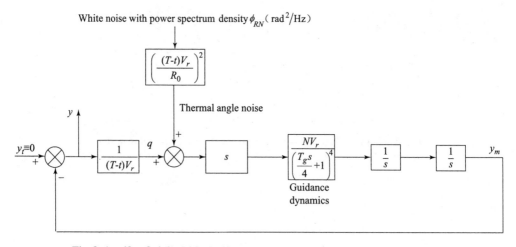

Fig. 8.1−48 Original block diagram under the thermal noise disturbance

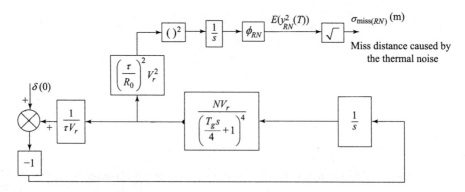

Fig. 8.1−49 Adjoint block diagram of the corresponding system under the thermal noise disturbance

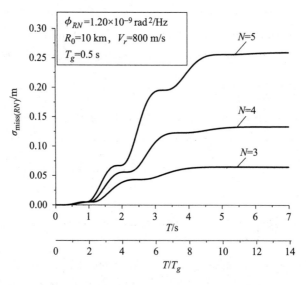

Fig. 8.1−50 Miss distance curves corresponding to different N for the proportional guidance system (under the thermal noise disturbance)

The adjoint method analysis result shows that thermal noise introduced miss distance is small and its value approaches constant for normal guidance time $T > 10T_g$.

§ 8.2 Optimal Proportional Navigation Guidance Laws OPN

8.2.1 OPN1: Considering the Missile Guidance Dynamics

It should be noticed that the proportional navigation law (Section 8.1) is derived under the assumption of zero guidance dynamics. But since it is a feedback guidance law with very good robustness, the guidance law can still be used when the guidance dynamics is not zero even though it is no longer optimal. The application restriction is normally $T/T_g > 10$. This will limit the missile's minimum operational range. When the real guidance dynamics model is known which is composed of a seeker, a guidance filter and autopilot dynamics, a modified proportional navigation guidance law could be derived. This modified guidance law can make the related guidance process optimum while greatly reducing the allowed missile minimum operation range.

Suppose that the total higher order guidance dynamics can be simplified to a first order model, i.e.,

$$\frac{a_m(s)}{a_c(s)} = \frac{1}{T_g s + 1}. \tag{8.2-1}$$

The extended state equations with this simplified guidance dynamics will be

$$\begin{bmatrix} \dot{y} \\ \dot{V} \\ \dot{a}_m \end{bmatrix} = \begin{bmatrix} 0 & 1 & 0 \\ 0 & 0 & -1 \\ 0 & 0 & -1/T_g \end{bmatrix} \begin{bmatrix} y \\ V \\ a_m \end{bmatrix} + \begin{bmatrix} 0 \\ 0 \\ 1/T_g \end{bmatrix} a_c. \tag{8.2-2}$$

The three state variables of this state equation are the relative distance y between the missile and the target perpendicular to the LOS, the relative velocity V, and the absolute acceleration a_m of the missile in inertial space. The control variable is the missile's autopilot command a_c. Take the objective function of this optimization problem as $J = S\frac{y(T)^2}{2} + \frac{1}{2}\int_0^T a_c^2(t) dt$, and take $S \to \infty$.

By solving the above optimal control problem, the optimal guidance law OPN1 is given as

$$a_c = N'\left[\frac{1}{t_{go}^2}y(t) + \frac{1}{t_{go}}V(t) + \frac{1}{t_{go}^2}(1 - e^{-\bar{t}_{go}} - \bar{t}_{go})a_m(t)\right]. \tag{8.2-3}$$

In equation (8.2-3), $t_{go} = T - t$ is the remaining engagement time, $\bar{t}_{go} = \frac{T-t}{T_g} = \frac{t_{go}}{T_g}$ is the non-dimensional remaining engagement time, N' is the effective navigation ratio and its expression is:

$$N' = \bar{t}_{go}^2(e^{-\bar{t}_{go}} + \bar{t}_{go} - 1)\left(-\frac{1}{2}e^{-2\bar{t}_{go}} - 2Te^{-\bar{t}_{go}} + \frac{1}{3}\bar{t}_{go}^3 - \bar{t}_{go}^2 + \bar{t}_{go} + \frac{1}{2}\right)^{-1}. \tag{8.2-4}$$

It is known that under the condition of small disturbance, the LOS angle can be approximated as $q = \dfrac{y(t)}{V_r \cdot (T-t)} = \dfrac{y(t)}{V_r \cdot t_{go}}$. Taking the derivative of q with respect to time t will give

$$\dot{q} = \dfrac{y(t)}{V_r \cdot t_{go}^2} + \dfrac{V(t)}{V_r \cdot t_{go}}. \quad (8.2-5)$$

A more practical proportional navigation law for engineering applications can be obtained as follows by substituting the expression of $\dot{q}$ into the above guidance law expression

$$a_c = N'V_r\dot{q} + C_1 a_m, \quad (8.2-6)$$

where

$$N' = \overline{t}_{go}^2 (e^{-\overline{t}_{go}} + \overline{t}_{go} - 1)\left(-\dfrac{1}{2}e^{-2\overline{t}_{go}} - 2\overline{t}_{go}e^{-\overline{t}_{go}} + \dfrac{1}{3}\overline{t}_{go}^3 - \overline{t}_{go}^2 + \overline{t}_{go} + \dfrac{1}{2}\right)^{-1}$$

and

$$C_1 = N'\dfrac{1}{\overline{t}_{go}^2}(1 - e^{-\overline{t}_{go}} - \overline{t}_{go}).$$

The differences between this extended proportional navigation guidance law OPN1 and the standard proportional navigation guidance law PN are:

(1) The OPN1 guidance law has one more feedback term (missile acceleration a_m).

(2) The feedback coefficients of the state $\dot{q}$ and state a_m have become time varying, which are related to $\overline{t}_{go} = \dfrac{t_{go}}{T_g} = \dfrac{T-t}{T_g}$.

When there is a heading error disturbance ε, the block diagram of the extended proportional navigation law is shown in Fig. 8.2-1.

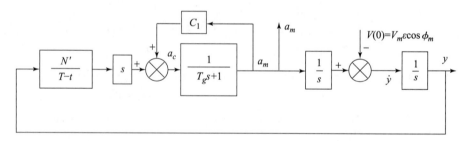

Fig. 8.2-1 Block diagram of the OPN1 extended proportional navigation guidance law under the initial heading disturbance ε

According to the proposition of this problem, when the guidance dynamics is a first order model, this extended proportional navigation guidance law will hit the target with minimum control costs and no limit on t_{go} as long as the required acceleration condition is not saturated. Next, the effect of initial heading error ε on this guidance law will be studied.

Make the non-dimensionalization of the loop with respect to the engagement time T. For this purpose, take $\overline{\overline{t}} = \dfrac{t}{T}$, $\overline{\overline{s}} = Ts$, and the non-dimensional time to go will be $\overline{t}_{go} = \dfrac{T-t}{T_g} = \dfrac{T}{T_g}(1 - \overline{\overline{t}})$. The block diagram after non-dimensionalization is shown in Fig. 8.2-2.

Fig. 8.2-3 shows the comparison between the extended proportional navigation OPN1 and the normal proportional navigation PN. It can be seen that in the initial stage of the guidance, OPN1

8 Proportional Navigation and Extended Proportional Navigation Guidance Laws

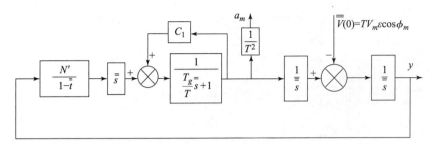

Fig. 8.2-2 Block diagram of the non-dimensional OPN1 guidance law under the initial heading error disturbance ε

needs more acceleration than PN. It is exactly because of the increase of the acceleration command in the early guidance stage that the effect of the lag of the guidance dynamics can be compensated. This ensures that the miss distance is zero even when there is the first order guidance dynamics. Similar to the proportional guidance, OPN1 also has the feature that the smaller the engagement time T/T_g, the greater the required acceleration is.

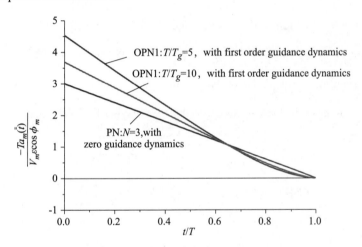

Fig. 8.2-3 Non-dimensional acceleration of proportional navigation and extended proportional guidance OPN1

Fig. 8.2-4 shows the acceleration command a_c and the acceleration output a_m for the PN and OPN1 guidance laws. It can be seen from the figure that the acceleration command a_c of the OPN1 guidance law at the end of the guidance is the same as the normal proportional navigation guidance. Both are zero, but the acceleration output a_m is not zero for OPN1 at the end of guidance. This is because a_m is the output of a_c through first order dynamics lag.

Next, the implementation approach of OPN1 in real engineering applications is studied. Suppose that the guidance dynamics model includes the dynamics of the seeker and the guidance filter $\left(\dfrac{T_g}{4}s + 1\right)^{-2}$, as well as the autopilot second order dynamics $\left(\dfrac{T_g}{4}s + 1\right)^{-2}$. In actual engineering applications, the feedback signal $\dot{q}$ can only be taken as the output of the guidance filter, and the feedback signal a_m taken as the output of the autopilot accelerometer. When the

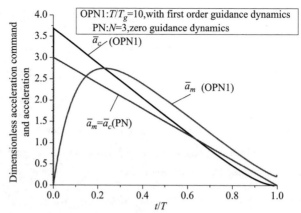

Fig. 8.2-4 Comparison between the non-dimensional acceleration command a_c, and the acceleration output a_m of the extended proportional navigation and the normal proportional navigation

feedback signals are taken accordingly, the guidance time constants in the calculation of the guidance law coefficient N', C_1 can be taken approximately as T_g (Model I) or $T_g/2$ (Model II), their corresponding block diagrams are shown in Fig. 8.2-5 and Fig. 8.2-6. In Fig. 8.2-5, $T_g = 4 \times \left(\frac{1}{4}T_g\right)$ is taken in the calculation of N', C_1. In Fig. 8.2-6, the second order dynamics lag of the autopilot alone is taken in the calculation of the guidance law coefficient N'^*, C_1^*, that is to say, $T_g^* = 2 \times \left(\frac{1}{4}T_g\right) = \frac{1}{2}T_g$ is taken in the calculation of the time-varying feedback coefficients N'^* and C_1^*.

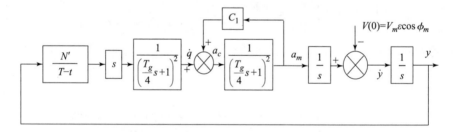

Fig. 8.2-5 Fourth order dynamics OPN1 guidance law model I in engineering applications (case ③)
(The guidance time constant is taken as T_g in the calculation of N' and C_1)

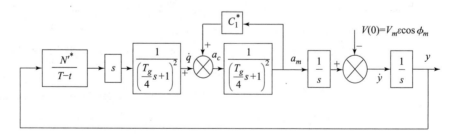

Fig. 8.2-6 Fourth order dynamics OPN1 guidance law model II in engineering applications (case ④)
(The guidance time constant is taken as $T_g^* = \frac{1}{2}T_g$ in the calculation of N'^* and C_1^*; that is, the seeker dynamics is ignored in the guidance law and only the pilot dynamics is taken into account)

Fig. 8.2-7 shows the non-dimensional miss distance curves of adopting the normal PN guidance law for the fourth order dynamics guidance system (taken as case ①), adopting the OPN1 guidance law with the first order dynamics system (taken as case ②) and adopting both engineering implementation approaches case ③ and ④ under the action of the initial heading error disturbance ε.

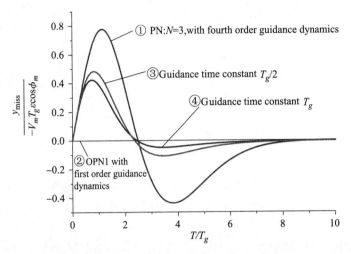

Fig. 8.2-7 Non-dimensional miss distance curves for different guidance laws under the action of the initial heading error disturbance ε

Obviously, under the action of the initial heading error disturbance ε, the OPN1 guidance law is optimal for first order guidance dynamics and its miss distance is zero for all engagement T/T_g. When this guidance law is applied with engineering implementation approaches case ③ and ④ with two engineering practical "second order seeker + second order pilot" dynamics models, the convergence time of the miss distance converges for $T/T_g > 6$ instead of $T/T_g > 10$ for normal PN, and its miss distance is smaller with the same T/T_g.

In the following, the acceleration of the OPN1 guidance law under the target maneuver a_t is analyzed and its block diagram is as shown in Fig. 8.2-8.

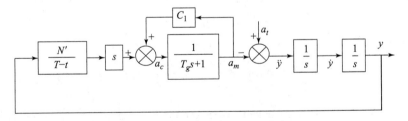

Fig. 8.2-8 Guidance block diagram of the guidance law OPN1 with constant target maneuver

The non-dimensional acceleration time history of the OPN1 guidance law with constant target maneuver is shown in Fig. 8.2-9. It can be known that the terminal acceleration of the guidance law diverges under the action of the target maneuver, so that the guidance law loses its effectiveness in the presence of the constant target maneuver. The reason is that this guidance law a_m feedback is used to compensate for the influence of the time constant T_g of the guidance dynamics, but not target maneuver. That is, this guidance law cannot be applied to the target maneuver case. The solution to this problem is to add the target maneuver compensation terms in the guidance law design (see the following OPN2 and OPN3).

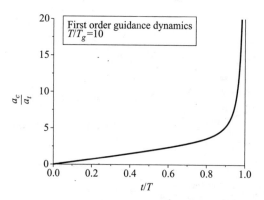

Fig. 8.2-9 Time history of the non-dimensional acceleration under constant target maneuver for guidance law OPN1

8.2.2 OPN2: Considering the Constant Target Maneuver

From the analysis of Section 8.1, it is known that when proportional guidance is used for target maneuvers, the required missile acceleration should be 3 - 5 times that of the target maneuver acceleration. If the available acceleration of the missile does not reach this value, the saturation of the missile's acceleration will cause excessive miss distance. When the missile has the ability to estimate the target maneuver acceleration, the estimated value of the target maneuver acceleration can be used to introduce compensation in the guidance law, so as to reduce the required missile acceleration and increase guidance accuracy.

When the input to the guidance system is constant target maneuver acceleration a_t alone, the dynamics model will be

$$\begin{cases} \dot{y} = V \\ \dot{V} = a_t - a_c \end{cases} \tag{8.2-7}$$

Take the objective function for deriving the optimal guidance law as

$$\min J = \min \left[S \frac{y(T)^2}{2} + \frac{1}{2} \int_0^T a_c^2(t) \, dt \right]. \tag{8.2-8}$$

In the above equation, T is the engagement time, and $\phi(T) = S \frac{y(T)^2}{2}$ is the penalty function of the miss distance at the time moment T. When the value of S approaches ∞, the miss distance is zero. The integration term of the objective function represents the minimum integral of the squared missile acceleration over the time interval T, which means that it tries to hit the target with minimum control costs.

By solving the above optimum control problem, the optimum control solution will be

8 Proportional Navigation and Extended Proportional Navigation Guidance Laws

$$a_c = N\left(\frac{y}{t_{go}^2} + \frac{\dot{y}}{t_{go}} + \frac{1}{2}a_t\right), \quad (8.2-9)$$

where t_{go} is the remaining flight time, and $t_{go} = T - t$.

Based on the previous knowledge, the above guidance law can be changed to a more practical form

$$a_c = N\left(V_r\dot{q} + \frac{1}{2}a_t\right). \quad (8.2-10)$$

Under the condition of there being no guidance dynamics, this guidance law guarantees that the miss distance is zero when a constant maneuver target is attacked. Its guidance block diagram is shown in Fig. 8.2 – 10.

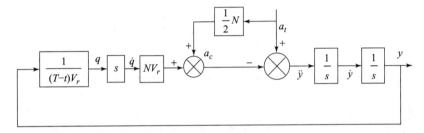

Fig. 8.2 – 10 Guidance loop of the enhanced proportional navigation law OPN2 with the effect of constant target maneuver

Make the non-dimensionalization of the loop with respect to the engagement time T. For this purpose, take $\bar{t} = \frac{t}{T}$, $\bar{s} = Ts$, so the non-dimensional remaining engagement time becomes $\bar{t}_{go} = \frac{T-t}{T_g} = \frac{T}{T_g}(1 - \bar{t})$. The block diagram after the non-dimensionalization is shown in Fig. 8.2 – 11.

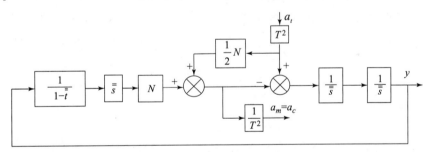

Fig. 8.2 – 11 Non-dimensional block diagram of OPN2 guidance law

From the analysis of the non-dimensional acceleration time history in Fig. 8.2 – 12, it is known that using this guidance law to compensate a_t can ensure that the required acceleration of the missile at the end of engagement is zero, and the maximum required acceleration in the initial stage of the guidance is only 1.5 times that of the target maneuver acceleration.

The guidance accuracy of this guidance law for a system with fourth order guidance dynamics will be analyzed in the following in this case. The guidance block diagram is shown in Fig. 8.2 – 13.

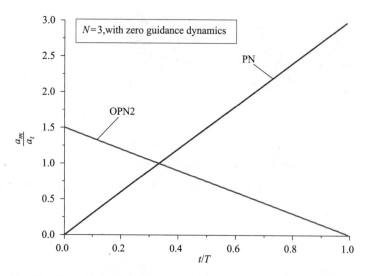

Fig. 8.2 – 12 Non-dimensional acceleration time history with PN and OPN2 guidance law with target maneuver and guidance dynamics ignored

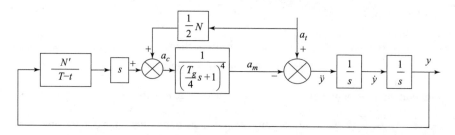

Fig. 8.2 – 13 Block diagram of the OPN2 guidance law with the fourth order guidance dynamics

Fig. 8.2 – 14 shows the miss distance over the engagement time when this guidance law is applied to an actual fourth order guidance system. It can be seen that this guidance law has no compensation for the influence of the guidance dynamics. Therefore, the required minimum

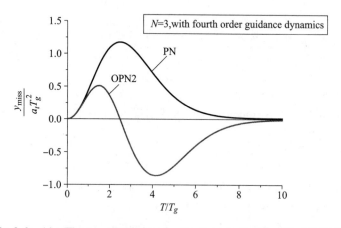

Fig. 8.2 – 14 The non-dimensional miss distance for the PN and OPN2 guidance law with constant target maneuver a_t

engagement time T is still 10 times that of the guidance dynamics time, but the required terminal acceleration becomes smaller and the miss distance is better. In order to reduce both the allowed minimum engagement time and the required terminal acceleration, it is necessary to compensate for both the target maneuver and the guidance dynamics in the guidance law (refer to the analysis of the guidance law OPN3 in the following section).

8.2.3 OPN3: Considering Both Constant Target Maneuvers and Missile Guidance Dynamics

According to the analysis in the previous section, in order to reduce the required terminal acceleration and further shorten the minimum allowable engagement time, it is necessary to introduce compensations for both the target maneuver a_t and the guidance dynamics T_g in the guidance law.

To accomplish this task, the required extended state equation is

$$\begin{bmatrix} \dot{y} \\ \dot{V} \\ \dot{a}_m \end{bmatrix} = \begin{bmatrix} 0 & 1 & 0 \\ 0 & 0 & -1 \\ 0 & 0 & -1/T_g \end{bmatrix} \begin{bmatrix} y \\ V \\ a_m \end{bmatrix} + \begin{bmatrix} 0 \\ 1 \\ 0 \end{bmatrix} a_t + \begin{bmatrix} 0 \\ 0 \\ 1/T_g \end{bmatrix} a_c . \quad (8.2-11)$$

The three states are the relative distance y between the missile and the target perpendicular to the LOS, the relative velocity V, and the absolute acceleration a_m of the missile in inertial space. The system disturbance input is the estimated target acceleration a_t, and its control variable is the acceleration autopilot command a_c.

The objective function for deriving the optimal guidance law is given in the same form as before

$$\min J = \min \left[S \frac{y(T)^2}{2} + \frac{1}{2} \int_0^T a_c^2(t) dt \right]. \quad (8.2-12)$$

By solving the optimal control problem, the optimal guidance law will be

$$a_c = N' \left[\frac{y}{t_{go}^2} + \frac{1}{t_{go}} V - \frac{1}{t_{go}^2} (e^{-\bar{t}_{go}} - 1 + \bar{t}_{go}) a_m + 0.5 a_t \right], \quad (8.2-13)$$

where $t_{go} = T - t$ is the remaining flight time, $\bar{t}_{go} = \frac{T-t}{T_g} = \frac{t_{go}}{T_g}$ is the non-dimensional remaining flight time, and N' is the effective navigation ratio.

$$N' = \bar{t}_{go}^2 (e^{-\bar{t}_{go}} - 1 + \bar{t}_{go}) \left(-\frac{1}{2} e^{-2\bar{t}_{go}} - 2Te^{-\bar{t}_{go}} + \frac{1}{3} \bar{t}_{go}^3 - \bar{t}_{go}^2 + \bar{t}_{go} + \frac{1}{2} \right)^{-1}. \quad (8.2-14)$$

Under the condition of small disturbance, the proportional navigation law in another form can be given by substituting the expression of $\dot{q}$ into the OPN3 guidance law, that is

$$a_c = N' V_r \dot{q} + C_1 a_m + 0.5 a_t , \quad (8.2-15)$$

where we define $C_1 = N' \frac{1}{\bar{t}_{go}^2} (1 - e^{-\bar{t}_{go}} - \bar{t}_{go})$.

This guidance law guarantees that in the presence of the first order guidance dynamics and constant target maneuver, the miss distances at different engagement time are all zero. The block diagram of the OPN3 guidance law with the constant target maneuver a_t is shown in Fig. 8.2-15.

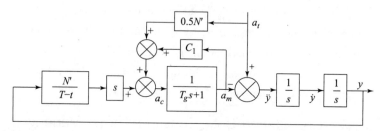

Fig. 8.2-15　Block diagram of the OPN3 guidance law with constant target maneuver a_t

From the analysis of the non-dimensional acceleration time history in Fig. 8.2-16, it is known that using the OPN3 guidance law ensures that the required acceleration of the missile at the end of the engagement is zero with the constant target maneuver. Although the maximum required acceleration in the early stage of the guidance is greater, it solves the problem of the acceleration divergence for the OPN1 guidance law. In addition, the required acceleration under the OPN3 guidance law decreases with the increase of the engagement time T/T_g. As shown in Fig. 8.2-16, the maximum required accelerations are respectively 2.25 and 1.8 times the target maneuver acceleration when $T/T_g = 5$ and 10.

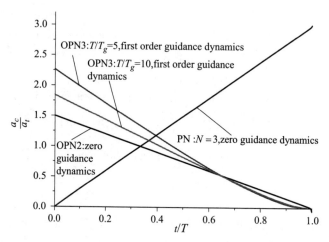

Fig. 8.2-16　Non-dimensional acceleration time history with the PN and OPN3 guidance law under constant target maneuver a_t

Fig. 8.2-17 shows the acceleration command a_c and missile acceleration a_m curves under the action of the PN and OPN3 guidance laws. It can be known from the figure that the acceleration command a_c under the OPN3 guidance law converges to zero at the end of the engagement, but its acceleration a_m is not zero. This is because a_m is the output of a_c when passing the first order guidance lag dynamics.

For practical applications, we can only take the guidance filter output $\dot{q}$ and autopilot accelerometer output as the feedback variables as in Section 8.2.1 and take T_g^* as $4 \times \left(\dfrac{1}{4}T_g\right) = T_g$ (cases ③) or $2 \times \left(\dfrac{1}{4}T_g\right) = \dfrac{T_g}{2}$ (cases ④), as seen in Fig. 8.2-19 and Fig. 8.2-20.

8 Proportional Navigation and Extended Proportional Navigation Guidance Laws

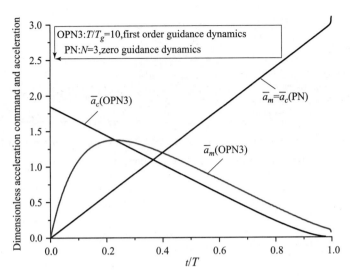

Fig. 8.2-17 Comparison between the non-dimensional acceleration command $\bar{a}_c$ and guidance dynamics output $\bar{a}_m$ of OPN3 and the PN when the target is under constant target maneuver a_t.

Fig. 8.2-21 shows the non-dimensional miss distance of the PN guidance law under the fourth order dynamics guidance system (Fig. 8.2-18 and case ①), the OPN3 guidance law under the first order dynamics guidance system (case ②), and the OPN3 guidance law adopting two engineering application schemes (cases ③ and ④) with constant target maneuver a_t.

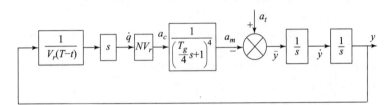

Fig. 8.2-18 Block diagram of the fourth order dynamics PN guidance law with constant target maneuver a_t (Case ①)

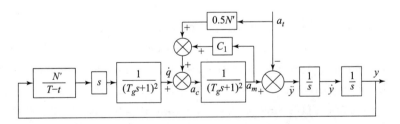

Fig. 8.2-19 Fourth order dynamics OPN3 guidance law in engineering applications with constant target maneuver a_t (case ③)

Obviously, under constant target maneuver a_t, the OPN3 guidance law is optimal when the first order guidance dynamics is taken into account, and the miss distance is zero. When the guidance law is applied to the dynamics model with two types of engineering implementation

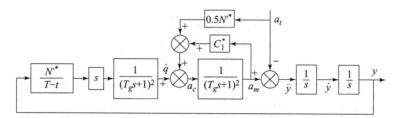

Fig. 8.2 – 20 Fourth order dynamics OPN3 guidance law in engineering applications with constant target maneuver a_t (case ④)

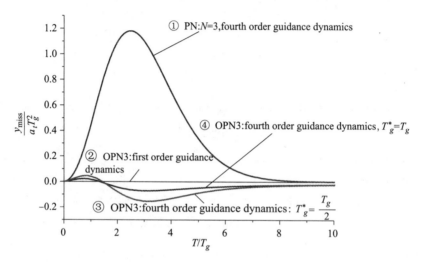

Fig. 8.2 – 21 Curves of the non-dimensional miss distance in different guidance law cases

approaches, the miss distance performance can be enhanced from the required $T/T_g > 10$ in the normal proportional guidance to $T/T_g > 7$. Moreover, since the target maneuver acceleration is compensated, this guarantees that the required acceleration at the end of the engagement is smaller and the miss distance is reduced with the same T/T_g.

8.2.4 Estimation of Target Maneuver Acceleration

An active radar seeker has the function of measuring the relative missile-target distance ΔR when working in the medium to low pulse repetition modes. The direction angle q of the target in inertial space can be acquired by adding together the missile attitude angle ϑ, the seeker gimbal angle ϕ, and the beam angle measurement error ε (Fig. 8.2 – 22).

In the condition that the LOS direction and the missile-target relative distance ΔR (the missile-target relative distance vector ΔR in the three-dimensional space) are given, the position vector $R_t = R_m + \Delta R$ of the target in the inertial space can be obtained from the position vector R_m of the missile in inertial space which is provided by the missile's navigation system and ΔR measured by the seeker (Fig. 8.2 – 23).

Obviously, the target maneuver acceleration $a_t(t)$ can be estimated by using the known target position vector $R_t(t)$ in inertial space with the help of Kalman filter technique. (refer to Section

8 Proportional Navigation and Extended Proportional Navigation Guidance Laws

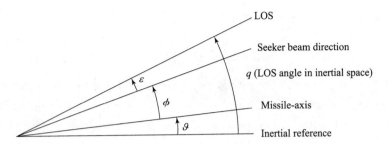

Fig. 8.2-22 Definition of related angles for the line of sight angle calculation

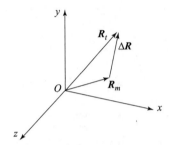

Fig. 8.2-23 Target position vector R_t in inertial space

8.4 for the Kalman filter theory on acceleration estimation.)

When the seeker works in the medium pulse repetition frequency mode, the missile can measure not only the relative missile-target distance, but also their radial relative velocity. The proper use of this additional information can further improve the target acceleration estimation accuracy.

With the improvement of the radar seeker velocity and distance measurement accuracy, it will be feasible for the missile to estimate the target maneuver acceleration in real time. Therefore, the proportional navigation guidance law mentioned above with target maneuver compensation can be applied to engineering practice in the future.

8.2.5 Estimation of t_{go}

(1) In practice, the required accuracy for t_{go} in the extended proportional navigation guidance law is not high. It can be estimated by dividing the missile-target relative distance measured by the seeker by the current missile velocity.

(2) When t_{go} is so small that the calculated acceleration command is close to the missile's acceleration capacity limit, it can be taken as a constant and does not have to be further decreased.

8.2.6 Extended Guidance Law with Impact Angle Constraint

In general, air-to-ground missiles often have a demand for a large impact angle when attacking a stationary target. For this reason, it is necessary to adopt a proportional navigation guidance law that can include an impact angle constraint. Assuming that the flight velocity of the missile V_{missile} is constant and the target has no maneuver, a_c is the acceleration autopilot command. Since the guidance dynamics is ignored, there is a missile lateral acceleration $a_m = a_c$. The state equation used to solve this problem remains

$$\begin{cases} \dot{y} = V \\ \dot{V} = -a_c \end{cases}, \quad (8.2-16)$$

where the state variable y is the relative distance of the missile and the target perpendicular to the LOS, V is the relative velocity ($V = V_t - V_m$) in the direction perpendicular to the LOS, and V_m is the missile velocity in the direction perpendicular to the missile-target line (Fig. 8.2 – 24). Since $V = V_t - V_m$, we will have $V_m = -V$ as $V_t = 0$.

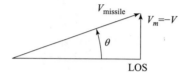

Fig. 8.2 – 24 Relationship between V and θ

Under the assumption of small disturbances and $V_t = 0$, it is known from the geometric relationship in Fig. 8.2 – 24 that

$$\theta = \frac{V_m}{V_{missile}} = \frac{-V}{V_r}. \quad (8.2-17)$$

Therefore, the expected impact angle θ_F can be given as

$$\theta_F = \frac{-V^*(T)}{V_r}, \quad (8.2-18)$$

where θ_F is the expected impact angle and $V^*(T)$ the expected terminal relative velocity.

The objective function for deriving the optimal guidance law is

$$\min J = \min \left[S_1 \frac{y(T)^2}{2} + S_2 \frac{(V(T) - V^*(T))^2}{2} + \frac{1}{2} \int_0^T a_c^2(t) \, dt \right], \quad (8.2-19)$$

where $V^*(T)$ is the expected value of $V(T)$; $\phi_1(T) = S_1 \frac{y(T)^2}{2}$ is the penalty function of the miss distance at the moment T; $\phi_2(T) = S_2 \frac{(V(T) - V^*(T))^2}{2}$ is the penalty function for the velocity perpendicular to the LOS at the moment T and S_1, S_2 are the related weighting coefficients. When S_1 approaches ∞, the miss distance is zero; when S_2 approaches ∞, the relative velocity $V(T)$ reaches the expected value $V^*(T)$, i.e., the terminal impact angle will be the expected impact angle $\theta_F \left(\theta_F = -\frac{V^*(T)}{V_r} \right)$.

By solving the above optimal problem, the optimal guidance law for the zero guidance dynamics system will be

$$a_c(t) = -\frac{1}{t_{go}^2}(6y + 4t_{go}V + 2t_{go}V(T)^*), \quad (8.2-20)$$

or

$$a_c(t) = 4V_r \left(\frac{y(t) + V(t)t_{go}}{V_r t_{go}^2} \right) + \frac{2V_r}{t_{go}} \left(\frac{y(t)}{V_r t_{go}} + \frac{V(T)^*}{V_r} \right). \quad (8.2-21)$$

In the above equations t_{go} is the remaining flight time. With small disturbance assumption, it is known that

$$q = \frac{y(t)}{V_r \cdot (T-t)}, \tag{8.2-22}$$

and

$$\dot{q} = \frac{y(t)}{V_r \cdot (T-t)^2} + \frac{V(t)}{V_r \cdot (T-t)}. \tag{8.2-23}$$

By using equation (8.2 – 22), equation (8.2 – 23) and $V^*(T) = -V_r \theta_F$, a more practical form of the proportional navigation guidance law with the impact angle constraint θ_F can be obtained as

$$a(t) = 4V_r \dot{q}(t) + \frac{2V_r}{t_{go}}(q(t) - \theta_F). \tag{8.2-24}$$

From the above equation, it is known that the proportional navigation guidance law with the impact angle constraint is composed of two terms: the proportional guidance $\dot{q}(t)$ feedback term ($4V_r \dot{q}(t)$) that ensures the hitting of the target and the feedback term ($2V_r(q(t) - \theta_F)/t_{go}$) of $(q(t) - \theta_F)$ that satisfies the impact angle constraint.

It should be noted that although the above guidance law is derived under the assumption of small disturbances, the result is a guidance law with both the miss distance feedback (feedback $\dot{q}$) and the impact angle control feedback (feedback $(q(t) - \theta_F)$). Therefore, this guidance law can be applied to a non-linear system that requires both target hitting and impact angle control. However, the absolute optimum cannot be guaranteed at this time.

In the following, a guided bomb example is used to verify the applicability of this guidance law in a nonlinear and high impact angle guidance problem. Suppose that the guidance dynamics of the missile is ignored and system in the pitch plane is given as follows

$$\begin{cases} \dot{V} = (-X - mg\sin\theta)/m \\ \dot{\theta} = (Y - mg\cos\theta)/(mV) \\ \dot{x} = V\cos\theta \\ \dot{y} = V\sin\theta \end{cases}, \tag{8.2-25}$$

where the drag is $X = c_x q S$, $c_x = 0.5$; the normal force is $Y = ma_c$, a_c is the above proportional navigation guidance law command; the reference area of the missile is taken as $S = 0.107, 5 \text{ m}^2$, the diameter is $D = 370$ mm and the mass is $m = 500$ kg.

Suppose that the guided bomb initial releasing condition is: altitude $H_0 = 1,000$ m, velocity $V_0 = 250$ m/s. The bomb is released horizontally. The target position is: $X_t = 5,000$ m, $Y_t = 0$ m and the expected impact angles θ_F are respectively $0°$, $-30°$, $-60°$ and $-90°$.

Fig. 8.2 – 25 shows the guided trajectory curves with different expected impact angles θ_F. It can be seen that the trajectory shaping guidance law is still useful even under large impact angle requirements and nonlinear conditions.

In engineering applications, a more general trajectory shaping guidance law with a weighting coefficient N_y and a weighting coefficient N_q is often used.

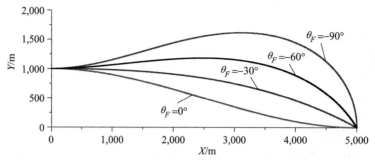

Fig. 8.2-25 Trajectory curves of the proportional navigation law with impact angle constraints under different expected impact angles

$$a_c(t) = N_y V_r \dot{q}(t) + N_q V_r (q(t) - \theta_F)/t_{go}, \qquad (8.2-26)$$

where the weighting coefficients N_y and N_q can be used to adjust the weight ratio of the impact position constraint and the impact angle constraint. Generally speaking, the constraint on impact position should be given higher weight than the impact angle constraint.

§ 8.3   Other Proportional Navigation Laws

8.3.1   Proportional Navigation Law with Gravity Over-compensation

Normal guidance bombs and some tactical air-to-ground missiles flying at subsonic speeds do not have high flight velocity in their terminal guidance phase, and their lateral acceleration capacity is limited (usually 2 ~ 4 g). When normal proportional navigation guidance law is adopted, an additional aerodynamic acceleration of about one g will have to be provided to counter balance the gravity. As this is done, the available missile lateral acceleration maneuvering upward will be $a_{\text{inertial available}} = a_{\text{aerodynamic available}} - g$, and the downward maneuvering capability will be $a_{\text{inertial available}} = a_{\text{aerodynamic available}} + g$. So the upward maneuverability of the missile is lower than its downward maneuverability. The total difference could be as high as 2 g. The gravity over-compensated proportional navigation guidance law studied below can make the missile lateral acceleration required at the end of engagement smaller than the normal value of one g, and the final maneuverability upward and downward more equal. Moreover, this guidance law can also increase the impact angle to increase the warhead effectiveness.

The general form of the gravity over-compensated proportional navigation guidance law is given as

$$a_c = NV_r\dot{q} + (c - 1)g. \qquad (8.3-1)$$

(normal gravity compensation: $c = 1$; over-gravity compensation: $c > 1$)

Fig. 8.3-1 shows its guidance block diagram, in which $a_{m\text{ aerodynamic}}$ represents the missile's aerodynamic lateral acceleration, and $a_{m\text{ inertial}}$ indicates the missile's maneuverability acceleration in inertial space (gravity force included). The relationship between the two is

$$a_{m\text{ inertial}} = a_{m\text{ aerodynamic}} - g \quad (a \text{ is positive for upward acceleration}). \qquad (8.3-2)$$

Therefore, as shown in Fig. 8.3 −2, the guidance law of the missile in the inertial space is

$$\begin{cases} a_{m\text{ inertial}} = NV_r\dot{q} + (c-1)g \\ a_{m\text{ aerodynamic}} = NV_r\dot{q} + cg \end{cases} \quad (8.3-3)$$

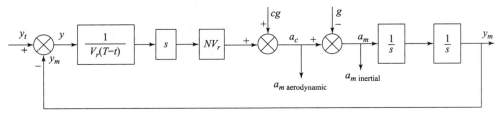

Fig. 8.3 −1 Block diagram of the gravity over-compensated proportional navigation guidance

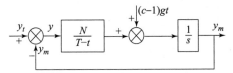

Fig. 8.3 −2 Simplified block diagram of the gravity over-compensated proportional navigation guidance law

It is known from Fig. 8.3 −2 that the linear time-varying differential equation of this guidance law is

$$\dot{y}_m + \frac{N}{T-t}y_m = (c-1)gt. \quad (8.3-4)$$

By solving this differential equation, the non-dimensional position, velocity and inertial acceleration under the guidance law can be obtained:

$$\frac{y_m}{(c-1)gT^2} = \frac{1-\bar{t}}{(N-1)(N-2)}[(N-1)\bar{t} - 1 + (\bar{t}_{go})^{N-1}]. \quad (8.3-5)$$

$$\frac{V_m}{(c-1)gT} = \frac{N}{(N-1)(N-2)} - \frac{2}{(N-2)}\bar{t} - \frac{N(\bar{t}_{go})^{N-1}}{(N-1)(N-2)}. \quad (8.3-6)$$

$$\frac{a_{m\text{ inertial}}}{(c-1)g} = \frac{\dot{V}_m}{(c-1)g} = \frac{1}{(N-2)}[N(\bar{t}_{go})^{N-2} - 2]. \quad (8.3-7)$$

where

$$\bar{t} = \frac{t}{T}, \quad \bar{t}_{go} = \frac{T-t}{T} = \frac{t_{go}}{T}.$$

Fig. 8.3 − 3, Fig. 8.3 − 4 and Fig. 8.3 − 5 respectively show the non-dimensional displacement, velocity and acceleration curves corresponding to the guidance law as initial heading error is zero.

The expression of the missile final velocity $V_m(T)$ when $t = T$ is

$$V_m(T) = -\frac{(c-1)gT}{(N-1)}. \quad (8.3-8)$$

It is clear that when c is larger than 1, the missile final velocity perpendicular to the ground is negative. That is, the gravity over-compensation can increase the impact angle of the missile.

The terminal acceleration values of the missile when $t = T$ are:

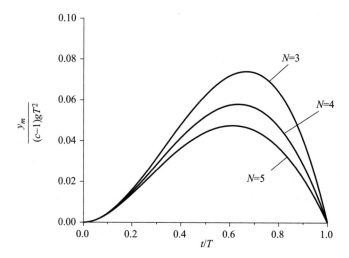

Fig. 8.3 – 3　Non-dimensional displacement curves

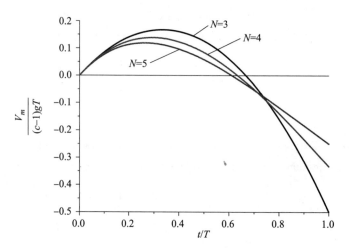

Fig. 8.3 – 4　Non-dimensional velocity curves

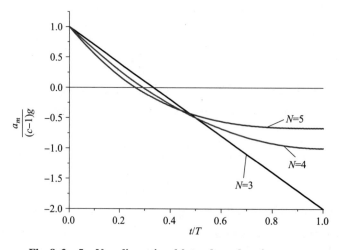

Fig. 8.3 – 5　Non-dimensional lateral acceleration curves

8 Proportional Navigation and Extended Proportional Navigation Guidance Laws

$$a_{m\ \text{inertial}} = \frac{-2(c-1)g}{(N-2)}, \qquad (8.3-9)$$

$$a_{m\ \text{aerodynamic}} = \frac{-2(c-1)g}{(N-2)} + g. \qquad (8.3-10)$$

Fig. 8.3-6 shows the aerodynamic acceleration values $a_{m\ \text{aerodynamic}}$ at the end of the engagement for gravity normal-compensation ($c = 1$) and gravity over-compensation ($c = 2$) with different proportional navigation constant $-\frac{a_c T}{V_m \varepsilon} = N\left(1 - \frac{t}{T}\right)^{N-2}$. It can be seen from the figure that for normal proportional navigation ($c = 1$), the missile always requires one g aerodynamic acceleration to balance the gravity force. If $c = 2$ and the proportional guidance coefficient $N = 4$, the aerodynamic acceleration required at the end of engagement could be zero. This solves the problem of asymmetry in the upward and downward acceleration capabilities of general proportional navigation guidance, and the lack of aerodynamic capability when the missile can be maneuvering upward has been effectively avoided.

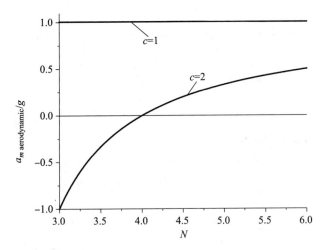

Fig. 8.3-6 Variation curve of $a_{m\ \text{aerodynamic}}$ over N ($c = 1, 2$)

8.3.2 Lead Angle Proportional Navigation Guidance Law

When a passive infrared guided missile is used to attack an aircraft target, the infrared seeker will track the infrared characteristic at the rear end of the target because the aircraft engine has the highest flame temperature. In order to damage the target to the maximum extent, it is better to modify the proportional navigation law according to the size of the target aircraft so that the seeker can trace the flame, but the missile will hit the vulnerable parts of the target body.

Based on the size of the target to be attacked, the forward displacement d of the attack point P relative to the engine flame can be set before the missile is launched (Fig. 8.3-7).

Suppose that point P is taken as the target, and an extended proportional navigation guidance law in the following form is introduced:

$$a_c = NV_r \dot{q}^*, \qquad (8.3-11)$$

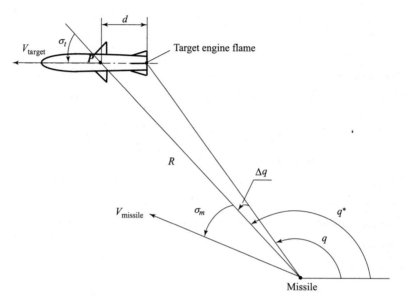

Fig. 8.3-7 Lead angle proportional navigation law

where q^* is the virtual missile-target line of sight, and its definition is

$$q^* = q + \Delta q, \qquad (8.3-12)$$
$$\dot{q}^* = \dot{q} + \Delta\dot{q}. \qquad (8.3-13)$$

Since the detector tracks the heat source at the end of the engine, its output is naturally the seeker output $\dot{q}$. The question now becomes how to get the value of $\Delta\dot{q}$. It is known from Fig. 8.3-7

$$\Delta q = \frac{d\sin\sigma_t}{R} = \frac{d\sin\sigma_t}{V_r(T-t)} = \frac{d\sin\sigma_t}{V_r t_{go}}, \qquad (8.3-14)$$

$$\Delta\dot{q} = \frac{d\sin\sigma_t}{V_r}\left(-\frac{1}{t_{go}^2}\right)(-1) = \frac{d\sin\sigma_t}{V_r t_{go}^2}. \qquad (8.3-15)$$

During the process of approaching the target, there is

$$V_{missile}\sin\sigma_m = V_{target}\sin\sigma_t. \qquad (8.3-16)$$

Therefore

$$\sin\sigma_t = \frac{V_{missile}}{V_{target}}\sin\sigma_m, \qquad (8.3-17)$$

$$\Delta\dot{q} = \frac{V_{missile} d}{V_{target} V_r}\left(\frac{1}{t_{go}^2}\right)\sin\sigma_m. \qquad (8.3-18)$$

It is known that the relationship between the infrared heat source energy W measured by the seeker detector and the missile-target relative distance R is as follows:

$$W = K\left(\frac{1}{R^\alpha}\right) \quad (\alpha \text{ is about } 2), \qquad (8.3-19)$$

$$\ln W = \ln K - \alpha \ln R, \qquad (8.3-20)$$

$$\frac{d}{dt}(\ln W) = -\alpha\frac{\dot{R}}{R} = -\alpha\frac{V_r}{V_r t_{go}} = -\alpha\frac{1}{t_{go}}, \qquad (8.3-21)$$

$$\frac{1}{t_{go}^2} = \frac{1}{\alpha^2}\left[\frac{\mathrm{d}}{\mathrm{d}t}(\ln W)\right]^2. \tag{8.3-22}$$

Therefore, the expression of $\Delta \dot{q}$ is

$$\Delta \dot{q} = \frac{V_{\mathrm{missile}} d}{V_{\mathrm{target}} V_r \alpha^2}\left[\frac{\mathrm{d}}{\mathrm{d}t}(\ln W)\right]^2 \sin\sigma_m. \tag{8.3-23}$$

Furthermore, the guidance equation of the extended proportional navigation law can be given as

$$a_c = NV_r\left\{\dot{q} + \frac{V_{\mathrm{missile}} d}{V_{\mathrm{target}} V_r \alpha^2}\left[\frac{\mathrm{d}}{\mathrm{d}t}(\ln W)\right]^2 \sin\sigma_m\right\}. \tag{8.3-24}$$

In engineering applications, V_{missile}, V_{target} and V_r can be approximately known. d and α are known, and σ_m can be acquired from the output of the seeker gimbal angle, and $[\mathrm{d}(\ln W)/\mathrm{d}t]^2$ can be obtained from the change rule of the infrared heat source energy W measured by the seeker detector. In this way, all the information required for applying the lead angle proportional navigation law has been obtained, so that the maximum damaging effect to the target can be achieved by using this extended proportional guidance law.

§ 8.4 Target Acceleration Estimation

It is known from the extended proportional navigation guidance law introduced in Section 8.2 that if the value and direction of the target maneuver acceleration $y_t = \frac{1}{2}at^2$ in the inertial space can be estimated, selecting the proportional navigation guidance law with target acceleration compensation will greatly reduce the required acceleration when the missile hits the target, thus ensuring higher guidance accuracy.

At present, the inertial navigation system of the air-to-air missile already can provide the coordinates and velocity of the missile at any given moment in inertial space. However, the target information can only be acquired in combination with the seeker information. The active radar seeker has different operation modes. When working in the medium distance and medium pulse repetition frequency mode, the active radar seeker can provide information on the missile-target distance and relative radial velocity, and the LOS direction. With this information, the three-dimensional information of the relative velocity and relative position of the missile-target can be calculated in real time. After that, combined with the missile position and the velocity information measured by the missile inertial navigation system, the motion information of the target in the inertial space can be acquired. When the missile approaches the target, the radar seeker can switch to the short distance and low pulse repetition frequency mode, and the position information of the target can be obtained. If the missile is far away from the target, the seeker will work in a long distance and high pulse repetition frequency Doppler velocity mode (table 8.4-1).

Table 8.4 – 1 Target information acquired by the seeker in different working modes

Working mode of the seeker	Seeker information	Target information acquired in combination with the missile inertial navigation system information
Long distance, high pulse repetition frequency	Velocity measurement, LOS direction measurement	Velocity
Medium distance, medium pulse repetition frequency	Distance measurement, velocity measurement, LOS direction measurement	Position, velocity
Short distance, low repetition frequency	Distance measurement, direction measurement	Position

It can be predicted that in the near future, with the continuous improvement of the accuracy of the radar seeker velocity and distance measurement, the target maneuver acceleration in the above three working modes can all be estimated by adopting the Kalman filter technique, so that the advanced proportional navigation law can be adopted to achieve a precise attack on a maneuvering target.

The method of estimating the target maneuver by using the Kalman filter technique will be briefly introduced in the following.

First, notice that the preconditions for estimating the system state using the Kalman filter are:

(1) The dynamic model of a certain linear time-varying system is $\dot{X}(t) = A(t)X(t) + B(t)u(t)$ and the output measurement model is $Z(t) = C(t)X(t)$, both are assumed known;

(2) A definitive input signal of the system $u(t)$ is given.

When there is a certain amount of disturbance and model parameters variation in the system, as long as the system can measure some state-related output $Z(t)$, the closed-loop state estimation method can be used to estimate the state $X(t)$ of the system. Define the estimated value of state $X(t)$ as $\hat{X}(t)$. The general estimator model is given as:

$$\dot{\hat{X}}(t) = A(t)\hat{X}(t) + B(t)u(t) + K(t)[Z(t) - C(t)\hat{X}(t)], \quad (8.4-1)$$

where $[Z(t) - C(t)\hat{X}(t)] = [Z(t) - \hat{Z}(t)]$ is the measurement estimation error, and $K(t)$ is the estimation feedback gain matrix.

The idea of this closed-loop estimation is that when there is an error between the measurement value $Z(t)$ and the estimated measurement value $\hat{Z}(t) = C(t)\hat{X}(t)$ of the system, these measurement estimation errors will be multiplied by a gain matrix and feedback to the state estimation equation. This feedback will force the state estimation $\hat{X}(t)$ to make proper adjustment to minimize the measurement estimation error. So, it is the minimization of the measurement estimation error which makes the estimation of the system state possible (Fig. 8.4 – 1).

The difference between different estimator models is that the models for the feedback gain matrix $K(t)$ are different. The model for the Kalman filter $K(t)$ is given as follows.

Suppose that there is white noise system input w_s and measured white noise input ν in the system, and the power spectrum of w_s is S_w and the power spectrum of ν is S_ν. That is, the system

8 Proportional Navigation and Extended Proportional Navigation Guidance Laws

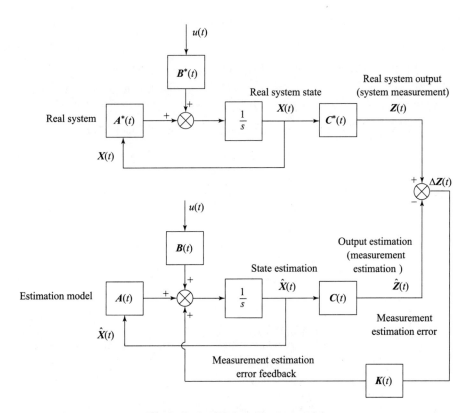

Fig. 8.4-1　Basic estimator model
(System theoretical models $A(t)$, $B(t)$, $C(t)$; system actual models $A^*(t)$, $B^*(t)$, $C^*(t)$)

model and measurement model are

$$\dot{X}(t) = A(t)X(t) + B(t)u(t) + G(t)w_s, \qquad (8.4-2)$$

$$Z(t) = C(t)X(t) + v. \qquad (8.4-3)$$

Obviously, in order to improve the estimation accuracy of the system response to the effect of the system input w_s, it is beneficial to widen the estimator bandwidth. However, in order to effectively filter out the negative effect of the measurement noise v, it is harmful to enlarge the bandwidth of the estimator. Therefore, if we want to not only estimate the system response under the action of w_s, but also reduce the negative effect of the measurement noise v, there has to be a compromised optimal estimator model according to the relative values of S_w and S_v.

The design idea of the Kalman filter is that with the given input power spectrum S_w, the measurement noise power spectrum S_v and the initial state estimation covariance matrix $P(0) = E[\{X(0) - \hat{X}(0)\}\{X(0) - \hat{X}(0)\}^T]$, a feedback matrix $K(t)$ can be derived to make the objective function $J(t) = E[\{X(t) - \hat{X}(t)\}^T\{X(t) - \hat{X}(t)\}]$ minimum at every moment of time t.

The algorithm to calculate $K(t)$ is first solve the nonlinear time-varying differential equations of the covariance matrix $P(t) = E[\{X(t) - \hat{X}(t)\}\{X(t) - \hat{X}(t)\}^T]$ for given initial value of the covariance matrix $P(0)$:

$$\dot{P}(t) = A(t)P(t) + P(t)A^{T}(t) + G(t)S_w G^{T}(t) - P(t)C^{T}(t)S_v^{-1}C(t)P(t). \tag{8.4-4}$$

After the estimated error covariance matrix $P(t)$ is acquired by solving the above differential equation, the following equation can be used to find the feedback gain matrix $K(t)$:

$$K(t) = P(t)C^{T}(t)S_v^{-1}. \tag{8.4-5}$$

Fig. 8.4-2 shows the flow chart for using continuous-time Kalman filter.

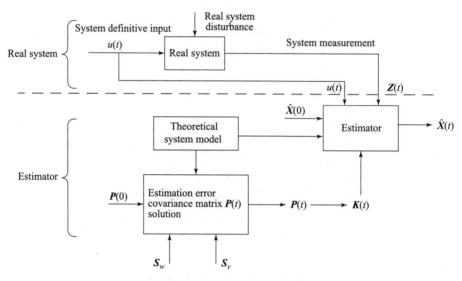

Fig. 8.4-2 Flow chart of the continuous-time Kalman filter

There will always be various external disturbances during the operation of a practical system, which will cause the system to deviate from its ideal operating state. The purpose of introducing the system noise w_s in the Kalman filter model is to use it to simulate the disturbance in the actual operation of a system.

When the system disturbance is relatively large or the system parameters have larger deviation from their designed values in applications, the Kalman filter bandwidth should be widened by increasing the value of S_w to improve the estimation accuracy. When the measurement noise of the system is larger, the filter bandwidth should be reduced by increasing the value of S_v to better filter out the noise disturbance. The specific values of S_w and S_v should be determined after iterative adjustments according to the actual operation of the system.

Generally, when selecting the initial value $P(0)$ of the state estimation error covariance matrix, it can be assumed that the estimation errors are statistically independent. That is, $E[(x_i - \hat{x}_i)(x_j - \hat{x}_j)] = 0 (i \neq j)$. Thus $P(0)$ can be simplified to a diagonal matrix:

$$P(0) = \begin{bmatrix} E[(x_1 - \hat{x}_1)^2] & 0 & 0 \\ 0 & \ddots & 0 \\ 0 & 0 & E[(x_n - \hat{x}_n)^2] \end{bmatrix}. \tag{8.4-6}$$

According to users' experience of the system and the disturbance during the initial operation, each element of $P(0)$ may be taken as the variance value σ^2 of the possible error of the state with respect to its theoretical value.

The use of Kalman filter in a linear time-varying system is introduced above. When the actual system is nonlinear, the system state estimation is performed by means of the extended Kalman filter method.

Consider the non-linear system model and measurement model:
$$\dot{X}(t) = f(X,t) + B(t)u(t) + G(t)w_s, \qquad (8.4-7)$$
$$Z(t) = f_z(X,t) + \nu. \qquad (8.4-8)$$

To use the Kalman filter method for a nonlinear system, first of all, the small disturbance linearization should be performed for the above nonlinear system model

$$\begin{cases} A(X,t) = \dfrac{\partial f(X)}{\partial X} \\ C(X,t) = \dfrac{\partial f_z(X)}{\partial X} \end{cases}. \qquad (8.4-9)$$

Therefore, the linearization model of the system will be given as
$$\begin{cases} \dot{X}(t) = A(X,t)X + B(t)u(t) + G(t)w_s \\ Z(t) = C(X,t)X + \nu \end{cases}. \qquad (8.4-10)$$

The difference from the standard Kalman filter system model is shown in equation 8.4-10, the A and C matrices are both not only a function of t but also X.

The extended Kalman filter model corresponding to the above linearized system model can be given as
$$\begin{cases} \dot{\hat{X}}(t) = f(\hat{X},t) + B(t)u(t) + K(t)[Z(t) - f_z(\hat{X},t)], \\ \dot{P}(t) = A(\hat{X},t)P(t) + P(t)A^T(\hat{X},t) + G(t)S_w G^T(t) - P(t)C^T(\hat{X},t)S_v^{-1}C(\hat{X},t)P(t). \\ K(t) = P(t)C^T(\hat{X},t)S_v^{-1}. \end{cases}$$
$$(8.4-11)$$

It should be noted that at this time, the linearized model is not used in the estimator $\dot{\hat{X}}(t)$ equations, and the nonlinear model is still used. In this way, a higher estimation accuracy can be achieved.

In engineering applications, the Kalman filter often uses its discrete form to facilitate the realization by the computer programs.

1) Discrete Kalman filter of the linear time-varying system

(1) System model:
$$\begin{cases} \hat{X}_k = \Phi_k X_{k-1} + B_k u_{k-1} \\ Z_k = C_k X_k \end{cases}. \qquad (8.4-12)$$

(2) Kalman filter model:
$$\begin{cases} \hat{X}_k = \Phi_k \hat{X}_{k-1} + B_k u_{k-1} + K_k(Z_k - C_k \Phi_k \hat{X}_{k-1} - C_k B_k u_{k-1}) \\ M_k = \Phi_k P_{k-1} \Phi_k^T + Q_k \\ K_k = M_k C_k^T [C_k M_k C_k^T + R_k]^{-1} \\ P_k = (I - K_k C_k) M_k \end{cases}, \qquad (8.4-13)$$

where Q_k is the white noise variance of the system input, and R_k is the measurement noise variance. As for the calculation flow, refer to Fig. 8.4 – 3.

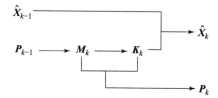

Fig. 8.4 – 3 Kalman filter calculation flow chart of the discrete linear time-varying system

Discrete algorithm of the extended Kalman filter

The discrete model of the system should be derived from the linearized model of the nonlinear system

$$\begin{cases} \hat{X}_k = \Phi_k(X)X_{k-1} + B_k u_{k-1} \\ Z_k = C_k(X)X_k \end{cases} \quad (8.4-14)$$

Its Kalman filter algorithm is

$$\begin{cases} \hat{X}_k^* = \int_{(k-1)T_s}^{kT_s} [f(\hat{X}_{k-1}) + B_k(\hat{X}_{k-1})u_k] dt + \hat{X}_{k-1} \\ \hat{X}_k = \hat{X}_k^* + K_k[Z_k - f_z(\hat{X}_k^*)] \\ M_k = \Phi_k(\hat{X}_{k-1})P_{k-1}\Phi_k^T(\hat{X}_{k-1}) + Q_k(\hat{X}_{k-1}) \\ K_k = M_k C_k^T(\hat{X}_{k-1})[C_k(\hat{X}_{k-1})M_k C_k^T(\hat{X}_{k-1}) + R_k]^{-1} \end{cases} \quad (8.4-15)$$

where Q_k is the white noise variance of the system input, and R_k is the measurement noise variance.

In the same way, the linearized model here is only used for the solution of M_k and K_k, while the estimation model still retains the nonlinear model in the solution of $\hat{X}_k^*$ and $\hat{X}_k$ in order to improve accuracy.

Refer to its calculation flow chart in Fig. 8.4 – 4.

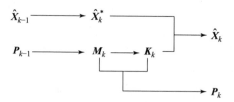

Fig. 8.4 – 4 Calculation flow chart of the Kalman filter of the discrete non-linear time-varying system

As described previously, it is known that to calculate the feedback matrix $K(t)$, three inputs are necessary in addition to the system model. They are the power spectrum S_w of the system noise (or Q_k in the discrete model), the power spectrum S_v of the measurement noise (or R_k in the discrete model) and the initial value $P(0)$ of the state estimation error covariance matrix. These three inputs are discussed in more detail in the following.

1) Initial value $P(0)$ of the state estimation error covariance matrix

Since the feedback gain matrix $K(t)$ in the state estimation equation is $K(t) = P(t)C^T(t)S_v^{-1}$, it can be seen that the higher the covariance matrix $P(t)$ value of the state estimation error, the higher the feedback matrix $K(t)$ value, that is, the wider the frequency bandwidth of the state estimator, the faster the estimation transient process.

For a linear time-varying system, $P(t)$ is obtained by solving the following nonlinear differential equation

$$\dot{P}(t) = AP(t) + P(t)A^{T} + GS_{w}G^{T} - P(t)C^{T}S_{v}^{-1}CP(t).$$

Since the system stimulation S_w and measurement error S_v in this equation are both white noise, with the increase of time, the optimal solution of the stationary random process should be independent of time after the transient process of $P(t)$ caused by the initial value $P(0)$ is over. Namely, the solution of min $E[\{X(\infty) - \hat{X}(\infty)\}^{T}\{X(\infty) - \hat{X}(\infty)\}]$ when $t = \infty$ of the Kalman filter is a constant value P_{∞}. Because $\dot{P} = 0$, the solution of P_{∞} is just the solution of the following Riccati non-linear algebraic equation

$$AP_{\infty} + P_{\infty}A^{T} + GS_{w}G^{T} - P_{\infty}C^{T}S_{v}^{-1}CP_{\infty} = 0. \qquad (8.4-16)$$

That is to say, the solution of $P(t)$ has the characteristics shown in Fig. 8.4 – 5.

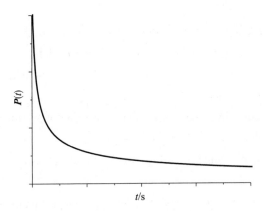

Fig. 8.4 – 5 Typical time characteristics of the covariance matrix solution $P(t)$ of the state estimation error

Therefore, a large value of $P(0)$ makes a large value of $P(t)$ in the initial phase, that is, the wider the Kalman filter bandwidth, the faster the estimation initial transient process in the initial phase. However, the value of $P(\infty)$ is independent of the value of $P(0)$, that is to say, $P(0)$ has no effect on the bandwidth and characteristics of the estimator after the filter enters the steady state.

Since the missile engagement time is not very long, there is a clear need for fast target acceleration estimation. Therefore, for the task of target acceleration estimation, it is very important to ensure that the choice of $P(0)$ value is appropriate to achieve a fast target maneuver acceleration estimation.

As the estimation begins, we do not know the target maneuver direction and its maneuver acceleration level, so the initial estimated value $\hat{a}(0)$ of the target acceleration of the estimator state equation can only be taken as 0, but the initial estimation error of $P(0)$ of the target acceleration should be taken as maximum possible target maneuver acceleration value a_t, that is, the corresponding element of $P(0)$ should be a_t^2. This is a large initial value for the estimation error. Widening the initial filter bandwidth is helpful to quickly eliminate the system state estimation error.

Generally, $P(0)$ is taken as a diagonal matrix. The initial value of each state estimation error can be taken as the estimated or known variance of each state.

2) Power spectrum matrix of the measurement noise S_v (or R_k matrix)

According to users' understanding of the selected measurement sensor, it should not be difficult to choose a reasonable S_v value. It is known from $K(t) = P(t)C^T(t)S_v^{-1}$ that S_v^{-1} and $K(t)$ will decrease with the increase of S_v. This will make the filter bandwidth narrower, which increases the weight of the estimator's filtering ability.

3) Power spectrum matrix S_w of the system noise (or Q_k matrix)

The rational choice of S_w is not so simple and its value will directly affect the bandwidth of the estimator in the initial stage as well as the steady state stage. The Kalman filter theory is to use the system white noise to simulate the phenomenon that the system could deviate from its ideal state by various factors, including

(a) Uncertain external disturbances in the operation of the system.

(b) The system parameters deviation from its designed model.

(c) With a given S_v, changing the filter bandwidth by changing the S_w matrix to quickly obtain the best estimate for the system state.

Due to the above reasons, in actual applications the value of S_w will be adjusted many times to determine its reasonable value with the consideration of the above requirements.

The following is a simple example to illustrate the ability of the Kalman filter to estimate the target maneuver acceleration.

Suppose that we have a one-dimensional missile-target engagement case and the target suddenly makes a 5g constant horizontal maneuver during the missile-target engagement to avoid the attack. If the missile's active seeker can measure the target maneuver distance y from the moment $t = 0$, then we can use the following simple Kalman filter state dynamics model and measurement model

$$\begin{cases} \dfrac{dy}{dt} = V \\ \dfrac{dV}{dt} = a \\ \dfrac{da}{dt} = 0 + w_s \end{cases}, \qquad (8.4-17)$$

$$z = y, \qquad (8.4-18)$$

where V is the target velocity, a is the target acceleration, z is the measurement and here z is the target distance y. Suppose that the distance measurement error is $\sigma_y = 20$ m, the initial value of the velocity estimation error is $\sigma_V = 50$ m/s and the initial value of the maneuver acceleration estimation error is $\sigma_a = 5g = 49$ m/s^2. Take the two estimator design parameters random input and measurement power spectrum as

$$S_v = 0.04 \ (\text{m}^2\text{s}), \ S_w = 0.01 \ (\text{m}^2/\text{s}^5),$$

and the initial value $\boldsymbol{P}(0)$ of the state estimation error covariance matrix as

$$\boldsymbol{P}(0) = \begin{bmatrix} 20^2 & 0 & 0 \\ 0 & 50^2 & 0 \\ 0 & 0 & (5 \times 9.81)^2 \end{bmatrix}.$$

The solutions to the nonlinear equation $\boldsymbol{P}(t)$ are shown in Fig. 8.4−6, Fig. 8.4−7 and Fig. 8.4−8.

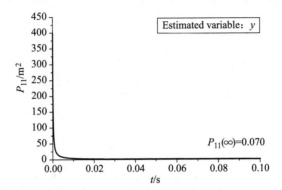

Fig. 8.4−6 $P_{11}(t)$ curve

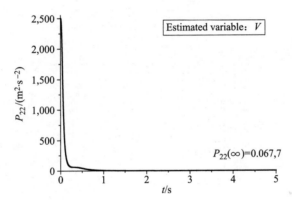

Fig. 8.4−7 $P_{22}(t)$ curve

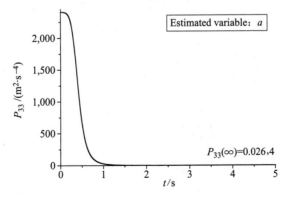

Fig. 8.4−8 $P_{33}(t)$ curve

As can be known from the figures, the initial values of P_{11}, P_{22} and P_{33} are the initial values of $P(0)$. After the transition process, $P(t)$ enters its respective steady state value P_∞ independent of $P(0)$.

The measurement estimation error feedback gain K_{11}, K_{21}, K_{31} can be obtained by using the equation $K(t) = P(t)C^T(t)S_v^{-1}$ (Fig. 8.4-9, Fig. 8.4-10 and Fig. 8.4-11). As can be seen from the figures, the feedback gain at the initial stage of each channel is high, which is conducive to quickly eliminate the initial estimation error. After the estimation transition process, the feedback gain becomes a steady state value. The magnitude of the steady state gain determines the ability of the system to resist various random disturbances and system parameter fluctuations during the steady state operation.

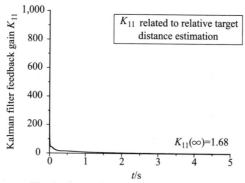

Fig. 8.4-9   Feedback gain K_{11} curve

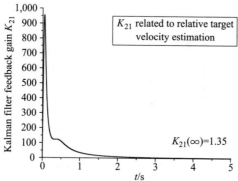

Fig. 8.4-10   Feedback gain K_{21} curve

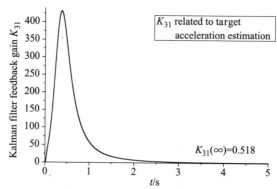

Fig. 8.4-11   Feedback gain K_{31} curve

Let us assume the following initial state deviations $y(0)$, $V(0)$ and $a(0)$:
$y(0) = 20$ m,
$V(0) = 50$ m/s,
$a(0) = 5g$ m/s^2.

Since the initial value of the estimator has no prior information, the estimated states $\hat{y}(0)$, $\hat{V}(0)$ and $\hat{a}(0)$ can only be taken as 0, that is,
$\hat{y}(0) = 0$ m,
$\hat{V}(0) = 0$ m/s,
$\hat{a}(0) = 0$ m/s^2.

The result of $\boldsymbol{K}(t)$ is substituted into the differential equations of the actual system and the estimator differential equations, and $y(t)$, $V(t)$, $a(t)$ curves of the actual system, and $\hat{y}(t)$, $\hat{y}(t)$, $\hat{a}(t)$ curves of the estimator output are shown in Fig. 8.4 – 12, Fig. 8.4 – 13, Fig. 8.4 – 14 and Fig. 8.4 – 15.

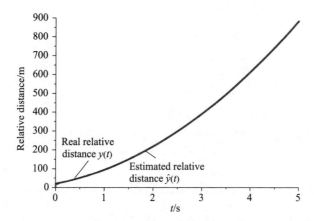

Fig. 8.4 – 12 Estimated and actual relative distance curves

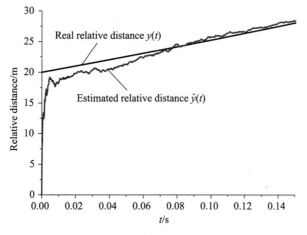

Fig. 8.4 – 13 Estimated and actual relative distance curves of the initial phase

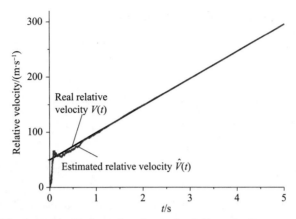

Fig. 8.4-14 Estimated and actual relative velocity curves

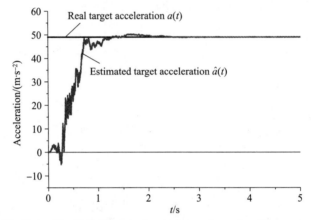

Fig. 8.4-15 Estimated and actual target acceleration curves

Fig. 8.4-16, Fig. 8.4-17 and Fig. 8.4-18 show the estimation errors $y - \hat{y}$, $V - \hat{V}$ and $a - \hat{a}$ curves. It can be known from the figures that even if we do not have any prior knowledge of the target maneuver ($\hat{a}(0) = 0$), as long as we can have some of the target motion measurement (here it is the relative distance $y(t)$), the Kalman filter technique can be used to quickly estimate the target maneuver acceleration.

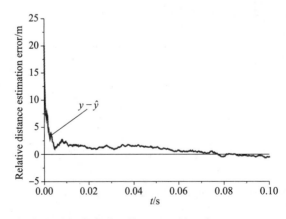

Fig. 8.4-16 Relative distance estimation error curve

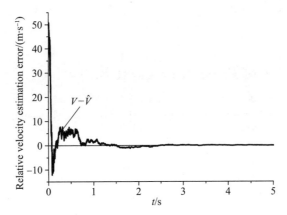

Fig. 8.4 – 17 Relative velocity estimation error curve

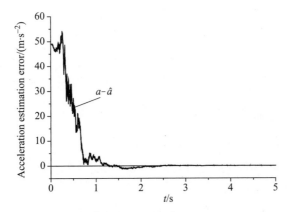

Fig. 8.4 – 18 Acceleration estimation error curve

When the seeker works at the medium pulse repetition frequency mode, and if the accuracy of the velocity measurement is acceptable, the velocity measurement can also be taken as the system measurement output along with the distance measurement. This will lead to a better state estimation accuracy.

When the seeker is operating at the high pulse repetition frequency mode, and if its velocity measurement is reliable, the velocity measurement alone can be used as the system measurement output to perform the system state estimation.

At present, the distance and velocity measurement accuracies of the current radar seeker are not very high. Therefore, it is still difficult to adopt the guidance law with target maneuver compensation, but with the improvement of the seeker performance, the application of this technology is still promising.

Finally, it should be noted that the perturbation model of a practical system is not completely consistent with the stochastic model set by Kalman filter, so it is not necessary to insist on obtaining accurate input values for S_w, S_v and $P(0)$ in practical applications. But the biggest advantage of Kalman filter is that it has pointed out the specific effects of the filter parameters S_w, S_v and $P(0)$ on the bandwidth of the estimator. When the actual estimation outcomes are found to be unsatisfactory, it is not difficult to know how to adjust the values of S_w, S_v or $P(0)$ to obtain better estimation results.

§ 8.5 Optimum Trajectory Control System Design

The mathematical model for the optimum trajectory control design is given as the system dynamics equation

$$\frac{dx}{dt} = f(x(t), u(t), t). \quad (8.5-1)$$

In the above equation, x is the system state variable and u is the system control variable.

The following design constraints may occur in the trajectory optimization problem.

(1) Initial state constraint:

$$\phi_{0min} < \phi_0(x_0, t_0) < \phi_{0max}. \quad (8.5-2)$$

(2) Terminal state constraint:

$$\phi_{fmin} < \phi_f(x_f, t_f) < \phi_{fmax}. \quad (8.5-3)$$

(3) State initial and terminal state value constraints and state variable constraints:

$$x_{0min} < x_0 < x_{0max},$$
$$x(t)_{min} < x(t) < x(t)_{max}, \quad (8.5-4)$$
$$x_{fmin} < x_f < x_{fmax}.$$

(4) Derived variable $C = C(x, u, t)$ constraints, where C is related to the state x and control u:

$$C(t)_{min} < C(x(t), u(t), t) \leqslant C(t)_{max}. \quad (8.5-5)$$

(5) Control constraints:

$$u(t)_{min} < u(t) < u(t)_{max}. \quad (8.5-6)$$

(6) System parameters constraints:

$$P_{min} < P < P_{max}. \quad (8.5-7)$$

It is required to obtain the optimal control $u(t)$ under the condition that all the above constraints are satisfied and at the same time the following objective function J is minimized:

$$J = \phi(x_0, x_f, t_0, t_f) + \int_{t_0}^{t_f} L(x(t), u(t), P, t) dt. \quad (8.5-8)$$

Because this optimal control problem is constrained at both ends of the trajectory, it is mathematically often called a two-point boundary value problem (TPBVP). If a continuous system model is used to solve such a problem, its computational cost may be so large that it is unacceptable for practical engineering applications.

It is noteworthy that the pseudo-spectral software currently available on the market can solve these problems through continuous function parameterization.

We know that using a polynomial interpolation function to fit a continuous function at finite number of nodes can parameterize the continuous function and approximate it with the polynomial function. However, in order to ensure interpolation accuracy, the number of interpolation nodes must be large, which means that the order of the interpolation polynomial must be high. When the equally spaced nodes are taken and the polynomial order is high, a very large interpolation error could occur. For example, taking the continuous function $f(x) = 1/(1 + 12x^2)$ as an example, when 25 equally spaced nodes are taken in the interval $[-1, 1]$ of the independent variable x, the polynomial fitting result and its error are shown in Fig. 8.5 – 1. It can be seen that the use of equally spaced mode polynomial functions cannot meet the task of approximating a continuous function.

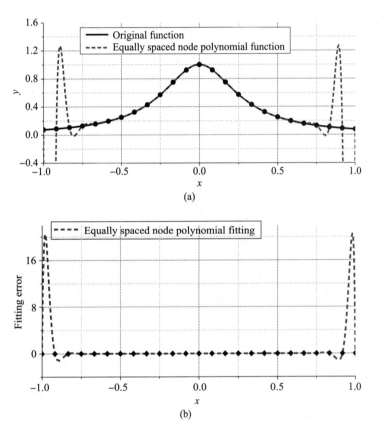

Fig. 8.5 – 1 Equally spaced node polynomial fitting with 25 nodes

(a) Equally spaced node polynomial fitting; (b) Equally spaced node polynomial fitting error

Orthogonal polynomial interpolation fitting uses variable spaced nodes to generate the interpolation fitting function, which can automatically narrow the node interval where there could be a large fitting error, thus ensuring consistent fitting accuracy in the entire fitting range. Fig. 8.5 – 2 uses orthogonal polynomials and takes 25 nodes to fit the aforementioned functions. It can be seen that when a proper number of nodes for orthogonal polynomials is chosen, a continuous function could be parameterized accurately.

Since the system state differential equation is an equality constraint that must be satisfied among the problems we are confronted with, the state derivative $\dot{x}(t)$ also has to be parameterized here. So, it is essential to ensure that the state derivative $\dot{x}(t)$ can be well approximated by the derivative of its

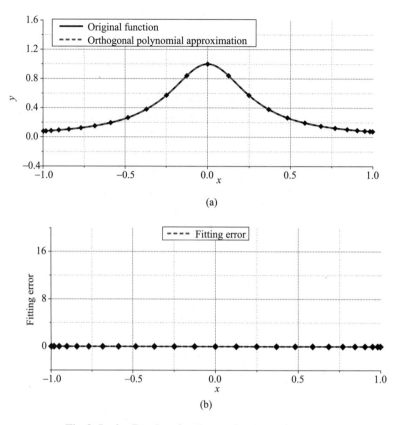

Fig. 8.5 − 2 Results of orthogonal polynomial fitting
(a) Orthogonal polynomial fitting; (b) Orthogonal polynomial fitting error

corresponding state polynomial approximation. Fig. 8.5 − 3 shows the difference of the function derivative mentioned above and its orthogonal polynomial derivative. It can be seen that the orthogonal polynomial not only guarantees the approximation of the function, but also guarantees the approximation of its derivative. With the above method of parameterization of continuous functions using orthogonal polynomials, the optimal trajectory problem mentioned above can be transformed into a standard constrained nonlinear programming problem.

The standard nonlinear programming problem has the following form.

(1) Objective function:
$$\min J(X);$$

(2) Subject to:
$$\text{equality constraints} f_E(X) = 0;$$
$$\text{inequality constraints} f_{NE}(X) \leqslant 0.$$

where X is the design variable vector of the nonlinear programming problem.

The trajectory optimization problem has the following nonlinear programming design variables after parameterization of the states and controls:

(1) Corresponding design variables of state functions parameterization each with n node polynomial parameters as the design variables

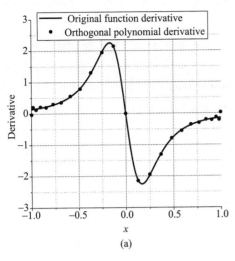
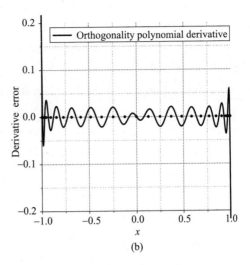

Fig. 8.5-3 Approximation result of the original function derivative and its orthogonal polynomial derivative

(a) Derivative curve; (b) Orthogonal polynomial derivative fitting error

$$x_i(t) \approx L_{xi}(x_{i1}, \cdots, x_{ik}) \ (i = 1, \cdots, n).$$

In the above, L_{xi} is the orthogonal polynomial corresponding to the ith state function $x_i(t)$, k is the number of nodes of the $(k-1)$ order orthogonal polynomial. So, the number of the nonlinear programming variables corresponding to the state function x is $n \cdot k$.

(2) Corresponding design variables of m control functions parameterization:

$$u_j(t) \approx L_{uj}(u_{j1}, \cdots, u_{jk}) \ (j = 1, \cdots, m).$$

Similarly, the number of nonlinear programming design variables corresponding to the parameterization of the control function is $m \cdot k$.

(3) The number of design variables corresponding to the system parameters $P_1, \cdots, P_s$ is s.

(4) The number of design variables corresponding to the initial value of time t_0 is 1.

(5) The number of design variables corresponding to the final time value t_f is 1.

It is known from the above that the nonlinear programming design variable X is

$$X = \begin{bmatrix} x \\ u \\ P \\ t_0 \\ t_f \end{bmatrix}.$$

The total number of design variables is $(m+n)k + s + 2$.

- Equality constraint: The equality constraint at the orthogonal polynomial nodes (Note: The original n state equation has been converted into $n \cdot k$ equality constraint at k nodes).

- Inequality constraint: The corresponding inequality constraint at the initial time, the final time and all orthogonal polynomial nodes.

- Objective function: The discrete integral result of the trajectory optimization objective function with J as the equation (8.5 – 8).

The orthogonal polynomial approximate solution of the optimal problem control function $u(t)$ can be obtained by solving the above nonlinear programming problem. It is noteworthy that the design inputs for using this software are still the continuous design inputs given as equations (8.5 – 1) – (8.5 – 8). The above discretization and nonlinear programming processes are the functions of the software which users do not need to be concerned with.

The scope of use of this software is very flexible. For example, the control variables can be taken as missile lateral acceleration $a(t)$, angle of attack $\alpha(t)$, sideslip angle $\beta(t)$, etc.; the objective function can be chosen as maximum range, maximum final velocity, minimum control integral, etc. The choices of the constraint are even flexible, for example, there may be dynamic pressure constraint $q = \frac{1}{2}\rho V^2 \leq q_{max}$, normal acceleration constraint $n(t) \leq n_{max}$, heat flow constraint $Q = \frac{c}{\sqrt{R_d}}\rho^{0.5}V^{3.08}$ (of which c is the constant associated with the aircraft characteristics, R_d is the radius of curvature of the aircraft stagnation point), angle of attack constraints, control surface deflection constraints and so on. The software can even be used for trajectory optimization of multi-stage rockets.

Next is a simple example presented to illustrate the importance of trajectory optimization. Suppose that there is a guided rocket, the burn-out velocity of the rocket is 1,800 m/s, the control is constrained by maximum lateral acceleration, the terminal inequality constraints are final velocity $V_f \geq 800$ m/s, the impact angle $|\theta_k| \geq 60°$, and the optimum missile lateral acceleration function $a(t)$ which can give the maximum range needs to be solved. The dashed trajectory in Fig. 8.5 – 4 gives the optimized missile flight trajectory, and the maximum range which satisfies the final velocity and impact angle constraints is 525 km. The results indicate that:

(1) When the guided rocket has a range of 400 – 800 km, the maximum height which corresponds to the optimal trajectory generally does not exceed 50 km, that is, the rocket can be considered as an air-guidance rocket, and the air rudder can be used to control the guidance rocket.

(2) Guidance rockets should adopt low-altitude leaping trajectories to increase long-range and low-air-density flight duration to obtain a larger range.

The maximum range available for this example with this optimal trajectory is approximately 525 km. If a general ballistic missile scheme is used and the angle of departure θ_0 is taken as an optimization parameter, the range of the missile can only reach 380 km and the maximum trajectory altitude should reach 104 km (see the solid line trajectory in Fig. 8.5 – 4) under the optimal angle of departure. It can be seen that the use of a general long-range ballistic missile program in this range is not reasonable. In short, even if the software is used during the concept design stage, it can help the designer to choose a reasonable control scheme.

8 Proportional Navigation and Extended Proportional Navigation Guidance Laws

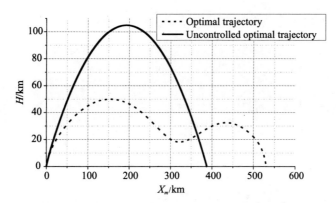

Fig. 8.5 – 4 Curves of the optimal trajectory and the uncontrolled optimal trajectory

9

Optimal Guidance for Trajectory Shaping

§ 9.1　Optimality of Error Dynamics in Missile Guidance

In essence, missile guidance law design is a kind of finite-time tracking problem depending on the operational objective. If only target interception is considered, the tracking error is defined as the ZEM distance. Nullifying the ZEM results in perfect interception with zero miss distance. For some types of tactical missiles, constraining the final impact angle or intercept angle is helpful to enhance the kill probability of the warhead or maintain advantageous homing engagement. To satisfy this requirement, the terminal impact angle error is considered as the tracking error in guidance law design. In order to enhance the survivability of anti-ship missiles against advanced close-in weapon system of battleships, the concept of salvo attack (reference [14]) is introduced to achieve simultaneous attack among all interceptors.

One typical implementation of salvo attack is the impact time guidance, in which the missile is forced to intercept the target at a desired time instance and hence the final impact time error is considered as the tracking error in guidance law design.

The guidance command is derived to force the system trajectory to follow the desired error dynamics, which provides desired convergence pattern of the tracking error. This is a general procedure to establish new missile guidance laws using nonlinear control methodologies. However, most previous studies only focus on how to force the tracking error to converge to zero and disregard what is the optimal error dynamics in terms of a meaningful performance index.

Inspired by the preceding observations, this chapter attempts to investigate the optimal convergence pattern of the tracking error and propose an optimal error dynamics with a meaningful cost function for missile guidance problems. To realize this, we solve the linear quadratic optimal control problem for a generalized tracking problem that is often observed in missile guidance law design using Schwarz's inequality approach.

For illustration, we present two examples to show how to apply the proposed optimal error dynamics to missile guidance law design. The presented examples include impact angle guidance and impact time guidance. By utilizing the proposed optimal error dynamics, the physical meaning of these guidance laws developed is theoretically analyzed and the cost functions provide useful guidelines on how to choose proper gains to guarantee bounded terminal guidance command. Theoretical analysis also reveals that some previous optimal guidance laws are special cases of the proposed optimal solutions. Hence, the proposed optimal error dynamics provides a unique and generalized way in establishing optimal guidance laws.

9.1.1 Optimal Error Dynamics

As stated earlier, missile guidance law design is a kind of finite-time tracking problem and the objective is to regulate a specific mission-dependent tracking error to zero in finite time. The general form of the tracking problem that is often observed in missile guidance law design is

$$\dot{\varepsilon}(t) = g(t)u(t), \qquad (9.1-1)$$

where $\varepsilon(t)$ represents the tracking error; $g(t)$ stands for a known time-varying function; and $u(t)$ denotes the control input.

According to a specific guidance problem, the tracking error can be ZEM, impact angle error, impact time error, heading error, etc.

In many previous works, one widely-accepted error dynamics for guidance law design is given by

$$\dot{\varepsilon}(t) + k\varepsilon(t) = 0, \qquad (9.1-2)$$

where $k > 0$ is the guidance gain to regulate the convergence rate of tracking error.

It is easy to verify that the closed-form solution of differential equation (9.1-2) is determined as

$$\varepsilon(t) = \varepsilon(t_0) e^{-kt}, \qquad (9.1-3)$$

where $\varepsilon(t_0)$ denotes the initial tracking error.

The preceding equation reveals that the tracking error converges to zero asymptotically with an exponential rate governed by the guidance gain k. Accordingly, this desired error dynamics has two major drawbacks:

(1) The finite-time convergence is not strictly guaranteed.

(2) It only focuses on how to drive the tracking error to zero and never considers what the optimal error dynamics in terms of a meaningful performance index is.

Motivated by these observations, this chapter aims to investigate the optimal convergence pattern of the tracking error and propose an optimal error dynamics that achieves this optimal pattern with guaranteed finite-time convergence for missile guidance law design.

Here we discuss the optimal error dynamics for missile guidance law design. The main results are presented in Theorem 9.1.

Theorem 9.1 Suppose that the system equation is given by equation (6.7-20) and the desired error dynamics is chosen as

$$\dot{\varepsilon}(t) + \frac{\Gamma(t)}{t_{go}} \varepsilon(t) = 0, \qquad (9.1-4)$$

where

$$\Gamma(t) = \frac{t_{go} R^{-1}(t) g^2(t)}{\int_t^{t_f} R^{-1}(\tau) g^2(\tau) d\tau}, \qquad (9.1-5)$$

with $R(t) > 0$ being an arbitrary weighting function.

Then, the resulting control input can be obtained as

$$u(t) = -\frac{\Gamma(t)}{g(t)t_{go}}\varepsilon(t) = -\frac{R^{-1}(t)g(t)}{\int_t^{t_f} R^{-1}(\tau)g^2(\tau)d\tau}\varepsilon(t), \quad (9.1-6)$$

to minimize the performance index

$$\min_u J = \frac{1}{2}\int_t^{t_f} R(\tau)u^2(\tau)d\tau \quad (9.1-7)$$

Proof The control input that enables system, the is, equation (6.7-20) to follow error dynamics that is, equation (9.1-4) is determined by substituting equation (9.1-4) into equation (9.1-4) as equation (9.1-6).

Next, we seek to prove that control input equation (9.1-6) is the optimal solution of the following optimization problem. The index of the problem is given as equation (9.1-7), while the constrains are

$$\dot{\varepsilon}(t) = g(t)u(t), \varepsilon(t_f) = 0. \quad (9.1-8)$$

Integrating from t to t_f on both sides of equation (9.1-8) gives

$$\varepsilon(t_f) - \varepsilon(t) = \int_t^{t_f} g(\tau)u(\tau)d\tau. \quad (9.1-9)$$

Imposing terminal constraint on equation (9.1-9) gives

$$-\varepsilon(t) = \int_t^{t_f} g(\tau)u(\tau)d\tau, \quad (9.1-10)$$

Introducing a slack variable $R(t)$, equation (9.1-10) can be transformed as

$$-\varepsilon(t) = \int_t^{t_f} g(\tau)R^{-1/2}(\tau)R^{1/2}(\tau)u(\tau)d\tau. \quad (9.1-11)$$

Applying Schwarz's inequality to the preceding equation yields

$$[-\varepsilon(t)]^2 \leq \left[\int_t^{t_f} R^{-1}(\tau)g^2(\tau)d\tau\right]\left[\int_t^{t_f} R(\tau)u^2(\tau)d\tau\right]. \quad (9.1-12)$$

Inequality equation (9.1-12) can be rewritten as

$$\frac{1}{2}\int_t^{t_f} R(\tau)u^2(\tau)d\tau \geq \frac{[-\varepsilon(t)]^2}{2\left[\int_t^{t_f} R^{-1}(\tau)g^2(\tau)d\tau\right]}. \quad (9.1-13)$$

As known, the equality of equation (9.1-13) holds if and only if there exists a constant C such that

$$u(t) = CR^{-1}(t)g(t). \quad (9.1-14)$$

Substituting equation (9.1-14) into equation (9.1-10), we have

$$-\varepsilon(t) = C\int_t^{t_f} R^{-1}(t)g^2(\tau)d\tau. \quad (9.1-15)$$

Solving equation (9.1-15) for C gives

$$C = \frac{-\varepsilon(t)}{\int_t^{t_f} R^{-1}(\tau)g^2(\tau)d\tau}. \quad (9.1-16)$$

Substituting equation (9.1-16) into equation (9.1-14) gives the optimal control input as

$$u(t) = -\frac{R^{-1}(t)g(t)}{\int_t^{t_f} R^{-1}(\tau)g^2(\tau)d\tau}\varepsilon(t), \quad (9.1-17)$$

which is identical with equation (9.1-6).

9.1.2 Analysis of Optimal Error Dynamics

One of the interesting points of the proposed optimal error dynamics is that it guarantees finite-time convergence since it is obtained by directly solving the finite-time optimal tracking problem. Also, the optimal error dynamics is given in a similar form as the existing desired error dynamics. The only difference lies in the proportional gain: a constant term k in the existing method and a time-varying term in the proposed method. Unlike the usual one, the proposed proportional gain changes from an initial small value to a final infinite value as the time-to-go goes to zero, namely

$$\lim_{t_{go} \to 0} \frac{\Gamma(t)}{t_{go}} = \infty. \qquad (9.1-18)$$

According to the value of $\Gamma(t)$, the evolving pattern of the proposed proportional gain further changes.

Let us discuss the characteristics of $\Gamma(t)$. From equation (9.1-5), for convenience, $\Gamma(t)$ can be rearranged as

$$\Gamma(t) = \frac{\varphi(t)}{\left(\int_t^{t_f} \varphi(\tau) \, d\tau / t_{go}\right)}, \qquad (9.1-19)$$

where $\varphi(t) = R^{-1}(t) g^2(t)$.

The function of $g(t)$ is given by the missile guidance problem under consideration and the weighting function $R(t)$ is the design parameter. According to different selections of $R(t)$, the time-varying term $\Gamma(t)$ changes differently. In the following, $\Gamma(t) > 0$, $\forall t > 0$ during the homing engagement in proposition 9.1.

Proposition 9.1 For given $g(t)$ and $R(t)$, $\Gamma(t)$ is always greater than zero, i.e., $\Gamma(t) > 0$.

Proof By definition, we have $R(t) > 0$ and $g(t) \neq 0$. From equation (9.1-19), it is obvious that the numerator is positive since $\varphi(t) = R^{-1}(t) g^2(t) > 0$. Also, the denominator is positive since the integration of positive function gives positive value. Accordingly, we have $\Gamma(t) > 0$.

Proposition 9.2 When function $\varphi(t)$ keeps constant, the equality $\Gamma(t) = 1$ holds. If $\varphi(t)$ decreases as $t \to t_f$, then $\Gamma(t) > 1$. If $\varphi(t)$ increases as $t \to t_f$, one has $\Gamma(t) < 1$.

Proof In equation (9.1-19), the denominator can be considered as the average value of $\varphi(t)$, denoted by $\bar{\phi}(t)$, during the remaining time of interception. Accordingly, the time-varying term $\Gamma(t)$ is the ratio of the current value $\varphi(t)$ to the average value $\bar{\phi}(t)$. If the term of $\varphi(t)$ is given by a constant value, then the $\Gamma(t)$ is unity as $\Gamma(t) = 1$ since $\varphi(t) = \bar{\phi}(t)$. If the term $\varphi(t)$ decreases as $t \to t_f$, $\Gamma(t)$ is greater than unity due to $\varphi(t) > \bar{\phi}(t)$. Conversely, if the term of $\varphi(t)$ has an increasing pattern, then $\Gamma(t)$ is less than unity because of $\varphi(t) < \bar{\phi}(t)$.

Remark 9.1 It follows from equation (9.1-4) that $\Gamma(t) \geq 1$ is desirable to ensure a stable and a fast convergence rate at the initial time. Accordingly, Proposition 9.2 reveals that it is desirable to impose a constant value or a decreasing pattern on the function $\varphi(t)$.

Remark 9.2 As shown in equation (9.1−5), the computation of $\Gamma(t)$ contains the integration of $\varphi(t)$, which is given as a function of $g(t)$ and $R(t)$. If $\varphi(t)$ is given by a closed-form function, $\Gamma(t)$ can be obtained analytically or numerically. Even though $g(t)$ is not given by a closed-form function due to the nature of the guidance problem, we can find closed-form solution of $\varphi(t)$ through appropriate choice of the weighting function $R(t)$. Under this condition, the chosen optimal error dynamics only considers the minimization of a specific performance index since the physical meaning of the performance index shown in equation (9.1−7) changes according to the choice of $R(t)$.

For a specific mission, the function $g(t)$ is fixed. Thus, by properly choosing the weighting function $R(t)$ for different objectives, the error dynamics is determined by theorem 9.1. Here, we provide two special cases. In the case of $R(t)=1$, the term $\Gamma(t)$ is given by

$$\Gamma(t) = \frac{t_{go} g^2(t)}{\int_t^{t_f} g^2(\tau) d\tau}. \qquad (9.1-20)$$

The optimal error dynamics with $\Gamma(t)$ shown in equation (9.1−5) minimizes the performance index

$$J = \frac{1}{2}\int_t^{t_f} u^2(\tau) d\tau, \qquad (9.1-21)$$

which provides an energy optimal guidance law.

In addition, if $K \geqslant 1$, we choose the weighting function as

$$R(t) = \frac{g^2(t)}{t_{go}^{K-1}}. \qquad (9.1-22)$$

Then $\varphi(t)$ is given by a function of time-to-go, regardless of $g(t)$, as

$$\varphi(t) = t_{go}^{K-1}. \qquad (9.1-23)$$

In this case, $\Gamma(t)$ is a constant as

$$\Gamma(t) = K. \qquad (9.1-24)$$

Then, the desired error dynamics is given by

$$\dot{\varepsilon}(t) + \frac{K}{t_{go}}\varepsilon(t) = 0. \qquad (9.1-25)$$

Solving equation (9.1−25) gives the closed-form of the tracking error as

$$\varepsilon(t) = \varepsilon(t_0)\left(\frac{t_{go}}{t_f}\right)^K. \qquad (9.1-26)$$

In practice, this optimal error dynamics, shown in equation (9.1−25), is useful due to the fact that the decreasing pattern of the tracking error is predictable as a function of time-to-go and the desired error dynamics is given by a simplified form. Therefore, in the next section, we will apply this desired error dynamics to various missile guidance problems for simplicity and practicality. Following theorem 9.1, the desired error dynamics equation (9.1−25) minimizes the performance index

$$J = \frac{1}{2}\int_t^{t_f} \frac{g^2(\tau)}{(t_f-\tau)^{K-1}} u^2(\tau) d\tau. \qquad (9.1-27)$$

§9.2 Optimal Predictor-corrector Guidance

9.2.1 General Approach for Guidance Law Design

This subsection presents the details of a general approach of how to utilize the proposed optimal error dynamics in guidance law design. To this end, let q be the variable that we would like to control for a specific mission objective. Suppose that the dynamics of q is given by a general nonlinear differential equation as

$$\dot{q}(t) = f(t) + b(t)u(t), \qquad (9.2-1)$$

where $f(t)$, $b(t) \neq 0$ are time-varying functions and $u(t)$ denotes the command input.

Formulate the guidance command in a composite form as

$$u(t) = u_0(t) + u_b(t), \qquad (9.2-2)$$

where $u_0(t)$ denotes the nominal control part and $u_b(t)$ stands for a biased term that can be designed by using the proposed error dynamics.

Based on the general command equation (9.2-2), the systematic way to utilize the proposed optimal error dynamics in guidance law design is to follow a prediction-correction manner:

(1) Predict the terminal error of the variable that we would like to control for a specific mission objective with a nominal trajectory, i.e., only controlled by u_0, as

$$q(t_f) = q(t) + \int_t^{t_f} f(\tau) + b(\tau)u_0(\tau)\,d\tau. \qquad (9.2-3)$$

(2) Define $\varepsilon(t) = q(t_f) - q_d$ as the tracking error with q_d being the desired state. Since the nominal control part u_0 imposes no effect on $q(t_f)$, the dynamics of tracking error $\varepsilon(t)$ under composite command equation (9.2-2) can then be readily determined as

$$\dot{\varepsilon}(t) = -b(t)u_b(t). \qquad (9.2-4)$$

(3) Correct the tracking error by using a biased term u_b based on the proposed error dynamics.

9.2.2 Impact Angle Control

Constraining the impact angle is often desirable in terms of increasing the warhead effectiveness as well as inflicting the maximum damage of the target for both anti-ship and anti-tank missiles since it enables exploiting the structural weakness of the target. This subsection will show the details of how to utilize the proposed optimal error dynamics in trajectory shaping for impact angle control. For simplicity, a stationary target interception scenario is considered in this subsection. Following the guideline of the general prediction-correction concept, the guidance command is formulated in a composite form as

$$a_M = a_{Nor} + a_{IA}. \qquad (9.2-5)$$

where a_{Nor} denotes the nominal command to guarantee zero ZEM for target interception and a_{IA} represents the biased command that is utilized to control the impact angle.

For the purpose of illustration, we leverage the PNG as the nominal guidance command, i.e.,

$$a_{Nor} = NV_M\dot{\sigma}. \qquad (9.2-6)$$

According to reference [16], the final flight path angle governed by PNG is given by

$$\gamma_{M_f} = \frac{N}{N-1}\sigma - \frac{1}{N-1}\gamma_M. \quad (9.2-7)$$

Let γ_f be the desired final flight path angle. Then, the impact angle error can be defined as $\varepsilon_\gamma = \gamma_f - \gamma_{M_f}$. In order to nullify the impact angle error, consider ε_γ as the tracking error, which gives the error dynamics as

$$\dot{\varepsilon}_\gamma = -\dot{\gamma}_{M_f} = -\frac{N}{N-1}\dot{\sigma} + \frac{1}{N-1}\frac{a_M}{V_M}. \quad (9.2-8)$$

Substituting $a_M = a_{Nor} + a_{IA}$ into equation (9.2-8), we have

$$\dot{\varepsilon}_\gamma = \frac{a_{IA}}{(N-1)V_M}. \quad (9.2-9)$$

For the impact angle error ε_γ, the optimal error dynamics is selected in a similar way as

$$\dot{\varepsilon}_\gamma + \frac{K}{t_{go}}\varepsilon_\gamma = 0, \quad (9.2-10)$$

where $K \geq 1$ is the guidance gain to be designed.

The corresponding performance index is obtained as

$$J = \frac{1}{2}\int_t^{t_f} \frac{1}{(N-1)^2 V_M^2 (t_f - \tau)^{K-1}} u^2(\tau) d\tau. \quad (9.2-11)$$

Since the constant terms in the performance index have no effect on the optimal pattern, performance index equation (9.2-11) is identical to

$$J = \frac{1}{2}\int_t^{t_f} \frac{1}{(t_f - \tau)^{K-1}} u^2(\tau) d\tau. \quad (9.2-12)$$

It can be clearly observed from the preceding performance index that the weighting function with $K > 1$ becomes infinite as $t_{go} \to 0$. Therefore, one can imply that the optimal error dynamics with $K > 1$ guarantees zero impact angle guidance command at the final time. Additionally, if $K = 1$, the above performance index coincides with the energy optimal case.

From equation (9.2-9), equation (9.2-10), the optimal solution of a_{IA} is easily obtained as

$$a_{IA} = -\frac{K(N-1)V_M}{t_{go}}\varepsilon_\gamma. \quad (9.2-13)$$

Combining equation (9.2-5) with equation (9.2-13), we can derive the final guidance command for impact angle control as

$$a_M = NV_M\dot{\sigma} - \frac{K(N-1)V_M}{t_{go}}\varepsilon_\gamma. \quad (9.2-14)$$

If we choose $N=3$ and $K=1$, guidance law equation (9.2-14) becomes optimal guidance law (OGL) for impact angle control (reference [15]). If the guidance gains satisfy $N \geq 3$ and $K = 1$, guidance law equation (9.2-14) reduces to interception angle control guidance (IACG) law [16]. If one enforces $N = K + 2$ and $K \geq 1$, guidance law equation (9.2-14) is identical with time-to-go weighted optimal impact angle guidance (TWOIAG) law (reference [17]). Finally, if one selects $N > K + 1$ and $K \geq 1$, guidance law equation (9.2-14) turns out to be time-to-go polynomial guidance (TPG) law (reference [18]).

Consequently, using the proposed approach generates a generalized optimal impact angle control guidance law that covers several previous results. From guidance command equation (9.2 – 14), the optimal impact angle guidance law can also be interpreted as a predictor-corrector guidance law: the optimal impact angle guidance law first predicts the terminal flight path angle by using the optimal PNG as its control input and corrects the terminal flight path angle error by using desired error dynamics equation (9.2 – 10).

Without loss of generality, the desired impact angle is set as $\gamma_f = -90°$. Fig. 9.2 – 1 presents the simulation results obtained from guidance law equation (9.2 – 14) with various guidance gains. From Fig. 9.2 – 1 (a) and (b), we can clearly note that the proposed optimal guidance law successfully guides the missile to intercept the target with a desired impact angle. Increasing the guidance gains N and K results in more curved trajectory and hence requires more control energy. The terminal guidance command converges to zero once the guidance gains satisfy condition $N > 3$, $K > 1$, as confirmed by Fig. 9.2 – 1 (c). The control effort consumption, demonstrated in Fig. 9.2 – 1 (d) indicates that guidance law equation (9.2 – 14) with $N = 3$ and $K = 1$ provides energy optimal interception. These results clearly conform with our analytical analysis.

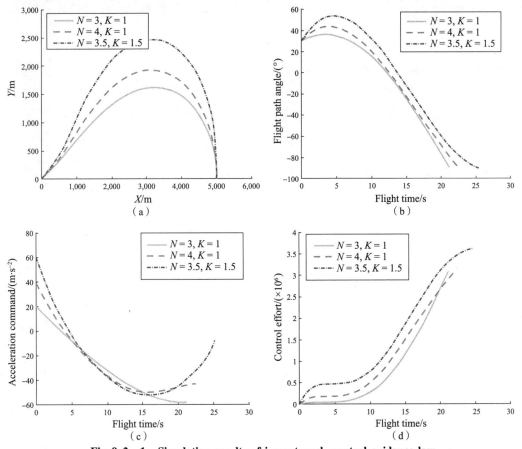

Fig. 9.2 – 1 Simulation results of impact angle control guidance law

(a) Interception trajectory; (b) Flight path angle; (c) Acceleration command; (d) Control effort

9.2.3 Impact Time Control

The ability to constrain the final impact time of anti-ship missiles is often desirable for increasing the survivability since this strategy enables penetrating ship-board self-defense systems by using multiple missiles to intercept a target simultaneously (reference [14]). One typical solution to realize simultaneous interception is the impact time control guidance, in which the missile is required to intercept the target at a specified time instant. In this subsection, we will show how to utilize the proposed optimal error dynamics to design impact time control guidance law. For simplicity, we only consider stationary target interception in this subsection.

Following the systematic prediction-correction guideline, we formulate the guidance command as

$$a_M = a_{Nor} + a_{IT}, \qquad (9.2-15)$$

where a_{Nor} denotes the nominal command to guarantee zero ZEM for target interception and a_{IT} stands for the biased correction part to nullify the impact time error.

Similarly, we choose the optimal PNG as the nominal guidance command, i.e.,

$$a_{Nor} = NV_M \dot{\sigma}. \qquad (9.2-16)$$

As derived in reference [14], the estimated final interception time under PNG is given by

$$t_f = t + \frac{r}{V_M}\left[1 + \frac{\sin^2\theta}{2(2N-1)}\right], \qquad (9.2-17)$$

where $\theta = \gamma_M - \sigma$ is the heading error.

Denote t_d as the desired impact time. Then, the impact time error can be defined as $\varepsilon_t = t_d - t_f$. In order to regulate the impact time error, consider ε_t as the tracking error, which gives the error dynamics as

$$\dot{\varepsilon}_t = -\dot{t}_f = -\frac{\dot{r}}{V_M}\left(1 + \frac{\sin^2\theta}{2(2N-1)}\right) - \frac{r(\sin\theta\cos\theta)\dot{\theta}}{(2N-1)V_M} - 1$$

$$= \cos\theta\left(1 + \frac{\sin^2\theta}{2(2N-1)}\right) - \frac{r\sin\theta\cos\theta\left(\frac{a_{M_0} + a_{IT}}{V_M} - \dot{\sigma}\right)}{(2N-1)V_M} - 1$$

$$= \cos\theta\left(1 + \frac{\sin^2\theta}{2(2N-1)}\right) + \frac{(N-1)\sin^2\theta\cos\theta}{2N-1} - \frac{r\sin\theta\cos\theta}{(2N-1)V^2}a_{IT} - 1. \qquad (9.2-18)$$

In practical flight, the lead angle $\theta = \gamma_M - \sigma$ is small and thus $\sin\theta \approx \theta$, $\cos\theta \approx 1 - \theta^2/2$. Using these approximations and neglecting the higher order term of θ, equation (9.2-18) can be reduced to

$$\dot{\varepsilon}_t = -\frac{r\sin\theta}{(2N-1)V_M^2}a_{IT}. \qquad (9.2-19)$$

In this example, we have $u(t) = a_{IT}$, then

$$g(t) = -\frac{r\sin\theta}{(2N-1)V_M^2}. \qquad (9.2-20)$$

The optimal error dynamics with respect to the impact time error ε_t is selected as

$$\dot{\varepsilon}_t + \frac{K}{t_{go}}\varepsilon_t = 0. \tag{9.2-21}$$

Substituting equation (9.2-21) into equation (9.2-19) gives the guidance command to nullify the impact time error as

$$a_{IT} = \frac{K(2N-1)V_M^2}{r\sin\theta t_{go}}\varepsilon_t. \tag{9.2-22}$$

Combining equation (9.2-16) with equation (9.2-22) leads to the impact time control guidance law as

$$a_M = NV_M\dot{\sigma} + \frac{K(2N-1)V_M^2}{r\sin\theta t_{go}}\varepsilon_t. \tag{9.2-23}$$

Note that if we choose $K = 4$ and $N = 3$, then guidance command equation (9.2-23) is identical to the impact time guidance law (reference [14]). Also, in the case of $K = N + 1$, the obtained impact time guidance law becomes the guidance law. Therefore, the obtained result is a generalized form of previous impact time guidance laws (reference [14]).

In addition, it follows from theorem 9.1 that the corresponding performance index of dynamics equation (9.2-21) is given by

$$J = \frac{1}{2}\int_t^{t_f} \frac{r^2\sin\theta^2}{(2N-1)^2 V_M^4(t_f - \tau)^{K-1}}u^2(\tau)\,\mathrm{d}\tau, \tag{9.2-24}$$

By neglecting the constant terms in the performance index, equation (9.2-24) further reduces to

$$J = \frac{1}{2}\int_t^{t_f} \frac{r^2\sin\theta^2}{(t_f - \tau)^{K-1}}u^2(\tau)\,\mathrm{d}\tau. \tag{9.2-25}$$

The performance index, given in equation (9.2-25), gives us general insights into the command pattern of the proposed guidance law and the behavior of guidance laws under the circumstance of impact time control. From equation (9.2-25), we can readily observe that the magnitude of the weighting function decreases as r and θ decrease for a given time-to-go. Therefore, the resultant guidance command of impact time control tends to increase as r and θ decrease. This is a general phenomenon of impact time-control guidance laws. For a stationary target, the parameter θ can be considered as a heading error. Therefore, decreasing r and θ means that the missile approaches a target, converging to a collision course. For the impact time control, the missile needs to make a detour away from the desired collision triangle, hence adjusting the flight time as desired. By the geometric rule, we readily know that changing the flight trajectory is getting easy as the missile deviates from the collision course. Namely, once the missile makes the collision course to a target, more control efforts are required to correct the flight time as desired. This is a general characteristic of impact time control guidance laws, and performance index equation (9.2-25) implies this fact.

Based on performance index equation (9.2-25), we can also determine the command pattern of the proposed guidance law. It follows from equation (9.2-25) that the weighting function $r^2\sin^2\theta/t_{go}^{K-1}$ gradually decreases with the decrease of r and θ. For this reason, it is recommended to choose relatively large guidance gain K such that t_{go}^{K-1} has faster decreasing rate than $r^2\sin4\theta$ to compensate for the decreasing of the weighting function. Notice that the decreasing rates of both relative range and velocity lead angle are governed by the PNG term. Since $r \approx V_M t_{go}$ when the interceptor approaches the target, the decreasing rate of r is proportional to t_{go}. In the case of PNG with constant navigation gain N, the closed-form solution of the velocity lead angle is $\theta = C t_{go}^{K-1}$ with C being a constant determined by the initial condition. Considering the fact that θ is small, the decreasing rate of $r^2 \sin^2 \theta$ is, therefore, proportional to t_{go}^{2N}. With this in mind, a suitable choice of K that guarantees a zero final guidance command is $K - 1 > 2N$.

The performance of generalized optimal impact time guidance law with $t_d = 20$ s is numerically analyzed in this subsection. The simulation results, including interception trajectory, impact time error, acceleration command and control effort, obtained by the proposed guidance law with various guidance gains are shown in Fig. 9.2 - 2. The results indicate that the proposed guidance law successfully guides the missile to intercept the target with a desired impact time. Increasing the value of the biased gain K, the convergence rate of impact time error becomes faster at the sacrifice of more energy consumption, as shown in Fig. 9.2 - 2 (b) and (d). We can also note from Fig. 9.2 - 2 (c) that the guidance gain under condition $K > 2N + 1$ ensures theoretical zero terminal guidance command and this fact supports the analytical findings.

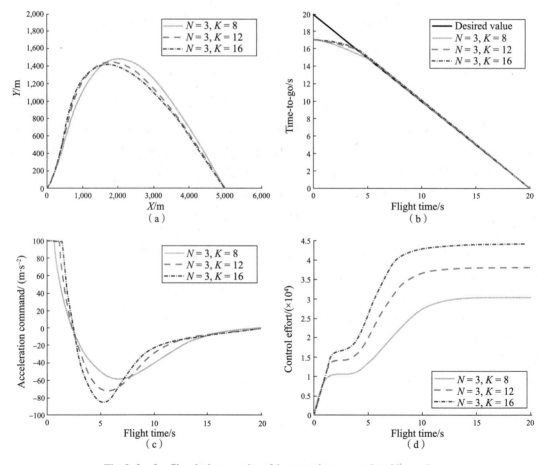

Fig. 9.2 - 2 Simulation results of impact time control guidance law

(a) Interception trajectory; (b) Time-to-go error; (c) Acceleration command; (d) Control effort

§9.3 Gravity-turn-assisted Optimal Guidance Law

The derivation of classical PNG assumes the vehicle is moving with a constant speed without the effect of gravitational acceleration. For endo-atmospheric scenarios, the vehicle's speed is only influenced by the aerodynamic forces and hence can be considered as a slowly varying variable. This means that the moving speed can be assumed as a piece-wise constant and justify the constant-speed assumption. However, exo-atmospheric interceptors, i.e., kinematic kill vehicles, require high impact energy to increase the effectiveness of direct hit and hence are usually subject to large accelerations. In this case, PNG cannot drive the missile to follow the desired straight line collision course and is far away from energy optimal.

To address this issue, the concept of guidance-to-collision (G2C) is introduced to design guidance laws for accelerating missiles. The basic idea of G2C is to consider the effect of axial acceleration in deriving the collision triangle. This strategy was realized by both SMC (reference [19]) and optimal control (reference [20]) to guide the interceptor on a straight line collision course to approach the predicted interception point (PIP).

The issue associated with the G2C law is that it is derived under the gravity-free assumption and hence requires additional compensation term to counteract the effect of gravity in implementation. Due to the existence of the additional term, the direct compensation strategy cannot guarantee zero terminal guidance command, thereby leading to the sacrifice of operational margins to cope with undesired disturbances. Notice that direct gravity compensation might require extra energy and hence result in the decrease of impact energy. Motivated by these observations, this chapter aims to propose a new gravity-turn-assisted optimal guidance law that automatically utilizes the gravity, rather than compensating it, for accelerating missiles. The main idea to solve this problem is to regulate the ZEM that considers both axial acceleration and gravity, to zero in a finite time, thus guiding the missile to follow a desired curved path to intercept the target. The key features of the proposed algorithm are twofold:

(1) Unlike PNG and G2C, the resulting desired interception path defined by the collision triangle that considers both axial acceleration and gravity is a curved trajectory instead of a straight line. This fact makes analytic guidance command derivation intractable. To address this problem, a new concept, called instantaneous ZEM, is introduced and the proposed optimal guidance law is then derived by using the optimal error dynamics proposed in section. 9.1.

(2) Detailed analysis reveals that both PNG and G2C are special cases of the proposed guidance law. The advantages of the proposed approach are clear: guaranteeing zero final guidance command and saving energy without requiring extra control effort to compensate for the gravity.

Due to the existence of axial acceleration and gravity, the flight path angle and speed evolve according to

$$\dot{\gamma}_M = \frac{a_x \sin\alpha_M - g\cos\gamma_M}{V_M}, \qquad (9.3-1)$$

$$\dot{V}_M = a_x \cos\alpha_M - g\sin\gamma_M. \qquad (9.3-2)$$

where g stands for the gravitational acceleration, which is assumed to be constant in guidance law design since the duration of terminal guidance phase is typically very short.

The control interest of exo-atmospheric interception is to design a guidance law to nullify the initial heading error such that the interceptor maintains its acceleration vector in the direction of its velocity vector thereafter. Consequently, if the gravitational effect is ignored, the missile will fly along a straight line that requires no extra control effort to the expected impact point. This characteristic is of great importance to kinematic kill vehicles since this strategy can reduce the magnitude of interceptor's angle-of-attack and hence increase the kill probability (reference [21]). To realize the concept of direct hit, the key part is to find the closed-form solution of an ideal collision triangle for the missile that requires no extra control effort, i.e., $\alpha_M = 0$. Once we have the closed-form solution, we can easily design a guidance law that drives the missile trajectory to converge to the ideal collision triangle in finite time by utilizing the optimal error dynamics proposed in section 9.1. However, due to the effect of gravity, the zero-control trajectory is no longer a straight line and hence most previous guidance laws were developed under the gravity-free assumption for simplicity. In practical applications, an additional gravity-compensation term $g\cos\gamma_M$ is then leveraged to reject the effect of gravity in implementation. This straightforward approach obviously has two main drawbacks:

(1) Compensating gravity using extra term $g\cos\gamma_M$ cannot guarantee zero terminal guidance command, leading to the sacrifice of operational margins.

(2) The additional term might require additional energy consumption for gravity compensation and hence impose adverse effects on the effectiveness of direct hit due to energy loss.

In order to address these two problems, the objective of this chapter is to propose a new gravity-turn-assisted optimal guidance law that automatically utilizes the gravity, rather than compensating it, for accelerating missiles.

9.3.1 Zero-control-effort Trajectory Considering Gravity

Once we obtain the ideal collision triangle, the optimal control theory can be utilized to design a guidance law that forces the missile to fly along the collision triangle to achieve the designed goal.

The ideal motion of the interceptor is defined as the missile kinematics with zero control input. And the instantaneous ZEM is defined as the final distance the missile would miss the target if the target continues along its present course and the missile follows a straight line along its current flight path angle with no further corrective maneuvers (Fig 9.3-1).

In the derivation of the desired collision triangle, it is natural to enforce the condition of ideal motion of the interceptor, that is, $\alpha_M = 0$ for exo-atmospheric case as our goal is to make the terminal guidance command converge to zero. Under this condition, the missile's kinematics is formulated as

$$\dot{x}_M = V_M \cos\gamma_M, \qquad (9.3-3)$$
$$\dot{y}_M = V_M \sin\gamma_M, \qquad (9.3-4)$$

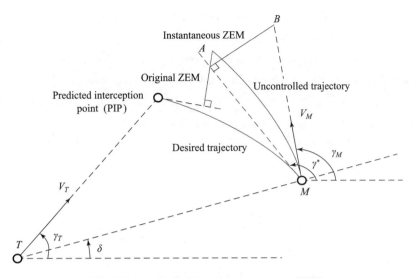

Fig 9.3-1 Definition of instantaneous ZEM

$$\dot{\gamma}_M = -\frac{g\cos\gamma_M}{V_M}, \tag{9.3-5}$$

$$\dot{V}_M = a_x - g\sin\gamma_M, \tag{9.3-6}$$

where (x_M, y_M) represents the inertial position of the interceptor. It follows from equation (9.3-3) – equation (9.3-6) that these four differential equations are dependent on the flight path angle γ_M and the direct integration seems to be intractable. From the point of problem dimension reduction, we manually change the argument from time t to flight path angle γ_M. Then, equation (9.3-3) – equation (9.3-6) can be reformulated as

$$\frac{\mathrm{d}t}{\mathrm{d}\gamma_M} = -\frac{V_M}{g\cos\gamma_M}, \tag{9.3-7}$$

$$\frac{\mathrm{d}x_M}{\mathrm{d}\gamma_M} = \frac{\mathrm{d}x_M}{\mathrm{d}t} \cdot \frac{\mathrm{d}t}{\mathrm{d}\gamma_M} = -\frac{V_M^2}{g}, \tag{9.3-8}$$

$$\frac{\mathrm{d}y_M}{\mathrm{d}\gamma_M} = \frac{\mathrm{d}y_M}{\mathrm{d}t} \cdot \frac{\mathrm{d}t}{\mathrm{d}\gamma_M} = -\frac{V_M^2}{g}\tan\gamma_M, \tag{9.3-9}$$

$$\frac{\mathrm{d}V_M}{\mathrm{d}\gamma_M} = \frac{\mathrm{d}V_M}{\mathrm{d}t} \cdot \frac{\mathrm{d}t}{\mathrm{d}\gamma_M} = V_M\tan\gamma_M - \frac{a_x V_M}{g\cos\gamma_M}. \tag{9.3-10}$$

Through this argument changing, we only need to solve three independent differential equations to find the analytical solution. Let $\kappa = -a_x/g$, then, equation (9.3-10) can be rewritten as

$$\frac{\mathrm{d}V_M}{\mathrm{d}\gamma_M} = V_M\tan\gamma_M + \frac{\kappa V_M}{\cos\gamma_M}. \tag{9.3-11}$$

Rearrange equation (9.3-11) as

$$\frac{\mathrm{d}V_M}{V_M} = \left(\tan\gamma_M + \frac{\kappa}{\cos\gamma_M}\right)\mathrm{d}\gamma_M. \tag{9.3-12}$$

Integrating both sides of equation (9.3-12) gives

$$\ln V_M(\gamma_M)\big|_{\gamma_{M0}}^{\gamma_{Mf}} = -\ln|\cos\gamma_M|\big|_{\gamma_{M0}}^{\gamma_{Mf}} + \kappa\ln|\sec\gamma_M + \tan\gamma_M|\big|_{\gamma_{M0}}^{\gamma_{Mf}}, \tag{9.3-13}$$

where γ_{Mf} and γ_{M0} denote the final and initial flight path angles, respectively.

Solving equation (9.3-13) for $V_M(\gamma_{Mf})$ yields

$$V_M(\gamma_{Mf}) = C|\sec\gamma_{Mf}||\sec\gamma_{Mf} + \tan\gamma_{Mf}|^\kappa, \qquad (9.3-14)$$

where C is the integration constant, determined by the initial conditions as

$$C = \frac{V_M(\gamma_{M0})}{|\sec\gamma_{M0}||\sec\gamma_{M0} + \tan\gamma_{M0}|^\kappa}. \qquad (9.3-15)$$

Setting γ_{Mf} as γ_M and substituting equation (9.3-14) into equation (9.3-7) – equation (9.3-9) result in

$$\frac{dt}{d\gamma_M} = -\operatorname{sgn}(\cos\gamma_M)\frac{C}{g}\sec\gamma_M|\sec\gamma_M + \tan\gamma_M|^\kappa, \qquad (9.3-16)$$

$$\frac{dx_M}{d\gamma_M} = -\frac{C^2}{g}\sec^2\gamma_M|\sec\gamma_M + \tan\gamma_M|^{2\kappa}, \qquad (9.3-17)$$

$$\frac{dy_M}{d\gamma_M} = -\frac{C^2}{g}\sec^2\gamma_M\tan\gamma_M|\sec\gamma_M + \tan\gamma_M|^{2\kappa}. \qquad (9.3-18)$$

After derivation, we have the closed-form solution as

$$t(\gamma_M) = t(\gamma_{M0}) - \operatorname{sgn}(\cos\gamma_M)\frac{C}{g}[f_t(\gamma_M) - f_t(\gamma_{M0})], \qquad (9.3-19)$$

$$x_M(\gamma_M) = x_M(\gamma_{M0}) - \frac{C^2}{g}[f_x(\gamma_M) - f_x(\gamma_{M0})], \qquad (9.3-20)$$

$$y_M(\gamma_M) = y_M(\gamma_{M0}) - \frac{C^2}{g}[f_y(\gamma_M) - f_y(\gamma_{M0})], \qquad (9.3-21)$$

where

$$f_t(\gamma_M) = \frac{1}{\kappa^2 - 1}(\kappa\sec\gamma_M - \tan\gamma_M)|\sec\gamma_M + \tan\gamma_M|^\kappa, \qquad (9.3-22)$$

$$f_x(\gamma_M) = \frac{1}{4\kappa^2 - 1}(2\kappa\sec\gamma_M - \tan\gamma_M)|\sec\gamma_M + \tan\gamma_M|^{2\kappa}, \qquad (9.3-23)$$

$$f_y(\gamma_M) = \frac{1}{4\kappa^2 - 4}(2\kappa\sec\gamma_M\tan\gamma_M - \tan^2\gamma_M - \sec^2\gamma_M)|\sec\gamma_M + \tan\gamma_M|^{2\kappa}. \qquad (9.3-24)$$

Setting $\gamma_{M0} = \gamma_M$ and $\gamma_M = \gamma_{Mf}$ in equation (9.3-19) – equation (9.3-21) provides the trajectory of the interceptor with zero control effort as

$$t(\gamma_{Mf}) = t(\gamma_M) - \frac{C}{g}[f_t(\gamma_{Mf}) - f_t(\gamma_M)], \qquad (9.3-25)$$

$$x_M(\gamma_{Mf}) = x_M(\gamma_M) - \frac{C^2}{g}[f_x(\gamma_{Mf}) - f_x(\gamma_M)], \qquad (9.3-26)$$

$$y_M(\gamma_{Mf}) = y_M(\gamma_M) - \frac{C^2}{g}[f_y(\gamma_{Mf}) - f_y(\gamma_M)]. \qquad (9.3-27)$$

Since the target acceleration is usually unavailable to the missile in practice, we assume that the target leverages a gravity compensation scheme with constant moving speed in the derivation of the collision triangle. With this in mind, the terminal position of the target after t_{go} is given by

$$x_T(t_f) = x_T + V_T\cos\gamma_T t_{go} = x_M + r\cos\sigma + V_T\cos\gamma_T t_{go}, \qquad (9.3-28)$$

$$y_T(t_f) = y_T + V_T\sin\gamma_T t_{go} = y_M + r\sin\sigma + V_T\sin\gamma_T t_{go}. \quad (9.3-29)$$

From equation (9.3 – 25), the time-to-go t_{go} can be formulated as

$$t_{go} = t(\gamma_{Mf}) - t(\gamma_M) = -\frac{C}{g}[f_t(\gamma_{Mf}) - f_t(\gamma_M)]. \quad (9.3-30)$$

A perfect interception requires

$$x_M(t_f) = x_T(t_f), y_M(t_f) = y_T(t_f). \quad (9.3-31)$$

Define the ideal instantaneous collision triangle considering gravity that requires no extra control effort for the missile to intercept the target.

Remark In previous derivations, we utilize the widely-accepted assumptions that the target adopts a gravity compensation scheme and maintains constant flying velocity. For ballistic targets with no gravity compensation, one can include the gravitational effect in target position prediction to obtain more accurate PIP. This can be easily achieved in a similar way as equation (9.3 – 20) – equation (9.3 – 21) by setting $\kappa = 0$.

9.3.2 Optimal Guidance Law Design and Analysis

One can directly derive the instantaneous ZEM as

$$z = -\sin(e_\gamma)\left(V_M t_{go} + \frac{1}{2}a_x t_{go}^2\right). \quad (9.3-32)$$

The first-order time derivative of the instantaneous ZEM can be approximated as

$$\dot{z} = -\dot{e}_\gamma\left(V_M t_{go} + \frac{1}{2}a_x t_{go}^2\right) = -a_M\left(t_{go} + \frac{1}{2V_M}a_x t_{go}^2\right). \quad (9.3-33)$$

9.3.2.1 Optimal guidance law design

In order to drive the instantaneous ZEM to converge to zero in finite time, various nonlinear control methods can be applied to the dynamics of ZEM as given in equation (9.3 – 33). Following the general prediction-correction concept established in section 9.2, a guidance correction command can be determined to make a designed error dynamics to nullify the instantaneous ZEM. To this end, we choose the optimal error dynamics, proposed in section 9.2, as

$$\dot{z} + \frac{Nz}{t_{go}} = 0, \quad (9.3-34)$$

where $N > 0$ is a constant guidance gain, which can be tuned to regulate the convergence rate of the instantaneous ZEM. Substituting equation (9.3 – 33) into equation (9.3 – 34) gives the guidance command as

$$a_M = \frac{Nz}{\left(1 + \frac{a_x}{2V_M}\right)t_{go}^2}. \quad (9.3-35)$$

The angle-of-attack command α_M is then obtained as

$$\alpha_M = \arcsin\left(\frac{a_M}{a_x}\right). \quad (9.3-36)$$

Guidance command in equation (9.3 – 35) is optimal in the sense of minimizing performance index

$$J = \int_t^{t_f} \frac{\left[1 + \frac{1}{2V_M}a_x(t_f - \tau)\right]^2}{(t_f - \tau)^{N-3}} a_M^2(\tau)\,d\tau. \quad (9.3-37)$$

For interceptors with fast moving velocity or short homing phase duration, the performance index can be approximated as

$$J \approx \int_t^{t_f} \frac{1}{(t_f - \tau)^{N-3}} a_M^2(\tau)\,d\tau, \quad (9.3-38)$$

which indicates that the proposed guidance law with $N = 3$ provides energy-optimal interception and it is wise to choose $N \geqslant 3$ to guarantee a zero final guidance command. When the interceptor converges to the ideal collision course, the terminal velocity can be obtained as

$$V_{Mf} = V_M + a_x t_{go}. \quad (9.3-39)$$

Following equation (9.3-39), the average velocity during the homing phase can be readily determined as

$$\bar{V}_M = V_M + \frac{1}{2}a_x t_{go}. \quad (9.3-40)$$

By using the concept of average velocity, the guidance command in equation (9.3-35) can be also rewritten as

$$a_M = \frac{\bar{N}z}{t_{go}^2}. \quad (9.3-41)$$

where the new guidance gain $\bar{N}$ is given by

$$\bar{N} = \frac{NV_M}{\bar{V}_M}, \quad (9.3-42)$$

which implies that the proposed guidance law is a PNG-type guidance law with a time-varying navigation gain $\bar{N}$. The initial and final values of the guidance gain are given by

$$\bar{N}_0 = \lim_{t \to t_f} N(t) = \frac{N}{1 + \frac{1}{2V_{M0}}a_x t_f},$$
$$\bar{N}_f = \lim_{t \to t_f} \bar{N}(t) = N \quad (9.3-43)$$

where V_{M0} denotes the initial vehicle speed.

9.3.2.2 Relationships with proportional navigation guidance

The behind idea of PNG is to generate a lateral acceleration to nullify the ZEM so as to follow a straight line interception course. To maintain the collision triangle, it is necessary to equalize between the distances travelled by the interceptor and the target perpendicular to the LOS. In PNG formulation, the assumptions regarding no axial acceleration and gravity are required. Under these two assumptions, one can imply that

$$L_M \sin(\gamma_{PNG}^* - \sigma) - L_T \sin(\gamma_T - \sigma) = 0, \quad (9.3-44)$$

where γ_{PNG}^* denotes the desired current flight path angle of PNG. L_M and L_T are the vehicles' travelled distances from the current point to the PIP point.

Solving γ_{PNG}^* using equation (9.3-44) gives

$$\gamma_{PNG}^* = \pi - \arcsin\left(\frac{L_T \sin(\gamma_T - \sigma)}{L_M}\right) + \sigma$$

$$= \pi - \arcsin\left(\frac{V_T \sin(\gamma_T - \sigma)}{V_M}\right) + \sigma. \tag{9.3-45}$$

Then, the ZEM dynamics of PNG can be obtained as

$$\begin{aligned} z_{PNG} &= -\sin(e_{\gamma,PNG}) V_M t_{go} \\ \dot{z}_{PNG} &= -\dot{e}_{\gamma,PNG} V_M t_{go} = -a_M t_{go} \end{aligned} \tag{9.3-46}$$

where $e_{\gamma,PNG} = \gamma_M - \gamma_{PNG}^*$ denotes the flight path angle tracking error. Note that the difference between the PNG ZEM and the proposed ZEM is that the desired flight path angle for the former is derived based on the assumption of constant flight velocity under the gravity-free condition, whereas the latter computes it under the influence of gravity for varying-speed missiles. The preceding PNG ZEM dynamics also reveals that if we remove the gravity and axial acceleration, the proposed instantaneous ZEM reduces to the PNG ZEM. As is well known, the guidance command of PNG is given by

$$a_M = \frac{N z_{PNG}}{t_{go}^2}. \tag{9.3-47}$$

Since classical PNG leverages the ZEM that is derived based on the gravity-free assumption, it requires extra term $g\cos\gamma_M$ to compensate the effect of gravity in practical implementation as

$$a_M = \frac{N z_{PNG}}{t_{go}^2} + g\cos\gamma_M, \tag{9.3-48}$$

which indicates that the terminal guidance command of PNG cannot converge to zero due to the extra term $g\cos\gamma_M$. As a comparison, the proposed instantaneous ZEM is derived by considering the gravity, and therefore the guidance command converges to zero once the interceptor is maintained on the collision triangle.

9.3.2.3 Relationships with guidance-to-collision

Similar to PNG, the objective of G2C is to maintain a straight line interception course to approach the target. This means that, the distances traveled by the interceptor and the target perpendicular to the LOS is equal at any time instant during the homing phase as

$$L_M \sin(\gamma_{G2C}^* - \sigma) - L_T \sin(\gamma_T - \sigma) = 0, \tag{9.3-49}$$

where γ_{G2C}^* denotes the desired current flight path angle of G2C. With constant axial acceleration, equation (9.3-49) can be reformulated as

$$\left(V_M + \frac{1}{2} a_M t_{go}\right)\sin(\gamma_{G2C}^* - \sigma) - V_T \sin(\gamma_T - \sigma) = 0. \tag{9.3-50}$$

Once a straight line collision course is reached and maintained after the heading error has been nulled, the time-to-go for G2C law can then be computed by (reference [19])

$$r = V_T t_{go} \cos(\gamma_T - \sigma) + V_M t_{go} \cos[\pi - (\gamma_{G2C}^* - \sigma)] + \frac{1}{2} a_M t_{go}^2 \cos[\pi - (\gamma_{G2C}^* - \sigma)]. \tag{9.3-51}$$

Solving equation (9.3-50) for γ_{G2C}^* gives

$$\gamma_{G2C}^{*} = \pi - \arcsin\left(\frac{V_T \sin(\gamma_T - \sigma)}{V_M + 0.5 a_M t_{go}}\right) + \sigma. \qquad (9.3-52)$$

Then, the G2C ZEM dynamics can be readily obtained as

$$z_{G2C} = -\sin(e_{\gamma,G2C})\left(V_M t_{go} + \frac{1}{2}a_x t_{go}^2\right), \qquad (9.3-53)$$

$$\dot{z}_{G2C} = -\dot{e}_{\gamma,G2C}\left(V_M t_{go} + \frac{1}{2}a_x t_{go}^2\right) = -a_M\left(t_{go} + \frac{a_x}{2V_M}t_{go}^2\right), \qquad (9.3-54)$$

where $e_{\gamma,PNG} = \gamma_M - \gamma_{G2C}^{*}$ denotes the flight path angle tracking error. Following the same line shown in the previous subsection, one can easily derive the optimal G2C command is given by

$$a_M = \frac{\bar{N} z_{G2C}}{t_{go}^2}. \qquad (9.3-55)$$

Similarly, as the ZEM of G2C is only valid without the gravitational effect, the practical implementation of G2C requires an additional compensation term as

$$a_M = \frac{\bar{N} z_{G2C}}{t_{go}^2} + g\cos\gamma_M. \qquad (9.3-56)$$

Since the proposed guidance law utilizes gravity instead of compensating it, one can safely predict that our approach can save energy in the case of instantaneous ZEM≤G2C ZEM.

Table 9.3-1 summarizes the relationships between the proposed guidance law and PNG as well as G2C. In conclusion, our approach is a generalized guidance law derived from the collision triangle and the gravity effect is automatically considered in the instantaneous ZEM. From the analysis, it can be noted that both PNG and G2C are special cases of the proposed guidance law. As we stated earlier, the advantage of using gravity-turn in guidance law design is to guarantee zero guidance command at the time of impact, thereby providing additional operation margins to cope with the undesired disturbances. Furthermore, by considering the gravitational effect in guidance law design, no extra energy for gravity compensation is required for guidance law implementation.

Table 9.3-1 Relationships between the proposed formulation and previous guidance laws

Guidance law	Guidance command	Terminal guidance command	Consideration
PNG	$a_M = \dfrac{\bar{N} z_{PNG}}{t_{go}^2} + g\cos\gamma_M$	Bounded	None
G2C	$a_M = \dfrac{\bar{N} z_{G2C}}{t_{go}^2} + g\cos\gamma_M$	Bounded	Speed variation
Proposed	$a_M = \dfrac{\bar{N} z_{proposed}}{t_{go}^2}$	Zero	Speed variation and gravity

9.3.3 Characteristics Analysis by Simulations

In this subsection, nonlinear simulations are performed to validate the proposed guidance law in an exo-atmospheric interception scenario. We assume the interceptor is equipped with a rocket motor that provides constant axial acceleration. The attitude of the interceptor can be adjusted to provide desired lateral acceleration to change the heading direction. The required initial conditions for a typical exo-atmospheric engagement, taken from reference [14] – reference [16], are summarized in table 9.3 – 2.

Table 9.3 – 2 Initial conditions for homing engagement

Parameters	Relative range	Initial LOS angle	Missile initial velocity	Target velocity	Axial acceleration
Values	50 km	0°	1,500 m/s	3,000 m/s	20g

9.3.3.1 Characteristics of the proposed guidance law

We first analyze the effect of guidance gain α on the guidance performance. In the simulations, the initial flight path angle is chosen as $\gamma_M(0) = 150°$ and $\gamma_T(0) = 20°$. The simulation results, including interception trajectory, angle-of-attack command, flight path angle error, control effort, time-to-go estimation and missile velocity, with various guidance gains $N = 2, 3, 4, 5, 6$ are presented in Fig. 9.3 – 2, where the control effort is defined as

$$E = \int_0^{t_f} a_M^2 d\tau. \qquad (9.3-57)$$

The results clearly reveal that guidance law with larger guidance gain N provides faster convergence speed of the flight path angle error. However, larger guidance gain also requires higher angle-of-attack command during the initial flight period, thereby generating more control efforts. When the flight path angle tracking error is close to zero, the control effort almost remains the same and the angle-of-attack commands are also close to zero for all guidance gain cases, meaning that the interceptor is maintained on the collision triangle. Since the gravity is considered in the derivation of the collision triangle, the guidance command converges to zero at the time of impact with the increasing of the guidance gain. This characteristic is totally different from previous guidance laws that used an additional term $g\cos\gamma_M$ to compensate the gravity.

From Fig. 9.3 – 2 (b) and (d), we can readily note that the proposed guidance law with $N \geqslant 3$ guarantees zero final guidance command and $N = 3$ provides energy-optimal interception. These results support the analytic findings of previous sections.

The results in Fig. 9.3 – 2 (e) gives accurate estimation of the time-to-go and the estimation error converges to zero once the interceptor is maintained on the desired trajectory. From the zoomed-in subfigure in Fig. 9.3 – 2 (f), one can observe that the missile velocity with smaller guidance gain N increases slightly faster than that with larger guidance gain N during the initial flight phase. This can be attributed to the fact that larger guidance gain requires more control efforts, i.e. larger magnitude of the angle-of-attack. Since the angle-of-attack remains very small during most of the flight period, the missile velocity in the considered scenario increases almost linearly, as shown in Fig. 9.3 – 2 (f).

Now let us investigate the performance of the proposed guidance law with various interceptor's initial flight path angles. The same scenario is simulated with four different initial flight path angles: $\gamma_M(0) = 90°, 120°, 150°, 180°$ and $\gamma_T(0) = 20°$. The simulation results with guidance gain

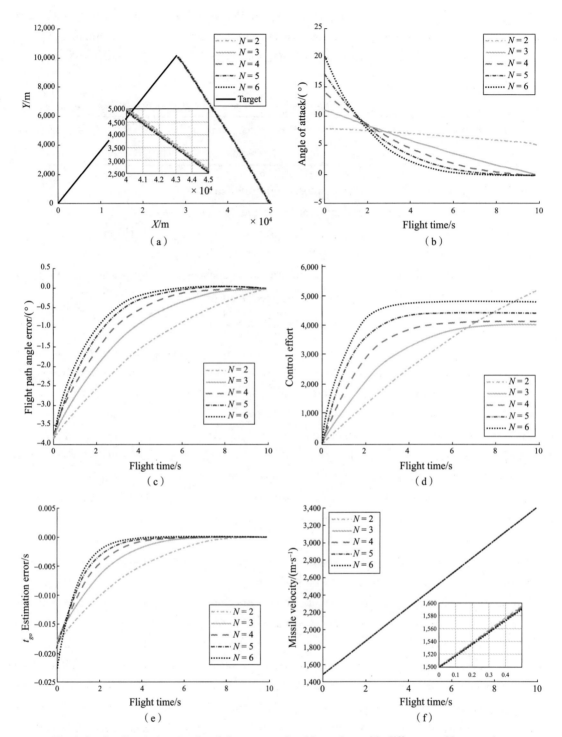

Fig. 9.3 − 2 Simulation results of the proposed guidance law with different guidance gains.

(a) Interception trajectory; (b) Angle-of-attack command; (c) Flight path angle error;
(d) Control effort; (e) Time-to-go estimation; (f) Missile velocity

$N=9$ are depicted in Fig. 9.3 − 3, which clearly shows that the proposed guidance law can guide the missile to successfully intercept the target in all tested scenarios.

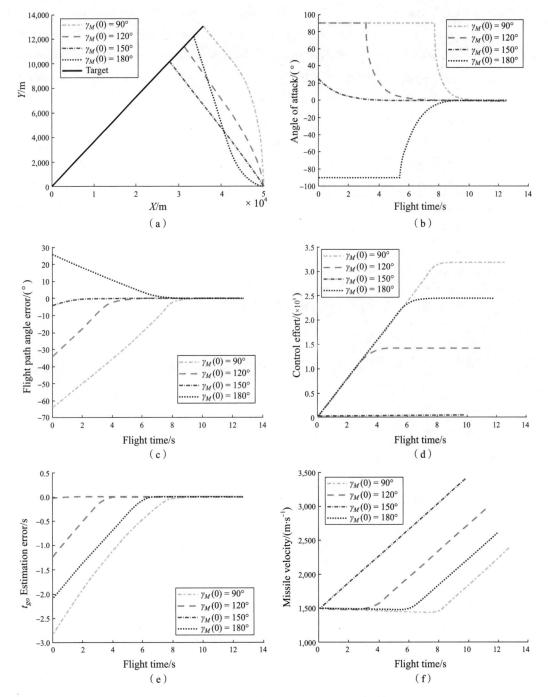

Fig. 9.3 − 3 Simulation results of the proposed guidance law with different initial flight path angles
(a) Interception trajectory; (b) Angle-of-attack command; (c) Flight path angle error;
(d) Control effort; (e) Time-to-go estimation; (f) Missile velocity

The interceptor with larger initial heading error experiences more curved trajectory to approach the target. For this reason, the duration of acceleration saturation of $\gamma_M(0) = 180°$ is longer than that of other cases. As the proposed time-to-go is calculated from the desired collision triangle, it gives an underestimation when the missile deviates from the zero control effort collision course. However, the simulation results indicate that the estimation accuracy is acceptable in most cases. From Fig. 9.3 – 3 (f), it can be noted that the missile's velocity slightly decreases due to the gravitational effect when the guidance command is saturated, i.e., all control effort is leveraged to nullify the instantaneous ZEM.

9.3.3.2 Comparison with other guidance laws

To further demonstrate the superiority of the proposed approach, the performance of the new guidance law is compared to that of classical PNG and G2C in this subsection. In the simulations, we utilize an additional term $g\cos\gamma_M$ in both PNG and G2C to compensate for the gravitational effect. In order to make fair comparisons, the guidance gain for all guidance laws is set as $N = 9$.

Fig. 9.3 – 4 compares the performance of these three guidance laws for intercepting a constant moving target. From the results, it can be noted that the axial acceleration of PNG does not align

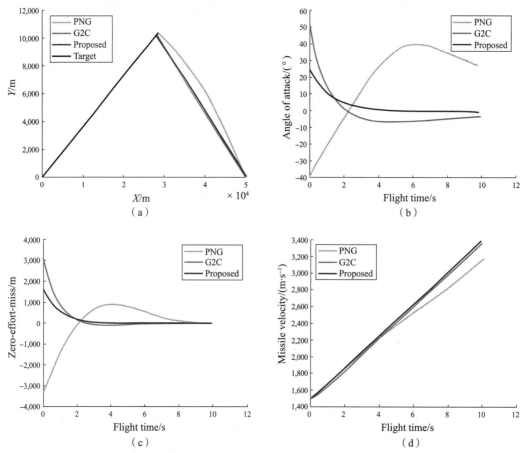

Fig. 9.3 – 4 Comparison results of exo-atmospheric interception with a non-maneuvering target
(a) Interception Trajectory; (b) Angle-of-attack command; (c) Zero-effort-miss; (d) Missile velocity

with the velocity vector, thereby forcing the missile to fly along a curved path to intercept the target. Unlike PNG, G2C guides the missile to intercept the target by following a straight line after the initial heading error is nullified. As a comparison, the proposed guidance law guides the missile to fly along a slightly curved trajectory as it exploits the gravity-turn concept. It is evident from Fig. 9. 3 – 4 (b) that the time evolutions of angle-of-attack of all guidance laws remain bounded and the proposed guidance law guarantees zero guidance command at the time of impact. Therefore, the proposed guidance law provides more operational margins than the G2C law when the missile is close to the target. Since the proposed guidance law requires no extra term to counteract the gravity, one can clearly observe from Fig. 9. 3 – 4 (b) that the proposed approach enables energy saving during the terminal guidance phase. The time history of ZEM, shown in Fig. 9. 3 – 4 (c), reveals that all guidance laws can nullify their corresponding ZEM to zero to guarantee target interception. From Fig. 9. 3 – 4 (d), one can clearly observe that the terminal missile velocity under both the proposed guidance law and the G2C law is larger than that of the PNG law, thereby enabling higher kill probability.

§ 9. 4   Three-dimensional Optimal Impact Time Guidance for Antiship Missiles

The primary objective of missile guidance laws is to drive the missile to intercept a specific target with zero miss distance. Proportional navigation guidance (PNG) has been proven to be an efficient and simple guidance algorithm for missile systems, thus showing wide applications in the past few decades (reference [22]). The optimality of PNG was analyzed in reference [23], and its extension to the three-dimensional (3D) scenario can be found in reference [24] . In the context of modern warfare, many high-value battleships, like destroyers and aircraft carriers, are equipped with powerful self-defense systems against antiship missiles (reference [25]) . To penetrate these formidable defensive systems, the concept of salvo attack or simultaneous attack was introduced: Many missiles are required to hit a battleship simultaneously, albeit their different initial locations. One typical solution of simultaneous attack is impact time control guidance.

Note that most previous impact time control guidance laws are dedicated for two-dimensional (2D) engagement scenarios, thus ignoring the cross-couplings between horizontal and vertical channels. It is well known that the application of 2D guidance laws in realistic 3D engagements is valid for roll-stabilized interceptors. However, designing 3D guidance law is meaningful since it can fully exploit the synergism effect of horizontal and vertical planes. Therefore, 3D guidance law is more beneficial if the effect of crosscouplings cannot be neglected. Although the Lyapunov-based guidance law (reference [28]) and the consensus-based guidance law (reference [31]) are directly derived using 3D kinematics, they are limited in addressing the optimality issue because they were formulated based on nonlinear control approaches rather than the optimal control framework. Furthermore, these two guidance laws only guarantee asymptotical convergence as proved by the authors.

Motivated by the aforementioned observations, this chapter aims to propose a generalized optimal 3D guidance law for antiship missiles to satisfy the impact time constraint. For this purpose, we use a composite guidance command, similar to reference [25], reference [26], reference [27], consisting of an optimal 3D PNG part and a feedback loop for regulating the impact time error. In determining the error feedback term, we first generalize the 2D PNG-based time-to-go estimation approach (reference [26]) to the 3D homing case. To drive the predicted time-to-go to its desired pattern, the optimal error dynamics method, developed in (reference [27]), was used to design the error feedback command. The finite-time convergence of the impact time error and the optimality of the proposed guidance law are also analyzed. Notice that the guidance algorithm developed can be viewed as an extension of (reference [27]) to a realistic 3D scenario with more rigorous analysis about guidance command and guidance gain selection. Compared with the 2D optimal impact time guidance law (reference [27]), the proposed guidance law provides more stable acceleration command and requires less energy consumption, especially when the cross-coupling effect is strong.

9.4.1 Problem Formulation

This section states the problem formulation. Before introducing the system kinematics, we make three basic assumptions as follows:

Assumption 1: The target is stationary.

Assumption 2: The missile is assumed as an ideal point-mass model.

Assumption 3: The missile is flying with constant velocity.

Note that these assumptions are widely accepted in impact time guidance law design for antiship missiles. In assumption 1, compared to antiship missiles, the target speed is ignorable. In assumption 2, typical philosophy treats the guidance and control loops separately by placing the guidance kinematics in an outer loop, generating guidance commands tracked by an inner dynamic control loop, also known as autopilot. In assumption 3, the vehicle's velocity is generally slowly varying and hence can be assumed as piecewise constant.

Under these assumptions, the 3D engagement geometry is shown in Fig. 9.4 – 1, in which

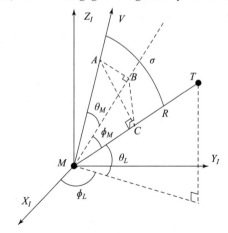

Fig. 9.4 – 1 Three-dimensional homing engagement geometry

(X_I, Y_I, Z_I) denotes the inertial reference coordinate system and V is the missile velocity. The missile-target relative range is denoted as R. The notations θ_M and ϕ_M stand for two velocity lead angles with respect to the LOS line in the pitch and yaw planes, respectively. Note that both θ_M and ϕ_M can be obtained indirectly from the onboard seeker's gimbal angles (reference [32]). The variables θ_L and ϕ_L represent the LOS angles in the azimuth and elevation directions, respectively. The angle σ is the missile velocity lead angle in the engagement plane, i.e., total velocity lead angle, also known as heading error. The differential equations describing the 3D kinematics can be formulated as (reference [32])

$$\dot{R} = -V\cos\theta_M\cos\phi_M, \tag{9.4-1}$$

$$\dot{\theta}_L = -\frac{V}{R}\sin\theta_M, \tag{9.4-2}$$

$$\dot{\phi}_L = -\frac{V}{R\cos\theta_L}\cos\theta_M\sin\phi_M, \tag{9.4-3}$$

$$\dot{\theta}_M = \frac{a_z}{V} + \frac{V}{R}\cos\theta_M\sin^2\phi_M\tan\theta_L + \frac{V}{R}\sin\theta_M\cos\phi_M. \tag{9.4-4}$$

$$\dot{\phi}_M = \frac{a_y}{V\cos\theta_M} - \frac{V}{R}\sin\theta_M\sin\phi_M\cos\phi_M\tan\theta_L + \frac{V}{R\cos\theta_M}\sin^2\theta_M\sin\phi_M + \frac{V}{R}\cos\theta_M\sin\phi_M \tag{9.4-5}$$

where a_y and a_z are missile accelerations in the yaw and pitch directions, respectively.

The complementary equation defining the relationship between the heading error and the projected velocity lead angles can be obtained from Fig. 9.4-1 as

$$\cos\sigma = \cos\theta_M\cos\phi_M. \tag{9.4-6}$$

The aim is to design a 3D optimal guidance law such that the missile can intercept a stationary target with a specific impact time t_d. Our solution to this problem is given by a composite guidance command, consisting of an optimal baseline 3D PNG and an optimal impact time error feedback term.

9.4.2 Three-dimensional Optimal Impact Time Guidance Law Design

This section will present the details of the proposed impact time guidance law. We will first predict the impact time under 3D PNG and then design an error feedback term using optimal error dynamics.

9.4.2.1 Impact time prediction in three-dimensional engagement

In impact time control, accurate impact time prediction is of paramount importance. For this reason, this subsection will generalize the 2D PNG-based time-to-go estimation (reference [26]), reference [30] to practical 3D scenarios. The classical PNG generates the commanded acceleration of the interceptor in proportion to the turning rate of LOS. In a 3D scenario, PNG is defined in a vector format as (reference [32])

$$\boldsymbol{a}^{\text{PNG}} = N\boldsymbol{\Omega}_L \times \boldsymbol{V} \tag{9.4-7}$$

where $N > 0$ denotes the navigation gain. The notations $\boldsymbol{\Omega}_L$ and $\boldsymbol{V}$ represent the LOS angular rate and missile velocity vectors, respectively. These two vectors determine the engagement plane of the 3D interception geometry (reference [33]). Since the relative range is usually not adjustable during terminal guidance phase, the 3D PNG is usually implemented in two planes in the velocity coordinate as (reference [32])

$$a_y^{PNG} = -NV\dot{\lambda}_y \sin\theta_M \sin\phi_M + NV\dot{\lambda}_z \cos\theta_M, \qquad (9.4-8)$$

$$a_z^{PNG} = -NV\dot{\lambda}_y \cos\phi_M, \qquad (9.4-9)$$

where $\dot{\lambda}_y$ and $\dot{\lambda}_z$ are LOS angular velocity vector components in the LOS coordinate, which can be directly measured using onboard seekers, and are determined as

$$\dot{\lambda}_y = \frac{V}{R}\sin\theta_M, \qquad (9.4-10)$$

$$\dot{\lambda}_z = -\frac{V}{R}\cos\theta_M \sin\phi_M. \qquad (9.4-11)$$

Substituting equation (9.4-10) and equation (9.4-11) into equation (9.4-8) and equation (9.4-9) yields

$$a_y^{PNG} = -\frac{NV^2}{R}\sin\phi_M, \qquad (9.4-12)$$

$$a_z^{PNG} = -\frac{NV^2}{R}\sin\theta_M \cos\phi_M. \qquad (9.4-13)$$

Differentiating equation (9.4-6) and substituting equation (9.4-4), equation (9.4-5), equation (9.4-12) and equation (9.4-13) into it results in

$$\begin{aligned}\dot{\sigma} &= \frac{1}{\sin\sigma}(\sin\theta_M \cos\phi_M \dot{\theta}_M + \cos\theta_M \sin\phi_M \dot{\phi}_M) \\ &= \frac{1}{\sin\sigma}\left[-\frac{(N-1)V}{R}\sin^2\theta_M \cos^2\phi_M - \frac{(N-1)V}{R}\sin^2\phi_M\right] \\ &= -\frac{(N-1)V}{R\sin\sigma}(\cos^2\phi_M - \cos^2\theta_M \cos^2\phi_M + \sin^2\phi_M) \\ &= -\frac{(N-1)V}{R}\sin\sigma. \end{aligned} \qquad (9.4-14)$$

Dividing equation (9.4-1) by equation (9.4-14) yields

$$\frac{dR}{d\sigma} = \frac{R\cot\sigma}{N-1}. \qquad (9.4-15)$$

Solving differential equation (9.4-15) in terms of σ gives

$$R = \frac{R_0}{(\sin\sigma_0)^{1/(N-1)}}(\sin\sigma)^{1/(N-1)}. \qquad (9.4-16)$$

where R_0 and σ_0 stand for the initial relative range and velocity lead angle.

Assume that the velocity leading angle satisfies $|\sigma| < \pi/2$, which implies that R is strictly decreasing from equation (9.4-1). Define an auxiliary variable $\eta = \sin\sigma$; then, equation (9.4-1) can be reformulated as

$$\frac{dt}{dR} = -\frac{1}{V\sqrt{1-\eta^2}}. \qquad (9.4-17)$$

Integrating the preceding expression using binomial series gives the predicted impact time t_f as

$$t_f = \frac{1}{V}\int_0^{R_0} \frac{1}{\sqrt{1-\eta^2}}dR$$
$$= \frac{1}{V}\int_0^{R_0}\left(1 + \frac{1}{2}\eta^2 + \frac{3}{8}\eta^4 + \frac{5}{16}\eta^6 + \cdots\right)dR. \qquad (9.4-18)$$

Substituting equation (9.4-16) into equation (9.4-18) and after integration, we have

$$t_f = \frac{R_0}{V}\left[1 + \frac{\eta_0^2}{2(2N-1)} + \frac{3\eta_0^4}{8(4N-3)} + \frac{5\eta_0^6}{16(6N-5)} + \cdots\right], \qquad (9.4-19)$$

where η_0 denotes the initial value of η.

Replacing R_0 and η_0 with R and η, respectively, gives the predicted time-to-go under 3D PNG as

$$t_{go} = \frac{R}{V}\left[1 + \frac{\eta^2}{2(2N-1)} + \frac{3\eta^4}{8(4N-3)} + \frac{5\eta^6}{16(6N-5)} + \cdots\right]. \qquad (9.4-20)$$

By neglecting the higher-order terms of η^2, we have the approximated time-to-go estimation

$$t_{go} = \frac{R}{V}\left[1 + \frac{\sin^2\sigma}{2(2N-1)}\right]. \qquad (9.4-21)$$

Remark 1. For practical interceptors that provide roll-stabilization capability, the 3D guidance problem can be treated in two separate channels. Accordingly, 3D homing guidance can also be achieved by constructing two separate 2D PNGs in the pitch and yaw planes for roll-stabilized airframes. The commanded accelerations in the two planes are defined as (reference [34])

$$a_y^{PNG} = -\frac{NV^2}{R}\sin\phi_M, \qquad (9.4-22)$$

$$a_z^{PNG} = -\frac{NV^2}{R}\sin\theta_M. \qquad (9.4-23)$$

It follows from equation (9.4-12) and equation (9.4-13) that separate 2D PNG is identical to 3D PNG if the cross-coupling between the pitch and the yaw planes is ignorable. However, if the relative motions in the two planes cannot be decoupled, performance degradation of separately implementing 2D PNG is inevitable due to the cross-coupling effect.

Remark 2. If we only consider the 2D engagement, the pitch plane, for example, we have $\sigma = \theta_M$ and $\phi_M = 0$. Then, the predicted time-to-go equation (9.4-20) reduces to

$$t_{go} = \frac{R}{V}\left[1 + \frac{\sin^2\theta_M}{2(2N-1)}\right], \qquad (9.4-24)$$

which coincides with the results, derived in reference [30], when the velocity leading angle is small, e.g., $\sin\theta_M \approx \theta_M$. Comparing equation (9.4-20) and equation (9.4-24), it can be concluded that the proposed time-to-go estimation extends the 2D algorithm to a projected plan containing the LOS vector and missile velocity vector in the 3D scenario.

Remark 3. In impact time guidance law design, the desired impact time t_d should be set to be achievable; i.e., the problem is well posed. From practical standpoint, the desired impact time t_d is required to be larger than the predicted impact time t_f. For this reason, a suitable choice of t_d is

$$t_d > \frac{R_0}{V}\left[1 + \frac{\sin^2\sigma_0}{2(2N-1)}\right]. \qquad (9.4-25)$$

9.4.2.2 Impact time guidance law design

To achieve impact time control, both target interception and zero impact time error are required to be satisfied. For this reason, we propose a 3D composite guidance law, which is composed of an optimal baseline 3D PNG and an optimal impact time error feedback term. Instead of using two different biased terms, the proposed guidance law only uses one unique feedback command, which is automatically allocated to both vertical and horizontal planes, as

$$a_y = a_y^{PNG} + a_b \sin\phi_M = \left(-\frac{NV^2}{R} + a_b\right)\sin\phi_M, \qquad (9.4-26)$$

$$a_z = a_z^{PNG} + a_b \sin\theta_M \cos\phi_M = \left(-\frac{NV^2}{R} + a_b\right)\sin\theta_M \cos\phi_M, \qquad (9.4-27)$$

where a_b denotes the error feedback term to be determined.

Define $e_t = t_d - t_{go} - t$ as the impact time error. Substituting equation (9.4-21) into e_t and taking the time derivative using equation (9.4-26) and equation (9.4-27) give

$$\dot{e}_t = -\dot{t}_{go} - 1$$

$$= -\frac{\dot{R}}{V} - \frac{\dot{R}}{V}\frac{\sin^2\sigma}{2(2N-1)} - \frac{R}{V}\cdot\frac{\sin\sigma\cos\sigma\dot{\sigma}}{2N-1} - 1$$

$$= \cos\sigma\left[1 + \frac{\sin^2\sigma}{2(2N-1)}\right] - \frac{R\cos\sigma}{(2N-1)V^2}\left[-\frac{(N-1)V^2}{R}\sin^2\sigma + a_b\sin^2\sigma\right] - 1. \qquad (9.4-28)$$

$$= \cos\sigma\left[1 + \frac{\sin^2\sigma}{2(2N-1)}\right] - \frac{R\cos\sigma\sin^2\sigma}{(2N-1)V^2}\left[-\frac{(N-1)V^2}{R} + a_b\right] - 1$$

$$= \cos\sigma\left[1 + \frac{\sin^2\sigma}{2(2N-1)}\right] + \frac{(N-1)\cos\sigma\sin^2\sigma}{2N-1} - \frac{R\cos\sigma\sin^2\sigma}{(2N-1)V^2}a_b - 1$$

Assume that the missile velocity lead angle σ is small. Then, we have $\sin\sigma \approx \sigma$ and $\cos\sigma \approx 1 - \sigma^2/2$. By using these two approximations and neglecting higher-order terms of σ, equation (9.4-28) reduces to

$$\dot{e}_t = -\frac{R\sin^2\sigma}{(2N-1)V^2}a_b. \qquad (9.4-29)$$

For system of equation (9.4-29), consider the optimal error dynamics (reference [21])

$$\dot{e}_t + \frac{K}{t_{go}}e_t = 0, \qquad (9.4-30)$$

where $K > 0$ is the guidance gain to be designed.

Combining equation (9.4-29) with equation (9.4-30) gives the guidance command to nullify the impact time error as

$$a_b = \frac{K(2N-1)V^2}{R\sin^2\sigma t_{go}}e_t. \qquad (9.4-31)$$

Substituting equation (9.4-31) into equation (9.4-26) and equation (9.4-27) yields the explicit guidance command as

$$a_y = \left[-\frac{NV^2}{R} + \frac{K(2N-1)V^2}{R\sin^2\sigma t_{go}}e_t\right]\sin\phi_M, \qquad (9.4-32)$$

$$a_z = \left[-\frac{NV^2}{R} + \frac{K(2N-1)V^2}{R\sin^2\sigma t_{go}}e_t\right]\sin\theta_M\cos\phi_M. \qquad (9.4-33)$$

Similar to 3D PNG, the proposed guidance law can be formulated as a vector, located in the engagement plane, as

$$\boldsymbol{a} = \left[-\frac{NV^2}{R}\sin\sigma + \frac{K(2N-1)V^2}{R\sin\sigma t_{go}} e_t \right] \boldsymbol{e}_a, \qquad (9.4-34)$$

where $\boldsymbol{e}_a = [0, \sin\phi_M/\sin\sigma, \sin\theta_M\cos\phi_M/\sin\sigma]^T$ denotes the unit vector that specifies the direction of the commanded acceleration in the velocity coordinate.

Remark 4. Although the proposed 3D guidance law is derived using stationary targets, the guidance law developed can be easily adapted to nonmaneuvering target scenarios through the well-known predicted interception point concept (reference [29]).

9.4.3 Analysis of Proposed Guidance Law

This section analyzes the properties of the proposed 3D optimal impact time guidance law in the following aspects.

9.4.3.1 Singularity issue

From equation (9.4-31), we can observe that $\sigma = 0$ is a singular point, which will result in infinite guidance command. However, it is easy to verify that this singular point is trivial since the velocity lead angle $\sigma \neq 0$ except for the final impact point. To see this, taking the time derivative of σ and substituting equation (9.4-32) and equation (9.4-33) into it yield

$$\dot{\sigma} = -\frac{(N-1)V}{R}\sin\sigma + \frac{K(2N-1)V}{R\sin\sigma t_{go}} e_t. \qquad (9.4-35)$$

By choosing the desired impact time t_d that satisfies condition equation (9.4-25), we can readily conclude that $e_{t,0} > 0$, where $e_{t,0}$ denotes the initial impact time error. With this in mind, one can imply that the term $K(2N-1)Ve_t/(R\sin\sigma t_{go})$ initially tries to increase the magnitude of the velocity lead angle for reducing the impact time error. Also note that the PNG term $-(N-1)V\sin\sigma/R$ is used to regulate the velocity lead angle to zero to guarantee target interception. It is well known that the velocity lead angle under PNG converges to zero only at the time of impact (reference [14]). Therefore, if the guidance gain K satisfies

$$\frac{K(2N-1)V}{R_0|\sin\sigma_0|t_f} e_{t,0} > \frac{(N-1)V}{R_0}|\sin\sigma_0|, \qquad (9.4-36)$$

or the equivalent form

$$K > \frac{(N-1)\sin^2\sigma_0 t_f}{(2N-1)e_{t,0}}, \qquad (9.4-37)$$

the error feedback term a_b will play a dominant role initially in the guidance command, thus forcing the magnitude of the velocity lead angle to increase until certain time instant t_1. When $t \geq t_1$, the PNG term will dominate over a_b, hence regulating the magnitude of the velocity lead angle to zero at the time of impact. Note that condition equation (9.4-37) is easily to be satisfied for practical scenarios. Therefore, the proposed guidance law is nonsingular. An example of the velocity leading angle response under the proposed guidance law is presented in Fig. 9.4-2.

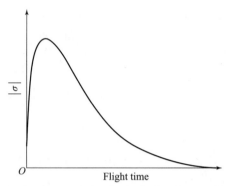

Fig. 9.4–2 An example of velocity leading angle profile

9.4.3.2 Finite-time convergence of impact time error

Under optimal error dynamics equation (9.4–30), it is easy to verify that the closed-form solution of the impact time error is determined as

$$e_t = e_{t,0} \left(\frac{t_{go}}{t_f} \right)^K, \tag{9.4-38}$$

which clearly reveals that the impact time error e_t will converge to zero at the time of impact if $K > 0$, thus satisfying the impact time control requirement. Furthermore, the convergence rate of the impact time error is determined by the guidance gain K: larger K results in faster convergence since $t_{go}/t_f \leqslant 1$.

9.4.3.3 Optimality of error feedback term

According to theorem 1 in reference [27], error dynamics equation (9.4–30) is optimal in terms of performance index

$$J = \frac{1}{2} \int_t^{t_f} \frac{R^2 \sin^4 \sigma}{(2N-1)^2 V^4 (t_f - \tau)^{K-1}} a_b(\tau) \, d\tau. \tag{9.4-39}$$

Since the constant terms in the performance index do not affect the optimal pattern, the previously mentioned performance index is identical to

$$J = \frac{1}{2} \int_t^{t_f} \frac{R^2 \sin^4 \sigma}{(t_f - \tau)^{K-1}} a_b(\tau) \, d\tau. \tag{9.4-40}$$

It follows from equation (9.4–40) that the weighting function $R^2 \sin^4 \sigma / (t_f - \tau)^{K-1}$ gradually decreases with the decrease of R and σ. This means that the magnitude of the error feedback term a_b tends to increase when the missile approaches the target. This property is, obviously, not desirable to guarantee the finite guidance command. For this reason, it is recommended to choose relatively large guidance gain K such that t_{go}^{K-1} is larger than $R^2 \sin^4 \sigma$ to compensate for the decreasing of the weighting function. Notice that the decreasing rates of both the relative range and velocity lead angle are governed by the PNG term; a suitable choice of K is $K > N$.

9.4.3.4 Relationship with 2D optimal impact time guidance law

When only considering the 2D homing engagement, e.g., the pitch plan, for example, we have $\sigma = \theta_M$ and $\phi_M = 0$. Then, the proposed 3D impact time guidance law, shown in equation (9.4–32) and equation (9.4–33), reduces to

$$a_y = 0, \quad (9.4-41)$$

$$a_z = -\frac{NV^2}{R}\sin\theta_M + \frac{K(2N-1)V^2}{R\sin\theta_M t_{go}}e_t. \quad (9.4-42)$$

which coincides with the generalized 2D optimal impact time guidance law proposed in reference [27].

It is well known that the 2D guidance law can be directly applied to 3D scenarios for roll-stabilized airframes by ignoring the crosscoupling effect between the horizontal and the vertical channels, i.e., assuming θ_M and ϕ_M are small. Under the condition that the relative motions in the two planes are decoupled, impact time control in a 3D scenario can be satisfied by using separate 2D guidance laws in the two planes. One feasible strategy to achieve this objective is to apply the 2D optimal impact time guidance law (reference [27]) in the vertical plane for impact time control and use the 2D PNG in the horizontal plane for the homing constraint. With this in mind, the individual 2D guidance command can be obtained as

$$a_y^{2D} = -\frac{NV^2}{R}\sin\phi_M, \quad (9.4-43)$$

$$a_z^{2D} = -\frac{NV^2}{R}\sin\theta_M + \frac{K(2N-1)V^2}{R\sin\theta_M t_{go}}e_t. \quad (9.4-44)$$

Comparing equation (9.4-32) and equation (9.4-33) with equation (9.4-43) and equation (9.4-44), one can observe that the proposed 3D guidance law automatically distributes the error feedback command term to both the horizontal and vertical planes while the 2D guidance law only leverages one channel in impact time control. This means that the proposed 3D guidance law fully exploits the synergism between these two channels and thus is beneficial when $\theta_M \neq 0$ and $\phi_M \neq 0$, especially when the effect of cross-coupling is strong. For example, if $\sin\theta_M\cos\phi_M > \sin\phi_M$, the proposed guidance law will mainly use the vertical plane for impact time control. Similarly, if $\sin\phi_M$ is dominant over $\sin\theta_M\cos\phi_M$, the horizontal plane will play an important role in impact time control. This property will be empirically evaluated in simulations. It is worth pointing out that the performance of 3D impact time guidance law is close to its 2D counterpart only when ϕ_M is small. As separately implementing the 2D guidance law ignores the cross-coupling effect, performance degradation is inevitable if this is a small angle approximation is violated. For example, if θ_M approaches to near zero before interception, the pitch guidance command equation (9.4-44) will suffer from a singular issue, as can be observed from the simulation studies.

Notice that the proposed 3D guidance leverages a kind of automatic command allocation; it is therefore helpful in saving energy consumption, compared to the separate 2D guidance law. To see this, define $E = a_y^2 + a_z^2$ as the quadratic energy consumption at each time instant; then, the required energy of the proposed 3D guidance law can be obtained as

$$E_{3D} = \left[-\frac{NV^2}{R} + \frac{K(2N-1)V^2}{R\sin^2\sigma t_{go}}e_t\right]^2(\sin^2\phi_M + \sin^2\theta_M\cos^2\phi_M)$$

$$= \left[-\frac{NV^2}{R} + \frac{K(2N-1)V^2}{R\sin^2\sigma t_{go}}e_t\right]^2\sin^2\sigma$$

$$= \frac{N^2 V^4}{R^2}\sin^2\sigma - \frac{2KN(2N-1)V^4 e_t}{R^2 t_{go}} + \frac{K^2(2N-1)^2 V^4}{R^2 \sin^2\sigma t_{go}^2}e_t^2. \qquad (9.4-45)$$

The required energy of using the separate 2D guidance law is given by

$$E_{2D} = \left(-\frac{NV^2}{R}\sin\phi_M\right)^2 + \left(-\frac{NV^2}{R}\sin\theta_M + \frac{K(2N-1)V^2}{R\sin\theta_M t_{go}}e_t\right)^2$$

$$= \frac{N^2 V^4}{R^2}(\sin^2\phi_M + \sin^2\theta_M) - \frac{2KN(2N-1)V^4 e_t}{R^2 t_{go}} + \frac{K^2(2N-1)^2 V^4}{R^2 \sin^2\theta_M t_{go}^2}e_t^2. \qquad (9.4-46)$$

Since $\cos^2\sigma = \cos^2\theta_M \cos^2\phi_M \le \cos^2\theta_M$, we have $\sin^2\sigma \ge \sin^2\theta_M$. Then, it follows from equation (9.4-47) that

$$E_{2D} \ge \frac{N^2 V^4}{R^2}(\sin^2\phi_M + \sin^2\theta_M \cos^2\phi_M) - \frac{2KN(2N-1)V^4 e_t}{R^2 t_{go}} + \frac{K^2(2N-1)^2 V^4}{R^2 \sin^2\sigma t_{go}^2}e_t^2$$

$$= \frac{N^2 V^4}{R^2}\sin^2\sigma - \frac{2KN(2N-1)V^4 e_t}{R^2 t_{go}} + \frac{K^2(2N-1)^2 V^4}{R^2 \sin^2\sigma t_{go}^2}e_t^2$$

$$= E_{3D}. \qquad (9.4-47)$$

where the equality holds if and only if $\phi_M = 0$. This expression clearly shows that the proposed 3D guidance law requires less energy consumption than separately implementing the 2D guidance law.

9.4.4 Numerical Simulations

In this section, the effectiveness of the proposed 3D optimal impact time guidance law is demonstrated through numerical simulations, in which an antiship missile is considered to intercept a stationary target. In the considered scenario, the target is located at (0 m, 0 m, 0 m). The interceptor is initially located at (6,000 m, 6,000 m, 0 m) with initial velocity lead angles $\theta_{M,0} = 10°$ and $\phi_{M,0} = 10°$. The missile flies with constant velocity $V = 250$ m/s. For implementing the proposed guidance law, the navigation gain of the baseline PNG is set as $N = 3$. In practice, the achieved acceleration of the missile is always bounded due to physical limits. For this reason, the magnitudes of both a_y and a_z are constrained by 100 m/s² in simulations.

9.4.4.1 Characteristics of proposed three-dimensional impact time guidance law

This subsection will empirically analyze the properties of the proposed 3D optimal impact time guidance law under various conditions. It is clear that the guidance gain K plays an important role in the proposed guidance law since it governs the convergence rate of the impact time error. For this reason, we first perform simulations with various guidance gains $K = 4, 8, 12$. In these simulations, the desired impact time is set as $t_d = 45$s, which satisfies condition equation (9.4-25). The simulation results, including the interception trajectories, history of the relative range, and acceleration command, are presented in Fig. 9.4-3. From this figure, it is clear that the proposed 3D guidance law successfully drives the missile to intercept the stationary target with the desired impact time. With higher guidance gain K, the response pattern of the relative range becomes more curved, and thus is closer to the desired pattern $(t_d - t)V$. That is, higher guidance gain K helps to increase the convergence speed of the impact time error. However, Fig. 9.4-3 (c) reveals that the proposed guidance law with higher guidance gain K requires more control energy during the initial flight phase. Furthermore, by choosing $K > N$, we can clearly note from Fig. 9.4-3 (d) that the guidance commands in both the vertical and horizontal planes converge to zero at the time of impact, demonstrating that the proposed guidance law has enough operational margins to cope with undesired disturbances when the missile approaches the target.

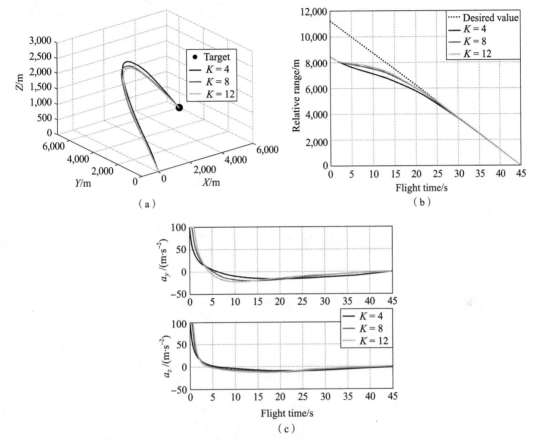

Fig. 9.4-3 Simulation results with respect to different guidance gains K
(a) Interception trajectories; (b) Relative range; (c) Acceleration command

Now, let us investigate the performance of the proposed 3D impact time guidance law with respect to different desired impact time t_d (40 s, 60 s and 80 s). For implementing the proposed guidance law, the guidance gain of the impact time error feedback term is chosen as $K = 12$. Fig. 9.4-4 (a) compares the interceptor trajectory for these three different impact time constraints. This figure clearly shows that the interceptor takes a longer flight path with a larger desired impact time. The history of the relative range with the three cases of desired impact time t_d is presented in Fig. 9.4-4 (b), which reveals that the proposed guidance law satisfies the impact time constraint precisely. The impact time error of the proposed guidance law turns out to be less than 0.01 s with different impact time constraints in our simulations. From this figure, it is obvious that the longer convergence phase is required to regulate the impact time error with a larger desired impact time t_d under the same initial conditions. The missile acceleration command produced by the proposed guidance law with different t_d is provided in Fig. 9.4 − 4 (c). Clearly, more energy consumption is required during the initial phase for a larger t_d. For this reason, the duration of the initial acceleration saturation of $t_d = 80$ s is longer than that of $t_d = 40$ s and $t_d = 60$ s.

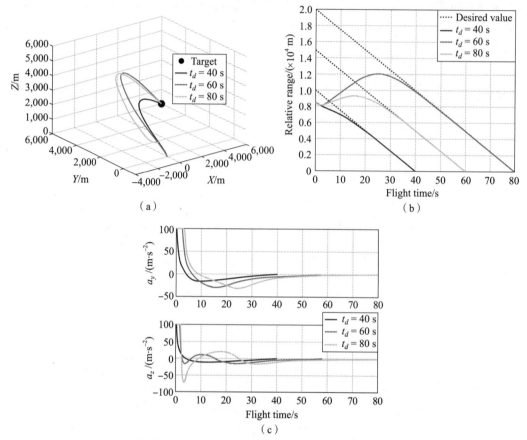

Fig. 9.4 − 4 Simulation results with respect to different desired impact time t_d
(a) Interception trajectories; (b) Relative range; (c) Acceleration command

9.4.4.2 Comparison with two-dimensional impact time guidance law

To further show the advantages of the proposed 3D impact time guidance law, comparisons with generalized 2D optimal impact time guidance law (reference [21]) are conducted in this subsection. For the purpose of comparison, three different initial conditions are considered as ① $\theta_{M,0} = 40°$, $\phi_{M,0} = 0°$; ② $\theta_{M,0} = 40°$, $\phi_{M,0} = 40°$; and ③ $\theta_{M,0} = 40°$, $\phi_{M,0} = 80°$. It is clear that case 1 is a 2D homing scenario that occurs in the vertical plane, case 2 considers a moderate level of the coupling effect between two planes, and case 3 corresponds to the strong cross-coupling scenario. For fair comparison, the guidance gain of the impact time error feedback term for both guidance laws is set as $K = 12$.

The simulation results, including the interception trajectories, relative range, and acceleration command, are shown in Fig. 9.4 − 5 with the desired impact time $t_d = 45$s. From the first row of Fig. 9.4 − 5, one can observe that these two guidance laws generate exactly the same results for case 1, as we expected. The reason is that the proposed 3D impact time guidance law reduces its 2D counterpart if we only consider the vertical plane as we discussed before. For case 2, since $\sin\theta_M\cos\phi_M < \sin\phi_M$, the proposed 3D guidance law will mainly use the horizontal plane to regulate the impact time error, as shown in the second row of Fig. 9.4 – 5. As we use the same guidance gain K for both guidance laws, the impact time error dynamics under the two guidance laws is the same. For this reason, the convergence patterns of the relative range under both guidance laws show similar characteristics, as can be confirmed from Fig. 9.4 – 5 (e). From Fig. 9.4 – 5 (f), we can observe that the proposed 3D guidance law leverages both the vertical and horizontal channels for impact time control and automatically distributes the error feedback command into these two channels. As a comparison, the 2D impact time guidance law only uses the vertical plane in impact time control. As for the strong crosscoupling engagement scenario in case 3, both guidance laws successfully drive the missile to intercept the target with the desired impact time constraint, as shown in the third row of Fig. 9.4 – 5. However, the 2D impact time guidance law shows oscillating patterns when the interceptor is close to the target, which is not desirable for onboard control systems. The control effort $\int_{t_0}^{t_f} [a_y^2(t) + a_z^2(t)] dt$ obtained from both the 2D and the 3D impact time guidance laws for these three different cases are summarized in table 9.4 – 1. From this table, we can readily note that the proposed 3D impact guidance law helps to reduce the energy consumption except for the particular case 1. This confirms the theoretical findings presented in the previous section.

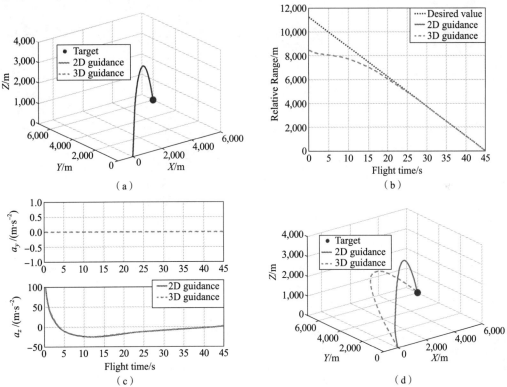

Fig. 9.4 – 5 Comparison results with 2D optimal impact time guidance law

(a) Interception trajectories for case 1; (b) Relative range for case 1; (c) Acceleration command for case 1;

(d) Interception trajectories for case 2

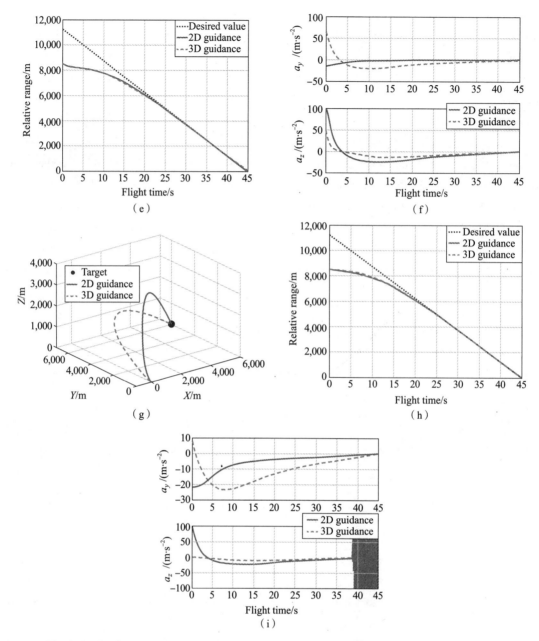

Fig. 9.4-5 Comparison results with 2D optimal impact time guidance law (Continued)

(e) Relative range for case 2; (f) Acceleration command for case 2; (g) Interception trajectories for case 3; (h) Relative range for case 3; (i) Acceleration command for case 3

Table 9.4-1 Control effort comparisons

Guidance	Case 1	Case 2	Case 3
2D	18,733	18,951	62,803
3D	18,733	13,080	9,926

References

[1] Bai W, Xue W, Huang Y, et. al. On extended state based Kalman filter design for a class of nonlinear time-varying uncertain systems [J]. Science China Information Sciences, 2018, 61 (4).

[2] S. N. Balakrishnan, Antonios Tsourdos, and Brian A White. Advances in missile guidance, control, and estimation [M]. Boca Raton: CRC Press, 2012.

[3] John H Blakelock. Automatic control of aircraft and missiles, 2nd edition [M]. New York: John Wiley & Sons, 1991.

[4] Голубев И. С., Светлов В. Г. ПРОЕКТИРОВАНИЕ ЗЕНИТНЫХ УПРАВЛЯЕМЫХ РАКЕТ [M]. Голубев И. С., Светлов В. Г., 2001.

[5] P. Garnell. Guided weapon control system, 2nd edition [M]. London: Royal Military College of Science, 1980.

[6] Ching-fang Lin. Modern navigation, guidance, and control processing, volume 2 [M]. Englewood Cliffs, NJ: Prentice Hall, 1991.

[7] Howard Musoff and Paul Zarchan. Fundamentals of Kalman filtering: a practical approach [M]. Virginia: American Institute of Aeronautics and Astronautics, 2009.

[8] Neryahu A Shneydor. Missiles guidance and pursuit: kinematics, dynamics and control [M]. Amsterdam: Elsevier Press, 1998.

[9] George M Siouris. Missile guidance and control systems [M]. Berlin: Springer Science & Business Media, 2004.

[10] Ashish Tewari. Advanced control of aircraft, spacecraft and rockets [M]. New York: John Wiley & Sons, 2011.

[11] Rafael Yanushevsky. Modern missile guidance [M]. Boca Raton: CRC Press, 2007.

[12] Paul Zarchan. Tactical and strategic missile guidance, 6th edition [M]. Virginia: American Institute of Aeronautics and Astronautics, Inc, 2012.

[13] Lin Defu, Wang Hui, Wang Jiang, Fan Junfang. Design and guidance law analysis of tactical missile autopilot [M]. Beijing: Beijing Institute of Technology Press, 2012.

[14] Jeon I-S, Lee J-I, Tahk M-J. Impact-time-control guidance law for anti-ship missiles [J]. IEEE Trans Control Syst Technol, 2006, 14 (2): 260-266.

[15] Ryoo C-K, Cho H, Tahk M-J. Optimal guidance laws with terminal impact angle constraint [J]. J Guid Control Dyn, 2005, 28 (4): 724-732.

[16] Lee C-H, Kim T-H, Tahk M-J. Interception angle control guidance using proportional navigation with error feedback [J]. J Guid Control Dyn, 2013, 36 (5): 1556-1561.

[17] Ohlmeyer E J, Phillips C A. Generalized vector explicit guidance [J]. J Guid Control Dyn, 2006, 29 (2): 261-268.

[18] Lee C-H, Kim T-H, Tahk M-J, Whang I-H. Polynomial guidance laws considering terminal impact angle and acceleration constraints [J]. IEEE Trans Aerosp Electron Syst, 2013, 49 (1): 74-92.

[19] Dwivedi P, Bhale P, Bhattacharyya A, Padhi R. Generalized estimation and predictive guidance for evasive targets [J]. IEEE Trans Aerosp Electron Syst, 2016, 52 (5): 2111-2122.

[20] He S, Lee C-H. Gravity-turn-assisted optimal guidance law [J]. J Guid Control Dyn, 2018, 41 (1): 171-183.

[21] Shima T, Golan OM. Exo-atmospheric guidance of an accelerating interceptor missile [J]. J Franklin Inst, 2012, 349 (2): 622-637.

[22] Zarchan, P. Tactical and strategic missile guidance [J]. AIAA, Reston, VA, 2012: 163-184.

[23] Jeon I-S, and Lee J-I. ATEGIC missile guidance, AIAA avigation based on nonlinear formulation [J]. IEEE Transactions on Aerospace and Electronic Systems, 2010, 46 (4): 2051-2055. (doi: 10.1109/TAES.2010.5595614)

[24] Cho N, and Kim Y, J-I. ATEGIC missile guidance, AIAA avigation based on nonlineeuvering target interception [J]. IEEE Transactions on Aerospace and Electronic Systems, 2016, 52 (2): 948-954. (doi: 10.1109/TAES.2015.140432)

[25] Jeon I-S, Lee J-I, and Tahk M-J. E guidance, AIAA avigation based on nonlineeuvering target [J]. IEEE Transactions on Control Systems Technology, 2006, 14 (2): 260-266. (doi: 10.1109/TCST.2005.863655)

[26] Jeon I-S, Lee J-I, and Tahk M-J. E guidance, AIAA avigation based on nonlineeuvering target [J]. IEEE Transactions on Control Systems Ttion, Journal of Guidance, Control, and Dynamics, 2016, 39 (8): 1887-1892. (doi: 10.2514/1.G001681)

[27] He S, and Lee C-H, and Tahk M-J. E guidance, AIAA avigation based on nonlineeuvering target [J]. IEEE Transactions on Control Systems, 2018, 7: 1620-1629. (doi: 10.2514/1.G003343)

[28] Kim M, Jung B, Han B, Lee S, and Kim Y, Lyapunov-based impact time control guidance laws against stationary targets [J]. IEEE Transactions on Aerospace and Electronic Systems, 2015, 51 (2): 1111-1122. (doi: 10.1109/TAES.2014.130717)

[29] Cho D, Kim H J, and Tahk M-J, Nonsingular sliding mode guidance for impact time control [J]. Journal of Guidance, Control, and Dynamics, 2016, 39 (1): 61-68. (doi: 10.2514/1.G001167)

[30] Jeon I-S, Lee J-I, and Tahk M-J. Homing guidance law for cooperative attack of multiple missiles [J]. Journal of Guidance, Control, and Dynamics, 2010, 33 (1): 275-280. (doi: 10.2514/1.40136)

[31] He S, Wang W, Lin D, and Lei H. Consensus-based two-stage salvo attack guidance [J]. IEEE Transactions on Aerospace and Electronic Systems, 2018, 54 (3): 1555 – 1566. (doi: 10.1109/TAES.2017.2773272)

[32] Song S – H, and Ha I – J. A lyapunov-like approach to performance analysis of 3 – dimensional pure PNG laws [J]. IEEE Transactions on Aerospace and Electronic Systems, 1994, 30 (1): 238 – 248. (doi: 10.1109/7.250424)

[33] Shin H – S, Tsourdos A, and Li K – B. A new three-dimensional sliding mode guidance law variation with finite time convergence [J]. IEEE Transactions on Aerospace and Electronic Systems, 2017, 53 (5): 2221 – 2232. (doi: 10.1109/TAES.2017.2689938)

[34] Zhou D, Sun S, and Teo K L. Guidance laws with finite time convergence [J]. Journal of Guidance, Control, and Dynamics, 2009, 32 (6): 1838 – 1846. (doi: 10.2514/1.42976)